Flexible Imputation of Missing Data

Second Edition

CHAPMAN & HALL/CRC
Interdisciplinary Statistics Series

Series editors: N. Keiding, B.J.T. Morgan, C.K. Wikle, P. van der Heijden

Recently Published Titles

For more information about this series, please visit: https://www.crcpress.com/go/ids

Flexible Imputation of Missing Data

Second Edition

Stef van Buuren

Netherlands Organization for Applied Scientific Research TNO
and
Utrecht University

CRC Press
Taylor & Francis Group
Boca Raton London New York

CRC Press is an imprint of the
Taylor & Francis Group, an **informa** business

A CHAPMAN & HALL BOOK

Cover design: Celine van Hoek

CRC Press
Taylor & Francis Group
6000 Broken Sound Parkway NW, Suite 300
Boca Raton, FL 33487-2742

First issued in paperback 2021

ISBN-13: 978-1-138-58831-8 (hbk)
ISBN-13: 978-1-03-217863-9 (pbk)
DOI: 10.1201/9780429492259

Library of Congress Cataloging-in-Publication Data

Names: Buuren, Stef van, author.
Title: Flexible imputation of missing data / by Stef van Buuren (TNO, Leiden, The Netherlands [and] University of Utrecht, The Netherlands).
Description: Second edition. | Boca Raton, Florida : CRC Press, [2019] | Includes bibliographical references and index.
Identifiers: LCCN 2018017122| ISBN 9781138588318 (hardback : alk. paper) | ISBN 9780429492259 (e-book : alk. paper) | ISBN 9780429960352 (web pdf : alk. paper) | ISBN 9780429960345 (epub : alk. paper) | ISBN 9780429960338 (mobi/kindle : alk. paper)
Subjects: LCSH: Multivariate analysis. | Multiple imputation (Statistics) | Missing observations (Statistics)
Classification: LCC QA278 .B88 2019 | DDC 519.5/35--dc23
LC record available at https://lccn.loc.gov/2018017122

Visit the Taylor & Francis Web site at
http://www.taylorandfrancis.com

and the CRC Press Web site at
http://www.crcpress.com

Voor Eveline, Guus, Otto en Maaike

Contents

II Advanced techniques 161

Foreword

I'm delighted to see this new book on multiple imputation by Stef van Buuren for several reasons. First, to me at least, having another book devoted to multiple imputation marks the maturing of the topic after an admittedly somewhat shaky initiation. Stef is certainly correct when he states in Section 2.1.2: "The idea to create multiple versions must have seemed outrageous at that time [late 1970s]. Drawing imputations from a distribution, instead of estimating the 'best' value, was a severe breach with everything that had been done before." I remember how this idea of multiple imputation was even ridiculed by some more traditional statisticians, sometimes for just being "silly" and sometimes for being hopelessly inefficient with respect to storage demands and outrageously expensive with respect to computational requirements.

Some others of us foresaw what was happening to both (a) computational storage (I just acquired a 64 GB flash drive the size of a small finger for under $60, whereas only a couple of decades ago I paid over $2500 for a 120 KB hard drive larger than a shoebox weighing about 10 kilos), and (b) computational speed and flexibility. To develop statistical methods for the future while being bound by computational limitations of the past was clearly inapposite. Multiple imputation's early survival was clearly due to the insight of a younger generation of statisticians, including many colleagues and former students, who realized future possibilities.

A second reason for my delight at the publication of this book is more personal and concerns the maturing of the author, Stef van Buuren. As he mentions, we first met through Jan van Rijckevorsel at TNO. Stef was a young and enthusiastic researcher there, who knew little about the kind of statistics that I felt was essential for making progress on the topic of dealing with missing data. But consider the progress over the decades starting with his earlier work on MICE! Stef has matured into an independent researcher making important and original contributions to the continued development of multiple imputation.

This book represents a 'no nonsense' straightforward approach to the application of multiple imputation. I particularly like Stef's use of graphical displays, which are badly needed in practice to supplement the more theoretical discussions of the general validity of multiple imputation methods. As I have said elsewhere, and as implied by much of what is written by Stef, "It's not that multiple imputation is so good; it's really that other methods for addressing missing data are so bad." It's great to have Stef's book on multiple imputation, and I look forward to seeing more editions as this rapidly developing methodology continues to become even more effective at handling missing data problems in practice.

Finally, I would like to say that this book reinforces the pride of an academic father who has watched one of his children grow and develop. This book is a step in the growing list of contributions that Stef has made, and, I am confident, will continue to make, in methodology, computational approaches and application of multiple imputation.

Donald B. Rubin

I am very pleased for the opportunity to add this short addendum to my preface to the first edition of Stef's wonderfully readable book on multiple imputation. Over the past few years, I've recommended that many check out Stef's first edition for excellent advice to practitioners of multiple imputation. The increased appreciation and use of multiple imputation between these two editions reflects, not only the growing maturity of computational statistics, but also the growing acceptance of multiple imputation as essentially being "the only game in town" for dealing generally with the problem of missing data, especially because it leads so naturally to visual displays of sensitivity of conclusions to differing assumptions about the reasons for the missing data. I am enthusiastic about the growing role of sensitivity analysis using visual displays, always prominent in Stef's contributions, but also now in other places, such as Liublinska and Rubin (2014) — some of us can be slow learners!

Donald B. Rubin, May 2018

Preface to second edition

Welcome to the second edition of *Flexible Imputation of Missing Data*, a book that can help you to solve missing data problems using `mice`. I am tremendously grateful for the success of the first edition. The `mice` community has been steadily growing over the last years, and the `mice` package is now downloaded from CRAN at a rate of about 750 downloads per day. My hope is that this book will sharpen your intuition on how to think about missing data, and provide you the tools to apply your ideas in practice.

Since the first edition was published in 2012, multiple imputation of missing data has become one of the great academic industries. Many analysts now employ multiple imputation on a regular basis as a generic solution to the omnipresent missing-data problem, and a substantial group of practitioners are doing the calculations in `mice`. This book aspires to combine a state-of-the-art overview of the field with a set of how-to instructions for practical data analysis.

Some sections of the first edition are still perfectly fine, but many others appear outdated. And of course, some of the newer developments are missing from the first edition. The second edition brings the text up-to-date. So what's new?

1. Multiple imputation of multilevel data has been a hot spot of statistical research. Multilevel data can arise from nested data collection designs, but also emerge when data are combined from multiple sources. Imputers and analysts now have a bewildering array of options. The three pages in the first edition have expanded into a full-blown new chapter on multilevel imputation. This chapter translates the current insights among the leading developers into practical advice for end users.

2. Another hot spot in statistics and data science is the creation of personalized estimates. Causal inference by multiple imputation of the potential outcomes is an innovative approach that attempts to answer "what if" questions on the level of the individual. This edition contains a short new chapter on individual causal effects that demonstrates how multiple imputation is applied to obtain well-grounded personalized estimates.

3. Data science has continued to grow at a phenomenal pace. The R language is now the dominant software for developing new statistical techniques. RStudio has successfully introduced the open tidyverse ecosystem for data acquisition, organization, analysis and visualization. This

edition targets this growing audience of data scientists by including new sections on parallel computation and MICE workflows using pipes within the `tidyverse` ecosystem.

4. New algorithms for creating imputations have appeared, in particular methodology based on predictive mean matching, for imputing binary and ordered variables, for interactions using classification and regression trees, and many types of machine learning methods. Chapters 3 and 4 incorporate these developments.

5. Important theoretical advances have been made on the convergence, compatibility, misspecification and stability of the simulation algorithms underlying MICE. Chapter 4 in this edition highlights these developments.

In parallel to the book, I worked on a significant update of the software: `mice 3.0`. The main MICE algorithm now iterates over blocks of variables instead of individual columns, so we may now easily combine univariate and multivariate imputation methods. In addition, it is now possible to specify exactly which cells in the data should be imputed. There are new functions for multivariate tests, the support for native formula's has improved, and, thanks to the `broom` package, parameter pooling is now available for a much wider selection of complete-data models. The calculations use better numerical algorithms for low-level imputation functions. I have tried hard to remain code-compatible with previous versions of `mice`. Existing code should run properly, but do not expect exact replication of the results. All code used in this book was tested with `mice 3.0`.

The previous edition had two colors, and some of the plots did not work as well as I had intended. I am very glad that this edition is in full color, so that the differences between the blue and red points stand out clearly and provide a unifying look to the book. There is also syntax coloring of the R code, which makes it very easy to distinguish the various language elements.

All data are incomplete, and so are all books. I had the luxury that I could devote my time during the period December 2017–March 2018 to this revision. A block of four months may seem like a formidable amount of time, but in retrospect it passed very quickly. While some topics I had planned have remained in the conceptual stage, overall I think that this edition covers the relevant developments in the field.

New statistical techniques will only be applied if there is high-quality and user-friendly software available. I would like to thank the following people for their contribution to the `mice` package over the years: Karin Groothuis-Oudshoorn, Gerko Vink, Lisa Doove, Shahab Jolani, Roel de Jong, Rianne Schouten, Florian Meinfelder, Philipp Gaffert, Alexander Robitzsch and Bernie Gray.

There is a growing ecosystem of related R packages that extend the functionality of `mice` in some way. Currently, these include `miceadds`, `mitml`, `micemd`, `countimp`, `CALIBERrfimpute`, `miceExt` and `ImputeRobust`. There is also a `Python` version in the works, which could result in an enormous expansion of the user base. I thank the authors of these packages for the time and effort they have put into creating these programs: Alexander Robitzsch, Simon Grund, Thorsten Henke, Oliver Lüdtke, Vincent Audigier, Matthieu Resche-Rigon, Kristian Kleinke, Jost Reinecke, Anoop Shah, Tobias Schumacher, Philipp Gaffert, Daniel Salfran, Martin Spiess, Sergey Feldman and Rianne Schouten.

I wish to thank Rob Calver, Statistics Editor at Chapman & Hall/CRC for his encouragement during both the first and second edition. Lara Spieker, Suzanne Lassandro and Shashi Kumar have been very helpful in meeting the ambitious production schedule. I thank Daan Kloet of TNO for his support for the idea of a mini-sabbatical, and his assistence in realizing the idea within TNO. I also wish to thank Peter van der Heijden of the University of Utrecht for his support over the years. Several people read and commented on parts of the manuscript. I thank Gerko Vink, Shahab Jolani, Iris Eekhout, Simon Grund, Tom Snijders and Joop Hox for their insightful and useful feedback. This has helped me a lot to understand the details much clearer, allowing me to improve my fumbled writings.

Last but not least, I thank my wife Eveline for her patience in living with an individual who can be so preoccupied with something else.

Stef van Buuren
May 2, 2018

Preface to first edition

We are surrounded by missing data. Problems created by missing data in statistical analysis have long been swept under the carpet. These times are now slowly coming to an end. The array of techniques for dealing with missing data has expanded considerably during the last decades. This book is about one such method: multiple imputation.

Multiple imputation is one of the great ideas in statistical science. The technique is simple, elegant and powerful. It is simple because it fills the holes in the data with plausible values. It is elegant because the uncertainty about the unknown data is coded in the data itself. And it is powerful because it can solve "other" problems that are actually missing data problems in disguise.

Over the last 20 years, I have applied multiple imputation in a wide variety of projects. I believe the time is ripe for multiple imputation to enter mainstream statistics. Computers and software are now potent enough to do the required calculations with little effort. What is still missing is a book that explains the basic ideas and that shows how these ideas can be put into practice. My hope is that this book can fill this gap.

The text assumes familiarity with basic statistical concepts and multivariate methods. The book is intended for two audiences:

- (bio)statisticians, epidemiologists and methodologists in the social and health sciences;

- substantive researchers who do not call themselves statisticians, but who possess the necessary skills to understand the principles and to follow the recipes.

In writing this text, I have tried to avoid mathematical and technical details as much as possible. Formulas are accompanied by a verbal statement that explains the formula in layperson terms. I hope that readers less concerned with the theoretical underpinnings will be able to pick up the general idea. The more technical material is marked by a club sign ♣, and can be skipped on first reading.

I used various parts of the book to teach a graduate course on imputation techniques at the University of Utrecht. The basics are in Chapters 1–4. Lecturing this material takes about 10 hours. The lectures were interspersed with sessions in which the students worked out the exercises from the book.

This book owes much to the ideas of Donald Rubin, the originator of multiple imputation. I had the privilege of being able to talk, meet and work with him on many occasions. His clear vision and deceptively simple ideas have been a tremendous source of inspiration. I am also indebted to Jan van Rijckevorsel for bringing me into contact with Donald Rubin, and for establishing the scientific climate at TNO in which our work on missing data techniques could prosper.

Many people have helped realize this project. I thank Nico van Meeteren and Michael Holewijn of TNO for their trust and support. I thank Peter van der Heijden of Utrecht University for his support. I thank Rob Calver and the staff at Chapman & Hall/CRC for their help and advice. Many colleagues have commented on part or all of the manuscript: Hendriek Boshuizen, Elise Dusseldorp, Karin Groothuis-Oudshoorn, Michael Hermanussen, Martijn Heymans, Nicholas Horton, Shahab Jolani, Gerko Vink, Ian White and the research master students of the Spring 2011 class. Their comments have been very valuable for detecting and eliminating quite a few glitches. I happily take the blame for the remaining errors and vagaries.

The major part of the manuscript was written during a six-month sabbatical leave. I spent four months in Krukö, Sweden, a small village of just eight houses. I thank Frank van den Nieuwenhuijzen and Ynske de Koning for making their wonderful green house available to me. It was the perfect tranquil environment that, apart from snowplowing, provided a minimum of distractions. I also spent two months at the residence of Michael Hermanussen and Beate Lohse-Hermanussen in Altenhof, Germany. I thank them for their hospitality, creativity and wit. It was a wonderful time.

Finally, I thank my family, in particular my beloved wife Eveline, for their warm and ongoing support, and for allowing me to devote time, often nights and weekends, to work on this book. Eveline liked to tease me by telling people that I was writing "a book that no one understands." I fear that her statement is accurate, at least for 99% of the people. My hope is that you, my dear reader, will belong to the remaining 1%.

Stef van Buuren

About the author

Stef van Buuren is a statistician at the Netherlands Organization for Applied Scientific Research TNO in Leiden, and professor of Statistical Analysis of Incomplete Data at Utrecht University. He is the originator of the MICE algorithm for multiple imputation of multivariate data, and co-developed the `mice` package in R. More information can be found at `www.stefvanbuuren.name`.

List of symbols

Y $n \times p$ matrix of partially observed sample data (2.2.3)

R $n \times p$ matrix, 0–1 response indicator of Y (2.2.3)

X $n \times q$ matrix of predictors, used for various purposes (2.2.3)

Y_{obs} observed sample data, values of Y with $R = 1$ (2.2.3)

Y_{mis} unobserved sample data, values of Y with $R = 0$ (2.2.3)

n sample size (2.2.3)

m number of multiple imputations (2.2.3)

ψ parameters of the missing data model that relates Y to R (2.2.4)

θ parameters of the scientifically interesting model for the full data Y (2.2.5)

Q $k \times 1$ vector with k scientific estimands (2.3.1)

$\hat{Q}$ $k \times 1$ vector, estimator of Q calculated from a hypothetically complete sample (2.3.1)

U $k \times k$ matrix, within-imputation variance due to sampling (2.3.1)

ℓ imputation number, where $\ell = 1, \ldots, m$ (2.3.2)

Y_ℓ ℓth imputed dataset, where $\ell = 1, \ldots, m$ (2.3.2)

$\bar{Q}$ $k \times 1$ vector, estimator of Q calculated from the incompletely observed sample (2.3.2)

$\bar{U}$ $k \times k$, estimator of U from the incomplete data (2.3.2)

B $k \times k$, between-imputation variance due to nonresponse (2.3.2)

T total variance of $(Q - \bar{Q})$, $k \times k$ matrix (2.3.2)

λ proportion of the variance attributable to the missing data for a scalar parameter (2.3.5)

γ fraction of information missing due to nonresponse (2.3.5)

r relative increase of variance due to nonresponse for a scalar parameter (2.3.5)

$\bar{\lambda}$ λ for multivariate Q (2.3.5)

$\bar{r}$ r for multivariate Q (2.3.5)

ν_{old} old degrees of freedom (2.3.6)

ν adjusted degrees of freedom (2.3.6)

y univariate Y (3.2.1)

y_{obs} vector with n_1 observed data values in y (3.2.1)

y_{mis} vector n_0 missing data values in y (3.2.1)

$\dot{y}$ vector n_0 imputed values in y (3.2.1)

X_{obs} subset of n_1 rows of X for which y is observed (3.2.1)

X_{mis} subset of n_0 rows of X for which y is missing (3.2.1)

$\hat{\beta}$ estimate of regression weight β (3.2)

$\dot{\beta}$ simulated regression weight for β (3.2.1)

$\hat{\sigma}^2$ estimate of residual variance σ^2 (3.2.1)

$\dot{\sigma}^2$ simulated residual variance for σ^2 (3.2.1)

κ ridge parameter (3.2.2)

η distance parameter in predictive mean matching (3.4.2)

$\hat{y}_i$ vector of n_1 predicted values given X_{obs} (3.4.2)

$\hat{y}_j$ vector of n_0 predicted values given X_{mis} (3.4.2)

δ shift parameter in nonignorable models (3.8.1)

Y_j jth column of Y (4.1.1)

Y_{-j} all columns of Y except Y_j (4.1.1)

I_{jk} proportion of usable cases for imputing Y_j from Y_k (4.1.2)

O_{jk} proportion of observed cases in Y_j to impute Y_k (4.1.2)

I_j influx statistic to impute Y_j from Y_{-j} (4.1.3)

O_j outflux statistic to impute Y_{-j} from Y_j (4.1.3)

ϕ parameters of the imputation model that models the distribution of Y (4.3.1)

M number of iterations (4.5)

D_1 test statistic of Wald test (5.3.1)

r_1 r for Wald test (5.3.1)

ν_1 ν for Wald test (5.3.1)

D_2 test statistic for χ^2-test (5.3.2)

r_2 r for χ^2-test (5.3.2)

ν_2 ν for χ^2-test (5.3.2)

D_3 test statistic for likelihood ratio test (5.3.3)

r_3 r for likelihood ratio test (5.3.3)

ν_3 ν for likelihood ratio test (5.3.3)

C number of classes (7.2)

c class index, $c = 1, \ldots, C$ (7.2)

y_c outcome vector for cluster c (7.2)

X_c design matrix, fixed effects (7.2)

Z_c design matrix, random effects (7.2)

Ω covariance matrix, random effects (7.2)

$Y_i(1)$ outcome of unit i under treatment (8.1)

$Y_i(0)$ outcome of unit i under control (8.1)

τ_i individual causal effect (8.1)

τ average causal effect (8.1)

$\dot{\tau}_{i\ell}$ simulated τ_i in ℓ'th imputed data set (8.3)

$\hat{\tau}_i$ estimate of τ_i (8.3)

$\hat{\sigma}_i^2$ variance estimate of $\hat{\tau}_i$ (8.3)

$\dot{\tau}_\ell$ simulated within-replication average causal effect (8.4.3)

$\dot{\sigma}_\ell^2$ variance of $\dot{\tau}_\ell$ (8.4.3)

List of algorithms

Part I

Basics

Chapter 1

Introduction

> *We should be suspicious of any dataset (large or small) which appears perfect.*
> David J. Hand

1.1 The problem of missing data

1.1.1 Current practice

The mean of the numbers 1, 2 and 4 can be calculated in R as

```
y <- c(1, 2, 4)
mean(y)
```

```
[1] 2.33
```

where y is a vector containing three numbers, and where mean(y) is the R expression that returns their mean. Now suppose that the last number is missing. R indicates this by the symbol NA, which stands for "not available":

```
y <- c(1, 2, NA)
mean(y)
```

```
[1] NA
```

The mean is now undefined, and R informs us about this outcome by setting the mean to NA. It is possible to add an extra argument na.rm = TRUE to the function call. This removes any missing data before calculating the mean:

```
mean(y, na.rm = TRUE)
```

```
[1] 1.5
```

This makes it possible to calculate a result, but of course the set of observations on which the calculations are based has changed. This may cause problems in statistical inference and interpretation.

It gets worse with multivariate analysis. For example, let us try to predict daily ozone concentration (ppb) from wind speed (mph) using the built-in `airquality` dataset. We fit a linear regression model by calling the `lm()` function to predict daily ozone levels, as follows:

```
fit <- lm(Ozone ~ Wind, data = airquality)
Error in na.fail.default: missing values in object}
```

Many R users have seen this message. The code cannot continue because there are missing values. One way to circumvent the problem is to omit any incomplete records by specifying the `na.action = na.omit` argument to `lm()`. The regression weights can now be obtained as

```
fit <- lm(Ozone ~ Wind, data = airquality, na.action = na.omit)
```

This works. For example, we may produce diagnostic plots by `plot(fit)` to study the quality of the model. In practice, it is cumbersome to supply the `na.action()` function each time. We can change the setting in `options` as

```
options(na.action = na.omit)
```

which eliminates the error message once and for all. Users of other software packages like `SPSS`, `SAS` and `Stata` enjoy the "luxury" that this deletion option has already been set for them, so the calculations can progress silently. Next, we wish to plot the predicted ozone levels against the observed data, so we use `predict()` to calculate the predicted values, and add these to the data to prepare for plotting.

```
airquality2 <- cbind(airquality, predict(fit))
Error: arguments imply differing number of rows: 153, 116
```

Argg... that doesn't work either. The error message tells us that the two datasets have a different number of rows. The `airquality` data has 153 rows, whereas there are only 116 predicted values. The problem, of course, is that there are missing data. The `lm()` function dropped any incomplete rows in the data. We find the indices of the first six cases by

```
head(na.action(fit))
```

```
 5 10 25 26 27 32
 5 10 25 26 27 32
```

The total number of deleted cases is found as

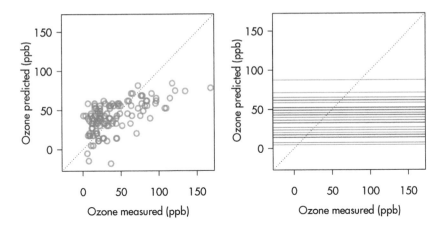

Figure 1.1: Predicted versus measured ozone levels for the observed (left, blue) and missing values (right, red).

```
naprint(na.action(fit))
```

```
[1] "37 observations deleted due to missingness"
```

The number of missing values per variable in the data is

```
colSums(is.na(airquality))
```

```
  Ozone Solar.R    Wind    Temp   Month     Day
     37       7       0       0       0       0
```

so in our regression model, all 37 deleted cases have missing ozone scores.

Removing the incomplete cases prior to analysis is known as *listwise deletion* or *complete-case analysis*. In R, there are two related functions for the subset of complete cases, `na.omit()` and `complete.cases()`.

Figure 1.1 plots the predicted against the observed values. Here we adopt the Abayomi convention for the colors (Abayomi et al., 2008): Blue refers to the observed part of the data, red to the synthetic part of the data (also called the *imputed values* or *imputations*), and black to the combined data (also called the *imputed data* or *completed data*). The printed version of the first edition of this book used gray instead of blue. The blue points on the left are all from the complete cases, whereas the figure on the right plots the points for the incomplete cases (in red). Since there are no measured ozone levels in that part of the data, the possible values are indicated by 37 horizontal lines.

Listwise deletion allows the calculations to proceed, but it may introduce additional complexities in interpretation. Let's try to find a better predictive model by including solar radiation (`Solar.R`) into the model as

```
fit2 <- lm(Ozone ~ Wind + Solar.R, data = airquality)
naprint(na.action(fit2))
```

```
[1] "42 observations deleted due to missingness"
```

Observe that the number of deleted days increased is now 42 since some rows had no value for `Solar.R`. Thus, changing the model altered the sample.

There are methodological and statistical issues associated with this procedure. Some questions that come to mind are:

- Can we compare the regression coefficients from both models?

- Should we attribute differences in the coefficients to changes in the model or to changes in the subsample?

- Do the estimated coefficients generalize to the study population?

- Do we have enough cases to detect the effect of interest?

- Are we making the best use of the costly collected data?

Getting the software to run is one thing, but this alone does not address the challenges posed by the missing data. Unless the analyst, or the software vendor, provides some way to work around the missing values, the analysis cannot continue because calculations on missing values are not possible. There are many approaches to circumvent this problem. Each of these affects the end result in a different way. Some solutions are clearly better than others, and there is no solution that will always work. This chapter reviews the major approaches, and discusses their advantages and limitations.

1.1.2   Changing perspective on missing data

The standard approach to missing data is to delete them. It is illustrative to search for missing values in published data. Hand et al. (1994) published a highly useful collection of small datasets across the statistical literature. The collection covers an impressive variety of topics. Only 13 out of the 510 datasets in the collection actually had a code for the missing data. In many cases, the missing data problem has probably been "solved" in some way, usually without telling us how many missing values there were originally. It is impossible to track down the original data for most datasets in Hand's book. However, we can easily do this for dataset number 357, a list of scores of 34 athletes in 10 sport events at the 1988 Olympic decathlon in Seoul. The table itself is complete, but a quick search on the Internet revealed that initially 39 instead of 34 athletes participated. Five of them did not finish for various reasons, including the dramatic disqualification of the German favorite Jürgen Hingsen because of three false starts in the 100-meter sprint.

It is probably fair to assume that deletion occurred silently in many of the other datasets.

The inclination to delete the missing data is understandable. Apart from the technical difficulties imposed by the missing data, the occurrence of missing data has long been considered a sign of sloppy research. It is all too easy for a referee to write:

> This study is weak because of the large amount of missing data.

Publication chances are likely to improve if there is no hint of missingness. Orchard and Woodbury (1972, p. 697) remarked:

> Obviously the best way to treat missing data is not to have them.

Though there is a lot of truth in this statement, Orchard and Woodbury realized the impossibility of attaining this ideal in practice.

The prevailing scientific practice is to downplay the missing data. Reviews on reporting practices are available in various fields: clinical trials (Wood et al., 2004; Powney et al., 2014; Díaz-Ordaz et al., 2014; Akl et al., 2015), cancer research (Burton and Altman, 2004), educational research (Peugh and Enders, 2004), epidemiology (Klebanoff and Cole, 2008; Karahalios et al., 2012), developmental psychology (Jeličić et al., 2009), general medicine (Mackinnon, 2010), developmental pediatrics (Aylward et al., 2010), and otorhinolaryngology, head and neck surgery (Netten et al., 2017). The picture that emerges from these studies is quite consistent:

- The presence of missing data is often not explicitly stated in the text;

- Default methods, such as listwise deletion are used without mentioning them;

- Different tables are based on different sample sizes;

- Model-based missing data methods, such as direct likelihood, full information maximum likelihood and multiple imputation, are notably underutilized.

Helpful resources include the STROBE (Vandenbroucke et al., 2007) and CONSORT checklists and flow charts (Schulz et al., 2010). Gomes et al. (2016) showed cases where the subset of full patient-reported outcomes is a selective, leading to misleading results. Palmer et al. (2018) suggested a classification scheme for the reasons of nonresponse in patient-reported outcomes.

Missing data are there, whether we like it or not. In the social sciences, it is nearly inevitable that some respondents will refuse to participate or to answer certain questions. In medical studies, attrition of patients is very common. The theory, methodology and software for handling incomplete data problems have been vastly expanded and refined over the last decades. The major statistical analysis packages now have facilities for performing the appropriate analyses.

This book aims to contribute to a better understanding of the issues involved, and provides a methodology for dealing with incomplete data problems in practice.

1.2 Concepts of MCAR, MAR and MNAR

Before we review a number of simple fixes for the missing data in Section 1.3 let us take a short look at the terms MCAR, MAR and MNAR. A more detailed definition of these concepts will be given later in Section 2.2.3. Rubin (1976) classified missing data problems into three categories. In his theory every data point has some likelihood of being missing. The process that governs these probabilities is called the *missing data mechanism* or *response mechanism*. The model for the process is called the *missing data model* or *response model*.

If the probability of being missing is the same for all cases, then the data are said to be missing completely at random (MCAR). This effectively implies that causes of the missing data are unrelated to the data. We may consequently ignore many of the complexities that arise because data are missing, apart from the obvious loss of information. An example of MCAR is a weighing scale that ran out of batteries. Some of the data will be missing simply because of bad luck. Another example is when we take a random sample of a population, where each member has the same chance of being included in the sample. The (unobserved) data of members in the population that were not included in the sample are MCAR. While convenient, MCAR is often unrealistic for the data at hand.

If the probability of being missing is the same only within groups defined by the *observed* data, then the data are missing at random (MAR). MAR is a much broader class than MCAR. For example, when placed on a soft surface, a weighing scale may produce more missing values than when placed on a hard surface. Such data are thus not MCAR. If, however, we know surface type and if we can assume MCAR *within* the type of surface, then the data are MAR. Another example of MAR is when we take a sample from a population, where the probability to be included depends on some known property. MAR is more general and more realistic than MCAR. Modern missing data methods generally start from the MAR assumption.

If neither MCAR nor MAR holds, then we speak of missing not at random (MNAR). In the literature one can also find the term NMAR (not missing at random) for the same concept. MNAR means that the probability of being missing varies for reasons that are unknown to us. For example, the weighing scale mechanism may wear out over time, producing more missing data as time progresses, but we may fail to note this. If the heavier objects are measured later in time, then we obtain a distribution of the measurements that will be distorted. MNAR includes the possibility that the scale produces more missing values for the heavier objects (as above), a situation that might be difficult to recognize and handle. An example of MNAR in public opinion research occurs if those with weaker opinions respond less often. MNAR is the most complex case. Strategies to handle MNAR are to find more data about the causes for the missingness, or to perform what-if analyses to see how sensitive the results are under various scenarios.

Rubin's distinction is important for understanding why some methods will work, and others not. His theory lays down the conditions under which a missing data method can provide valid statistical inferences. Most simple fixes only work under the restrictive and often unrealistic MCAR assumption. If MCAR is implausible, such methods can provide biased estimates.

1.3 Ad-hoc solutions

1.3.1 Listwise deletion

Complete-case analysis (listwise deletion) is the default way of handling incomplete data in many statistical packages, including SPSS, SAS and Stata. The function na.omit() does the same in S-PLUS and R. The procedure eliminates all cases with one or more missing values on the analysis variables.

An important advantage of complete-case analysis is convenience. If the data are MCAR, listwise deletion produces unbiased estimates of means, variances and regression weights. Under MCAR, listwise deletion produces standard errors and significance levels that are correct for the reduced subset of data, but that are often larger relative to all available data.

A disadvantage of listwise deletion is that it is potentially wasteful. It is not uncommon in real life applications that more than half of the original sample is lost, especially if the number of variables is large. King et al. (2001) estimated that the percentage of incomplete records in the political sciences exceeded 50% on average, with some studies having over 90% incomplete records. It will be clear that a smaller subsample could seriously degrade the ability to detect the effects of interest.

If the data are not MCAR, listwise deletion can severely bias estimates of means, regression coefficients and correlations. Little and Rubin (2002, pp. 41–44) showed that the bias in the estimated mean increases with the difference between means of the observed and missing cases, and with the proportion of the missing data. Schafer and Graham (2002) reported an elegant simulation study that demonstrates the bias of listwise deletion under MAR and MNAR. However, complete-case analysis is not always bad. The implications of the missing data are different depending on where they occur (outcomes or predictors), and the parameter and model form of the complete-data model.

In the context of regression analysis, listwise deletion possesses some unique properties that make it attractive in particular settings. There are cases in which listwise deletion can provide better estimates than even the most sophisticated procedures. Since their discussion requires a bit more background than can be given here, we defer the treatment to Section 2.7.

Listwise deletion can introduce inconsistencies in reporting. Since listwise deletion is automatically applied to the active set of variables, different analyses on the same data are often based on different subsamples. In principle, it is possible to produce one global subsample using all active variables. In practice, this is unattractive since the global subsample will always have fewer cases than each of the local subsamples, so it is common to create different subsets for different tables. It will be evident that this complicates their comparison and generalization to the study population.

In some cases, listwise deletion can lead to nonsensical subsamples. For example, the rows in the `airquality` dataset used in Section 1.1.1 correspond to 154 consecutive days between May 1, 1973 and September 30, 1973. Deleting days affects the time basis. It would be much harder, if not impossible, to perform analyses that involve time, e.g., to identify weekly patterns or to fit autoregressive models that predict from previous days.

The opinions on the merits of listwise deletion vary. Miettinen (1985, p. 231) described listwise deletion as

> ...the only approach that assures that no bias is introduced under any circumstances...

a bold statement, but incorrect. At the other end of the spectrum we find Enders (2010, p. 39):

> In most situations, the disadvantages of listwise deletion far outweigh its advantages.

Schafer and Graham (2002, p. 156) cover the middle ground:

> If a missing data problem can be resolved by discarding only a small part of the sample, then the method can be quite effective.

The leading authors in the field are, however, wary of providing advice about the percentage of missing cases below which it is still acceptable to do listwise deletion. Little and Rubin (2002) argue that it is difficult to formulate rules of thumb since the consequences of using listwise deletion depend on more than the missing data rate alone. Vach (1994, p. 113) expressed his dislike for simplistic rules as follows:

> It is often supposed that there exists something like a critical missing rate up to which missing values are not too dangerous. The belief in such a global missing rate is rather stupid.

1.3.2 Pairwise deletion

Pairwise deletion, also known as *available-case analysis*, attempts to remedy the data loss problem of listwise deletion. The method calculates the means and (co)variances on all observed data. Thus, the mean of variable X is based on all cases with observed data on X, the mean of variable Y uses all cases with observed Y-values, and so on. For the correlation and covariance, all data are taken on which both X and Y have non-missing scores. Subsequently, the matrix of summary statistics are fed into a program for regression analysis, factor analysis or other modeling procedures.

SPSS, SAS and Stata contain many procedures with an option for pairwise deletion. In R we can calculate the means and correlations of the `airquality` data under pairwise deletion in R as:

```
data <- airquality[, c("Ozone", "Solar.R", "Wind")]
mu <- colMeans(data, na.rm = TRUE)
cv <- cov(data, use = "pairwise")
```

The standard `lm()` function does not take means and covariances as input, but the `lavaan` package (Rosseel, 2012) provides this feature:

```
library(lavaan)
fit <- lavaan("Ozone ~ 1 + Wind + Solar.R
               Ozone ~~ Ozone",
               sample.mean = mu, sample.cov = cv,
               sample.nobs = sum(complete.cases(data)))
```

The method is simple, and appears to use all available information. Under MCAR, it produces consistent estimates of mean, correlations and covariances (Little and Rubin, 2002, p. 55). The method has also some shortcomings. First, the estimates can be biased if the data are not MCAR. Further, the covariance and/or correlation matrix may not be positive definite, which is requirement for most multivariate procedures. Problems are generally more severe for highly correlated variables (Little, 1992). It is not clear what sample size should be used for calculating standard errors. Taking the average sample size yields standard errors that are too small(Little, 1992). Also, pairwise deletion requires numerical data that follow an approximate normal distribution, whereas in practice we often have variables of mixed types.

The idea to use all available information is good, but the proper analysis of the pairwise matrix requires sophisticated optimization techniques and special formulas to calculate the standard errors (Van Praag et al., 1985; Marsh, 1998), which somewhat defeats its utility. Pairwise deletion works best used if the data approximate a multivariate normal distribution, if the correlations between the variables are low, and if the assumption of MCAR is plausible. It is not recommended for other cases.

1.3.3   Mean imputation

A quick fix for the missing data is to replace them by the mean. We may use the mode for categorical data. Suppose we want to impute the mean in Ozone and Solar.R of the airquality data. SPSS, SAS and Stata have pre-built functions that substitute the mean. This book uses the R package mice (Van Buuren and Groothuis-Oudshoorn, 2011). This software is a contributed package that extends the functionality of R. Before mice can be used, it must be installed. An easy way to do this is to type:

```
install.packages("mice")
```

which searches the Comprehensive R Archive Network (CRAN), and installs the requested package on the local computer. After succesful installation, the mice package can be loaded by

```
library("mice")
```

Imputing the mean in each variable can now be done by

```
imp <- mice(airquality, method = "mean", m = 1,
            maxit = 1)
```

```
iter imp variable
 1   1  Ozone  Solar.R
```

The argument method = "mean" specifies mean imputation, the argument m = 1 requests a single imputed dataset, and maxit = 1 sets the number of iterations to 1 (no iteration). The latter two options can be left to their defaults with essentially the same result.

Mean imputation distorts the distribution in several ways. Figure 1.2 displays the distribution of Ozone after imputation. In the figure on the left, the red bar at the mean stands out. Imputing the mean here actually creates a bimodal distribution. The standard deviation in the imputed data is equal to 28.7, much smaller than from the observed data alone, which is 33. The figure on the right-hand side shows that the relation between Ozone and Solar.R is distorted because of the imputations. The correlation drops from 0.35 in the blue points to 0.3 in the combined data.

Mean imputation is a fast and simple fix for the missing data. However, it will underestimate the variance, disturb the relations between variables, bias almost any estimate other than the mean and bias the estimate of the mean when data are not MCAR. Mean imputation should perhaps only be used as a rapid fix when a handful of values are missing, and it should be avoided in general.

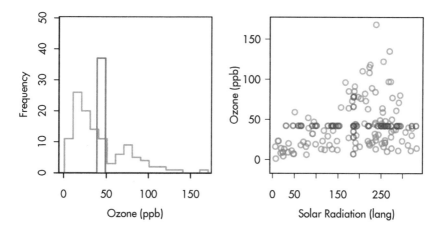

Figure 1.2: Mean imputation of Ozone. Blue indicates the observed data, red indicates the imputed values.

1.3.4 Regression imputation

Regression imputation incorporates knowledge of other variables with the idea of producing smarter imputations. The first step involves building a model from the observed data. Predictions for the incomplete cases are then calculated under the fitted model, and serve as replacements for the missing data. Suppose that we predict Ozone by linear regression from Solar.R.

```
fit <- lm(Ozone ~ Solar.R, data = airquality)
pred <- predict(fit, newdata = ic(airquality))
```

Another possibility for regression imputation uses mice:

```
data <- airquality[, c("Ozone", "Solar.R")]
imp <- mice(data, method = "norm.predict", seed = 1,
            m = 1, print = FALSE)
xyplot(imp, Ozone ~ Solar.R)
```

Figure 1.3 shows the result. The imputed values correspond to the most likely values under the model. However, the ensemble of imputed values vary less than the observed values. It may be that each of the individual points is the best under the model, but it is very unlikely that the real (but unobserved) values of Ozone would have had this distribution. Imputing predicted values also has an effect on the correlation. The red points have a correlation of 1 since they are located on a line. If the red and blue dots are combined, then the correlation increases from 0.35 to 0.39. Note that this upward bias grows with the percent missing ozone levels (here 24%).

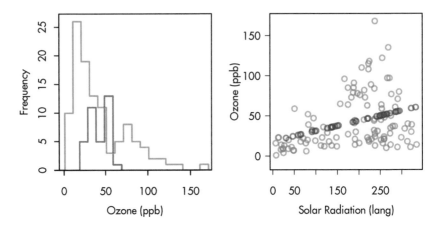

Figure 1.3: Regression imputation: Imputing Ozone from the regression line.

Regression imputation yields unbiased estimates of the means under MCAR, just like mean imputation, and of the regression weights of the imputation model if the explanatory variables are complete. Moreover, the regression weights are unbiased under MAR if the factors that influence the missingness are part of the regression model. In the example this corresponds to the situation where `Solar.R` would explain any differences in the probability that `Ozone` is missing. On the other hand, correlations are biased upwards, and the variability of the imputed data is systematically underestimated. The degree of underestimation depends on the explained variance and on the proportion of missing cases (Little and Rubin, 2002, p. 64).

Imputing predicted values can yield realistic imputations if the prediction is close to perfection. If so, the method reconstructs the missing parts from the available data. In essence, there was not really any information missing in the first place, it was only coded in a different form.

Regression imputation, as well as its modern incarnations in machine learning is probably the most dangerous of all methods described here. We may be led to believe that we're to do a good job by preserving the relations between the variables. In reality however, regression imputation artificially strengthens the relations in the data. Correlations are biased upwards. Variability is underestimated. Imputations are too good to be true. Regression imputation is a recipe for false positive and spurious relations.

1.3.5 Stochastic regression imputation

Stochastic regression imputation is a refinement of regression imputation attempts to address correlation bias by adding noise to the predictions. The following code imputes `Ozone` from `Solar.R` by stochastic regression imputation.

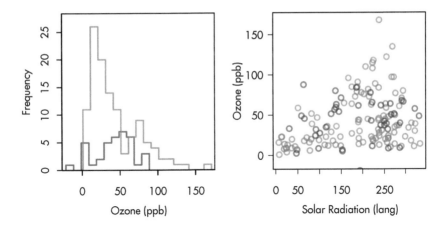

```
data <- airquality[, c("Ozone", "Solar.R")]
imp <- mice(data, method = "norm.nob", m = 1, maxit = 1,
            seed = 1, print = FALSE)
```

The `method = "norm.nob"` argument requests a plain, non-Bayesian, stochastic regression method. This method first estimates the intercept, slope and residual variance under the linear model, then calculates the predicted value for each missing value, and adds a random draw from the residual to the prediction. We will come back to the details in Section 3.2. The `seed` argument makes the solution reproducible. Figure 1.4 shows that the addition of noise to the predictions opens up the distribution of the imputed values, as intended.

Note that some new complexities arise. There is one imputation with a negative value. Such values need not be due to the draws from the residual distribution, but can also be a consequence of the use of a linear model for non-negative data. In fact, Figure 1.1 shows several negative predicted values in the observed data. Since negative `Ozone` concentrations do not exist in the real world, we cannot consider negative values as plausible imputations. Note also that the high end of the distribution is not well covered. The observed data form a cone, i.e., the data are heteroscedastic, but the imputation model assumes equal dispersion around the regression line. The variability of `Ozone` increases up to the solar radiation level of 250 langleys, and decreases after that. Though it is unclear whether this is a genuine meteorological phenomenon, the imputation model does not account for this feature.

Nevertheless, stochastic regression imputation represents a major conceptual advance. Some analysts may find it counterintuitive to "spoil" the best prediction by adding random noise, yet this is precisely what makes it suitable for imputation. A well-executed stochastic regression imputation preserves not

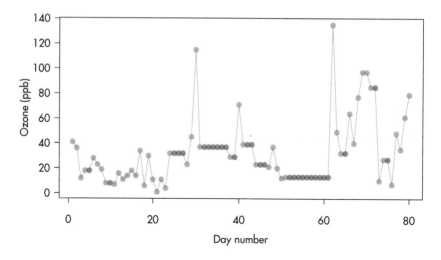

Figure 1.5: Imputation of Ozone by last observation carried forward (LOCF).

only the regression weights, but also the correlation between variables (cf. Exercise 3). The main idea to draw from the residuals is very powerful, and forms the basis of more advanced imputation techniques.

1.3.6 LOCF and BOCF

Last observation carried forward (LOCF) and baseline observation carried forward (BOCF) are ad-hoc imputation methods for longitudinal data. The idea is to take the previous observed value as a replacement for the missing data. When multiple values are missing in succession, the method searches for the last observed value.

The function `fill()` from the `tidyr` package applies LOCF by filling in the last known value. This is useful in situations where values are recorded only as they change, as in time-to-event data. For example, we may use LOCF to fill in `Ozone` by

```
airquality2 <- tidyr::fill(airquality, Ozone)
```

Figure 1.5 plots the results of the first 80 days of the `Ozone` series. The stretches of red dots indicate the imputations, and are constant within the same batch of missing ozone levels. The real, unseen values are likely to vary within these batches, so applying LOCF here gives implausible imputations.

LOCF is convenient because it generates a complete dataset. It can be applied with confidence in cases where we are certain what the missing values should be, for example, for administrative variables in longitudinal data. For outcomes, LOCF is dubious. The method has long been used in clinical trials. The U.S. Food and Drug Administration (FDA) has traditionally viewed

LOCF as the preferred method of analysis, considering it conservative and less prone to selection than listwise deletion. However, Molenberghs and Kenward (2007, pp. 47–50) show that the bias can operate in both directions, and that LOCF can yield biased estimates even under MCAR. LOCF needs to be followed by a proper statistical analysis method that distinguishes between the real and imputed data. This is typically not done however. Additional concerns about a reversal of the time direction are given in Kenward and Molenberghs (2009).

The Panel on Handling Missing Data in Clinical Trials recommends that LOCF and BOCF should not be used as the primary approach for handling missing data unless the assumptions that underlie them are scientifically justified (National Research Council, 2010, p. 77).

1.3.7 Indicator method

Suppose that we want to fit a regression, but there are missing values in one of the explanatory variables. The indicator method (Miettinen, 1985, p. 232) replaces each missing value by a zero and extends the regression model by the response indicator. The procedure is applied to each incomplete variable. The user analyzes the extended model instead of the original.

In R the indicator method can be coded as

```
imp <- mice(airquality, method = "mean", m = 1,
            maxit = 1, print = FALSE)
airquality2 <- cbind(complete(imp),
                r.Ozone = is.na(airquality[, "Ozone"]))
fit <- lm(Wind ~ Ozone + r.Ozone, data = airquality2)
```

Observe that since the missing data are in `Ozone` we needed to reverse the direction of the regression model.

The indicator method has been popular in public health and epidemiology. An advantage is that the indicator method retains the full dataset. Also, it allows for systematic differences between the observed and the unobserved data by inclusion of the response indicator, and could be more efficient. White and Thompson (2005) pointed out that the method can be useful to estimate the treatment effect in randomized trials when a baseline covariate is partially observed. If the missing data are restricted to the covariate, if the interest is solely restricted to estimation of the treatment effect, if compliance to the allocated treatment is perfect and if the model is linear without interactions, then using the indicator method for that covariate yields an unbiased estimate of the treatment effect. This is true even if the missingness depends on the covariate itself. Additional work can be found in Groenwold et al. (2012); Sullivan et al. (2018). It is not yet clear whether the coverage of the confidence interval around the treatment estimate will be satisfactory for multiple incomplete baseline covariates.

Table 1.1: Overview of assumptions made by ad-hoc methods.

	Mean	Unbiased Reg Weight	Correlation	Standard Error
Listwise	MCAR	MCAR	MCAR	Too large
Pairwise	MCAR	MCAR	MCAR	Complicated
Mean	MCAR	–	–	Too small
Regression	MAR	MAR[1]	–	Too small
Stochastic	MAR	MAR	MAR	Too small
LOCF	–	–	–	Too small
Indicator	–	–	–	Too small

The conditions under which the indicator method works may not be met in practice. For example, the method does not allow for missing data in the outcome, and generally fails in observational data. It has been shown that the method can yield severely biased regression estimates, even under MCAR and for low amounts of missing data (Vach and Blettner, 1991; Greenland and Finkle, 1995; Knol et al., 2010). The indicator method may have its uses in particular situations, but fails as a generic method to handle missing data.

1.3.8 Summary

Table 1.1 provides a summary of the methods discussed in this section. The table addresses two topics: whether the method yields the correct results on average (unbiasedness), and whether it produces the correct standard error. Unbiasedness is evaluated with respect to three types of estimates: the mean, the regression weight (with the incomplete variable as dependent) and the correlation.

The table identifies the assumptions on the missing data mechanism each method must make in order to produce unbiased estimates. The first line of the table should be read as follows:

1. *Listwise deletion* produces an unbiased estimate of the *mean* provided that the data are *MCAR*;

2. *Listwise deletion* produces an estimate of the standard error that is *too large*.

The interpretation of the other lines is similar. The "–" sign in some cells indicates that the method cannot produce unbiased estimates. Observe that both deletion methods require MCAR for all types. Regression imputation and stochastic regression imputation can yield unbiased estimates under MAR. In order to work, the model needs to be correctly specified. LOCF and the indicator method are incapable of providing consistent estimates, even under

---

[1]If the missing values occur in the outcome only. See Section 2.5.3 for another case.

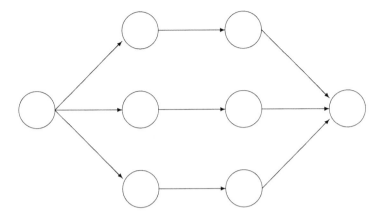

Incomplete data   Imputed data   Analysis results   Pooled result

Figure 1.6: Scheme of main steps in multiple imputation.

MCAR. Note that some special cases are not covered in Table 1.1. For example, listwise deletion is unbiased under two special MNAR scenarios (cf. Section 2.7).

Listwise deletion produces standard errors that are correct for the subset of complete cases, but in general too large for the entire dataset. Calculation of standard errors under pairwise deletion is complicated. The standard errors after single imputation are too small since the standard calculations make no distinction between the observed data and the imputed data. Correction factors for some situations have been developed (Schafer and Schenker, 2000), but a more convenient solution is multiple imputation.

1.4 Multiple imputation in a nutshell

1.4.1 Procedure

Multiple imputation creates $m > 1$ complete datasets. Each of these datasets is analyzed by standard analysis software. The m results are pooled into a final point estimate plus standard error by pooling rules ("Rubin's rules"). Figure 1.6 illustrates the three main steps in multiple imputation: imputation, analysis and pooling.

The analysis starts with observed, incomplete data. Multiple imputation creates several complete versions of the data by replacing the missing values by plausible data values. These plausible values are drawn from a distribution specifically modeled for each missing entry. Figure 1.6 portrays $m = 3$ imputed datasets. In practice, m is often taken larger (cf. Section 2.8). The number $m = 3$ is taken here just to make the point that the technique creates multiple versions of the imputed data. The three imputed datasets are identical for the observed data entries, but differ in the imputed values. The magnitude of these difference reflects our uncertainty about what value to impute.

The second step is to estimate the parameters of interest from each imputed dataset. This is typically done by applying the analytic method that we would have used had the data been complete. The results will differ because their input data differ. It is important to realize that these differences are caused only because of the uncertainty about what value to impute.

The last step is to pool the m parameter estimates into one estimate, and to estimate its variance. The variance combines the conventional sampling variance (within-imputation variance) and the extra variance caused by the missing data extra variance caused by the missing data (between-imputation variance). Under the appropriate conditions, the pooled estimates are unbiased and have the correct statistical properties.

1.4.2   Reasons to use multiple imputation

Multiple imputation (Rubin, 1987a, 1996) solves the problem of "too small" standard errors in Table 1.1. Multiple imputation is unique in the sense that it provides a mechanism for dealing with the inherent uncertainty of the imputations themselves.

Our level of confidence in a particular imputed value is expressed as the variation across the m completed datasets. For example, in a disability survey, suppose that the respondent answered the item whether he could walk, but did not provide an answer to the item whether he could get up from a chair. If the person can walk, then it is highly likely that the person will also be able to get up from the chair. Thus, for persons who can walk, we can draw a "yes" for missing "getting up from a chair" with a high probability, say 0.99, and use the drawn value as the imputed value. In the extreme, if we are really certain, we always impute the same value for that person. In general however, we are less confident about the true value. Suppose that, in a growth study, height is missing for a subject. If we only know that this person is a woman, this provides some information about likely values, but not so much. So the range of plausible values from which we draw is much larger here. The imputations for this woman will thus vary a lot over the different datasets. Multiple imputation is able to deal with both high-confidence and low-confidence situations equally well.

Another reason to use multiple imputation is that it separates the solution of the missing data problem from the solution of the complete-data problem. The missing-data problem is solved first, the complete-data problem next. Though these phases are not completely independent, the answer to the scientifically interesting question is not obscured anymore by the missing data. The ability to separate the two phases simplifies statistical modeling, and hence contributes to a better insight into the phenomenon of scientific study.

1.4.3 Example of multiple imputation

Continuing with the `airquality` dataset, it is straightforward to apply multiple imputation. The following code imputes the missing data twenty times, fits a linear regression model to predict `Ozone` in each of the imputed datasets, pools the twenty sets of estimated parameters, and calculates the Wald statistics for testing significance of the weights.

```
imp <- mice(airquality, seed = 1, m = 20, print = FALSE)
fit <- with(imp, lm(Ozone ~ Wind + Temp + Solar.R))
summary(pool(fit))
```

	estimate	std.error	statistic	df	p.value
(Intercept)	-62.7055	21.1973	-2.96	106.3	0.003755025718
Wind	-3.0839	0.6281	-4.91	91.7	0.000003024665
Temp	1.5988	0.2311	6.92	115.4	0.000000000271
Solar.R	0.0573	0.0217	2.64	112.8	0.009489765888

There is much more to say about each of these steps, but it shows that multiple imputation need not be a daunting task. Assuming we have set `options(na.action = na.omit)`, fitting the same model to the complete cases can be done by

```
fit <- lm(Ozone ~ Wind + Temp + Solar.R, data = airquality)
coef(summary(fit))
```

	Estimate	Std. Error	t value	Pr(>|t|)
(Intercept)	-64.3421	23.0547	-2.79	0.00622663809
Wind	-3.3336	0.6544	-5.09	0.00000151593
Temp	1.6521	0.2535	6.52	0.00000000242
Solar.R	0.0598	0.0232	2.58	0.01123663550

The solutions are nearly identical here, which is due to the fact that most missing values occur in the outcome variable. The standard errors of the multiple imputation solution are slightly smaller than in the complete-case analysis. Multiple imputation is often more efficient than complete-case analysis. Depending on the data and the model at hand, the differences can be dramatic.

Figure 1.7 shows the distribution and scattergram for the observed and imputed data combined. The imputations are taken from the first completed dataset. The blue and red distributions are quite similar. Problems with the negative values as in Figure 1.4 are now gone since the imputation method used observed data as donors to fill the missing data. Section 3.4 describes

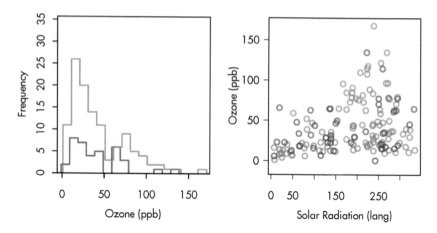

Figure 1.7: Multiple imputation of Ozone. Plotted are the imputed values from the first imputation.

the method in detail. Note that the red points respect the heteroscedastic nature of the relation between Ozone and Solar.R. All in all, the red points look as if they could have been measured if they had not been missing. The reader can easily recalculate the solution and inspect these plots for the other imputations.

Figure 1.8 plots the completed Ozone data. The imputed data of all five imputations are plotted for the days with missing Ozone scores. In order to avoid clutter, the lines that connect the dots are not drawn for the imputed values. Note that the pattern of imputed values varies substantially over the days. At the beginning of the series, the values are low and the spread is small, in particular for the cold and windy days 25–27. The small spread for days 25–27 indicates that the model is quite sure of these values. High imputed values are found around the hot and sunny days 35–42, whereas the imputations during the moderate days 52–61 are consistently in the moderate range. Note how the available information helps determine sensible imputed values that respect the relations between wind, temperature, sunshine and ozone.

One final point. The airquality data is a time series of 153 days. It is well known that the standard error of the ordinary least squares (OLS) estimate is inefficient (too large) if the residuals have positive serial correlation (Harvey, 1981). The first three autocorrelations of the Ozone are indeed large: 0.48, 0.31 and 0.29. The residual autocorrelations are however small and within the confidence interval: 0.13, −0.02 and 0.04. The inefficiency of OLS is thus negligible here.

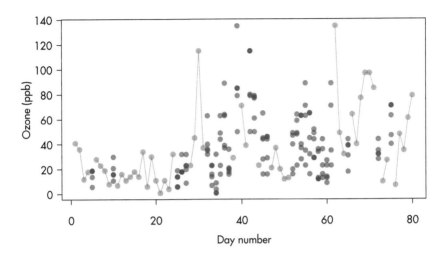

Figure 1.8: Multiple imputation of Ozone. Plotted are the observed values (in blue) and the multiply imputed values (in red).

1.5 Goal of the book

The main goal of this book is to add multiple imputation to the tool chest of practitioners. The text explains the ideas underlying multiple imputation, discusses when multiple imputation is useful, how to do it in practice and how to report the results of the steps taken.

The computations are done with the help of the R package mice, written by Karin Groothuis-Oudshoorn and myself (Van Buuren and Groothuis-Oudshoorn, 2011). The book thus also serves as an extended tutorial on the practical application of mice. Online materials that accompany the book can be found on www.multiple-imputation.com. My hope is that this hands-on approach will facilitate understanding of the key ideas in multiple imputation.

1.6 What the book does not cover

The field of missing data research is vast. This book focuses on multiple imputation. The book does not attempt cover the enormous body of literature on alternative approaches to incomplete data. This section briefly reviews three of these approaches.

1.6.1   Prevention

With the exception of McKnight et al. (2007, Chapter 4), books on missing data do not mention prevention. Yet, prevention of the missing data is the most direct attack on problems caused by the missing data. Prevention is fully in spirit with the quote of Orchard and Woodbury given on p. 7. There is a lot one could do to prevent missing data. The remainder of this section lists point-wise advice.

Minimize the use of intrusive measures, like blood samples. Visit the subject at home. Use incentives to stimulate response, and try to match up the interviewer and respondent on age and ethnicity. Adapt the mode of the study (telephone, face to face, web questionnaire, and so on) to the study population. Use a multi-mode design for different groups in your study. Quickly follow-up for people that do not respond, and where possible try to retrieve any missing data from other sources.

In experimental studies, try to minimize the treatment burden and intensity where possible. Prepare a well-thought-out flyer that explains the purpose and usefulness of your study. Try to organize data collection through an authority, e.g., the patient's own doctor. Conduct a pilot study to detect and smooth out any problems.

Economize on the number of variables collected. Only collect the information that is absolutely essential to your study. Use short forms of measurement instruments where possible. Eliminate vague or ambivalent questionnaire items. Use an attractive layout of the instruments. Refrain from using blocks of items that force the respondent to stay on a particular page for a long time. Use computerized adaptive testing where feasible. Do not allow other studies to piggy-back on your data collection efforts.

Do not overdo it. Many Internet questionnaires are annoying because they force the respondent to answer. Do not force your respondent. The result will be an apparently complete dataset with mediocre data. Respect the wish of your respondent to skip items. The end result will be more informative.

Use double coding in the data entry, and chase up any differences between the versions. Devise nonresponse forms in which you try to find out why people they did not respond, or why they dropped out.

Last but not least, consult experts. Many academic centers have departments that specialize in research methodology. Sound expert advice may turn out to be extremely valuable for keeping your missing data rate under control.

Most of this advice can be found in books on research methodology and data quality. Good books are Shadish et al. (2001), De Leeuw et al. (2008), Dillman et al. (2008) and Groves et al. (2009).

1.6.2   Weighting procedures

Weighting is a method to reduce bias when the probability to be selected in the survey differs between respondents. In sample surveys, the responders are weighted by design weights, which are inversely proportional to their probability of being selected in the survey. If there are missing data, the complete cases are re-weighted according to design weights that are adjusted to counter any selection effects produced by nonresponse. The method is widely used in official statistics. Relevant pointers include Cochran (1977) and Särndal et al. (1992) and Bethlehem (2002).

The method is relatively simple in that only one set of weights is needed for all incomplete variables. On the other hand, it discards data by listwise deletion, and it cannot handle partial response. Expressions for the variance of regression weights or correlations tend to be complex, or do not exist. The weights are estimated from the data, but are generally treated as fixed. The implications for this are unclear (Little and Rubin, 2002, p. 53).

There has been interest recently in improved weighting procedures that are "double robust" (Scharfstein et al., 1999; Bang and Robins, 2005). This estimation method requires specification of three models: Model A is the scientifically interesting model, Model B is the response model for the outcome, and model C is the joint model for the predictors and the outcome. The dual robustness property states that: if either Model B or Model C is wrong (but not both), the estimates under Model A are still consistent. This seems like a useful property, but the issue is not free of controversy (Kang and Schafer, 2007).

1.6.3 Likelihood-based approaches

Likelihood-based approaches define a model for the observed data. Since the model is specialized to the observed values, there is no need to impute missing data or to discard incomplete cases. The inferences are based on the likelihood or posterior distribution under the posited model. The parameters are estimated by maximum likelihood, the EM algorithm, the sweep operator, Newton–Raphson, Bayesian simulation and variants thereof. These methods are smart ways to skip over the missing data, and are known as direct likelihood, full information maximum likelihood (FIML), and more recently, pairwise likelihood estimation.

Likelihood-based methods are, in some sense, the "royal way" to treat missing data problems. The estimated parameters nicely summarize the available information under the assumed models for the complete data and the missing data. The model assumptions can be displayed and evaluated, and in many cases it is possible to estimate the standard error of the estimates.

Multiple imputation extends likelihood-based methods by adding an extra step in which imputed data values are drawn. An advantage of this is that it is generally easier to calculate the standard errors for a wider range of parameters. Moreover, the imputed values created by multiple imputation can be inspected and analyzed, which helps us to gauge the effect of the model assumptions on the inferences.

The likelihood-based approach receives an excellent treatment in the book by Little and Rubin (2002). A less technical account that should appeal to social scientists can be found in Enders (2010, chapters 3–5). Molenberghs and Kenward (2007) provide a hands-on approach of likelihood-based methods geared toward clinical studies, including extensions to data that are MNAR. The pairwise likelihood method was introduced by Katsikatsou et al. (2012) and has been implemented in `lavaan`.

1.7 Structure of the book

This book consists of three main parts: basics, case studies and extensions. Chapter 2 reviews the history of multiple imputation and introduces the notation and theory. Chapter 3 provides an overview of imputation methods for univariate missing data. Chapter 4 distinguishes three approaches to attack the problem of multivariate missing data. Chapter 5 reviews issues pertaining to the analysis of the imputed datasets.

Chapter 6 discusses practical issues for multivariate missing data. Chapter 7 discusses the problem how to impute for nested data so as to preserve the multilevel structure. Chapter 8 explores the use of multiple imputation to estimate individual causal effects.

Chapters 9–11 contain case studies of the techniques described in the previous chapters. Chapter 9 deals with "problems with the columns," while Chapter 10 addresses "problems with the rows". Chapter 11 discusses studies on problems with both rows and columns.

Chapter 12 concludes the main text with a discussion of limitations and pitfalls, reporting guidelines, alternative applications and future extensions.

1.8 Exercises

1. *Reporting practice.* What are the reporting practices in your field? Take a random sample of articles that have appeared during the last 10 years in the leading journal in your field. Select only those that present quantitative analyses, and address the following topics:

   (a) Did the authors report that there were missing data?

   (b) If not, can you infer from the text that there must have been missing data?

   (c) Did the authors discuss how they handled the missing data?

   (d) Were the missing data properly addressed?

(e) Can you detect a trend over time in reporting practice?

(f) Would the editors of the journal be interested in your findings?

2. *Loss of information.* Suppose that a dataset consists of 100 cases and 10 variables. Each variable contains 10% missing values. What is the largest possible subsample under listwise deletion? What is the smallest? If each variable is MCAR, how many cases will remain?

3. *Stochastic regression imputation.* The correlation of the data in Figure 1.4 is equal to 0.33. This is relatively low compared to the other correlations reported in Section 1.3. This seems to contradict the statement that stochastic regression imputation does not bias the correlation. Could this low correlation be due to random variation?

   (a) Rerun the code with a different `seed` value. What is the correlation now?

   (b) Write a loop to apply apply stochastic regression imputation with the `seed` increasing from 1 to 1000. Calculate the regression weight and the correlation for each solution, and plot the histogram. What are the mean, minimum and maximum values of the correlation?

   (c) Do your results indicate that stochastic regression imputation alters the correlation?

4. *Stochastic regression imputation (continued).* The largest correlation found in the previous exercise exceeds the value found in Section 1.3.4. This seems odd since the correlation of the imputed values under regression imputation is equal to 1, and hence the imputed data have a maximal contribution to the overall correlation.

   (a) Can you explain why this could happen?

   (b) Adapt the code from the previous exercise to test your explanation. Was your explanation satisfactory?

   (c) If not, can you think of another reason, and test that? Hint: Find out what is special about the solutions with the largest correlations.

5. *Nonlinear model.* The model fitted to the `airquality` data in Section 1.4.3 is a simple linear model. Inspection of the residuals reveals that there is a slight curvature in the average of the residuals.

   (a) Start from the completed cases, and use `plot(fit)` to obtain diagnostic plots. Can you explain why the curvature shows up?

   (b) Experiment with solutions, e.g., by transforming `Ozone` or by adding a quadratic term to the model. Can you make the curvature disappear? Does the amount of explained variance increase?

(c) Does the curvature also show up in the imputed data? If so, does the same solution work? Hint: You can assess the jth fitted model by `getfit(fit, j)`, where `fit` was created by `with(imp,...)`.

(d) Advanced: Do you think your solution would necessitate drawing new imputations?

Chapter 2

Multiple imputation

Imputing one value for a missing datum cannot be correct in general, because we don't know what value to impute with certainty (if we did, it wouldn't be missing).

Donald B. Rubin

2.1   Historic overview

2.1.1   Imputation

The English verb "to impute" comes from the Latin *imputo*, which means to reckon, attribute, make account of, charge, ascribe. In the Bible, the word "impute" is a translation of the Hebrew verb *hāshab*, which appears about 120 times in the Old Testament in various meanings (Renn, 2005). The noun "imputation" has a long history in taxation. The concept "imputed income" was used in the 19th century to denote income derived from property, such as land and housing. In the statistical literature, imputation means "filling in the data." Imputation in this sense is first mentioned in 1957 in the work of the U.S. Census Bureau (US Bureau of the Census, 1957).

Imputation is not alien to human nature. Yuval (2014) presented a world map, created in 1459 in Europe, that imputes fictitious continents in geographies that had yet to be discovered. One century later, the world map looked like a series of coastlines, with huge white spots for the inner lands, and these were all systematically explored during the later centuries. It's only when you can admit your own ignorance that you can start learning.

Allan and Wishart (1930) were the first to develop a statistical method to replace a missing value. They provided two formulae for estimating the value of a single missing observation, and advised filling in the estimate in the data. They would then proceed as usual, but deduct one degree of freedom to correct for the missing data. Yates (1933) generalized this work to more than one missing observation, and thus planted the seeds via a long and fruitful chain of intermediates that led up to the now classic EM algorithm (Dempster et al., 1977). Interestingly, the term "imputation" was not used by Dempster et al. or by any of their predecessors; it only gained widespread use after the monumental work of the Panel on Incomplete Data in 1983. Volume 2 devoted about 150 pages to an overview of the state-of-the-art of imputation technology (Madow et al., 1983). This work is not widely known, but it was the predecessor to the first edition of Little and Rubin (1987), a book that established the term firmly in the mainstream statistical literature.

2.1.2   Multiple imputation

Multiple imputation is now accepted as the best general method to deal with incomplete data in many fields, but this was not always the case. Multiple imputation was developed by Donald B. Rubin in the 1970's. It is useful to know a bit of its remarkable history, as some of the issues in multiple imputation may resurface in contemporary applications. This section details historical observations that provide the necessary background.

The birth of multiple imputation has been documented by Fritz Scheuren (Scheuren, 2005). Multiple imputation was developed as a solution to a practical problem with missing income data in the March Income Supplement to the Current Population Survey (CPS). In 1977, Scheuren was working on a joint project of the Social Security Administration and the U.S. Census Bureau. The Census Bureau was then using (and still does use) a *hot deck* imputation procedure. Scheuren signaled that the variance could not be properly calculated, and asked Rubin what might be done instead. Rubin came up with the idea of using multiple versions of the complete dataset, something he had already explored in the early 1970s (Rubin, 1994). The original 1977 report introducing the idea was published in 2004 in the history corner of the *American Statistician* (Rubin, 2004a). According to Scheuren: "The paper is the beginning point of a truly revolutionary change in our thinking on the topic of missingness" (Scheuren, 2004, p. 291).

Rubin observed that imputing *one* value (single imputation) for the missing value could not be correct in general. He needed a model to relate the unobserved data to the observed data, and noted that even for a given model the imputed values could not be calculated with certainty. His solution was simple and brilliant: create multiple imputations that reflect the uncertainty of the missing data. The 1977 report explains how to choose the models and how to derive the imputations. A low number of imputations, say five, would be enough.

The idea to create multiple versions of the data must have seemed outrageous at that time. Drawing imputations from a distribution, instead of estimating the "best" value, was a drastic departure from everything that had been done before. Rubin's original proposal did not include formulae for calculating combined estimates, but instead stressed the study of variation because of uncertainty in the imputed values. The idea was rooted in the Bayesian framework for inference, quite different from the dominant randomization-based framework in survey statistics. Moreover, there were practical issues involved in the technique, the larger datasets, the extra works to create the model and the repeated analysis, software issues, and so on. These issues have all been addressed by now, but in 1983 Dempster and Rubin wrote: "Practical implementation is still in the developmental state" (Dempster and Rubin, 1983, p. 8).

Rubin (1987a) provided the methodological and statistical footing for the method. Though several improvements have been made since 1987, the book was really ahead of its time and discusses the essentials of modern imputation technology. It provides the formulas needed to combine the repeated complete-data estimates (now called Rubin's rules), and outlines the conditions under which statistical inference under multiple imputation will be valid. Furthermore, pp. 166–170 provide a description of Bayesian sampling algorithms that could be used in practice.

Tests for combinations of parameters were developed by Li et al. (1991a), Li et al. (1991b) and Meng and Rubin (1992). Technical improvements for the degrees of freedom were suggested by Barnard and Rubin (1999) and Reiter (2007). Iterative algorithms for multivariate missing data with general missing data patterns were proposed by Rubin (1987, p. 192) Schafer (1997), Van Buuren et al. (1999), Raghunathan et al. (2001) and King et al. (2001).

In the 1990s, multiple imputation came under fire from various sides. The most severe criticism was voiced by Fay (1992). Fay pointed out that the validity of multiple imputation can depend on the form of subsequent analysis. He produced "counterexamples" in which multiple imputation systematically understated the true covariance, and concluded that "multiple imputation is inappropriate as a general purpose methodology." Meng (1994) pointed out that Fay's imputation models omitted important relations that were needed in the analysis model, an undesirable situation that he labeled *uncongenial*. Related issues on the interplay between the imputation model and the complete-data model have been discussed by Rubin (1996) and Schafer (2003).

Several authors have shown that Rubin's estimate of the variance can be biased (Wang and Robins, 1998; Robins and Wang, 2000; Nielsen, 2003; Kim et al., 2006). If there is bias, the estimate is usually too large. Rubin (2003) emphasized that variance estimation is only an intermediate goal for making confidence intervals, and generally not a parameter of substantive interest. He also noted that observed bias does not seem to affect the coverage of these intervals across a wide range of cases of practical interest.

The tide turned around 2005. Reviews started to appear that criticize insufficient reporting practice of the missing data in diverse fields (cf. Section 1.1.2). Nowadays multiple imputation is almost universally accepted, and in fact acts as the benchmark against which newer methods are being compared. The major statistical packages have all implemented modules for multiple imputation, so effectively the technology is implemented, almost three decades after Dempster and Rubin's remark.

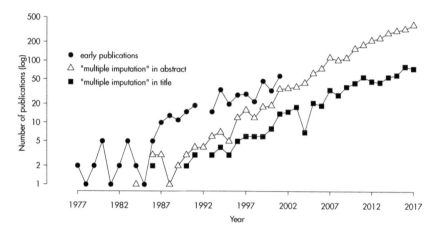

Figure 2.1: Multiple imputation at age 40. Number of publications (log) on multiple imputation during the period 1977–2017 according to three counting methods. Data source: https://www.scopus.com (accessed Jan 14, 2018).

2.1.3   The expanding literature on multiple imputation

Figure 2.1 contains three time series with counts on the number of publications on multiple imputation during the period 1977–2017. Counts were made in three ways. The rightmost series corresponds to the number of publications per year that featured the search term "multiple imputation" in the title. These are often methodological articles in which new adaptations are being developed. The series in the middle is the number of publication that featured "multiple imputation" in the title, abstract or key words in Scopus on the same search data. This set includes a growing group of papers that contain applications. The leftmost series is the number of publications in a collection of early publications available at http://www.multiple-imputation.com. This collection covers essentially everything related to multiple imputation from its inception in 1977 up to the year 2001. This group also includes chapters in books, dissertations, conference proceedings, technical reports and so on.

Note that the vertical axis is set in the logarithm. Perhaps the most interesting series is the middle series counting the applications. The pattern is approximately linear, meaning that the number of applications is growing at an exponential rate.

Several books devoted to missing data saw the light since the first edition of this book appeared in 2012. Building upon Schafer's work, Graham (2012) provides many insightful solutions for practical issues in imputation. Carpenter and Kenward (2013) propose methodological advances on important aspects of multiple imputation. Mallinckroth (2013) and O'Kelly and Ratitch (2014) concentrate on the missing data problem in clinical trials, Zhou et al. (2014) target health sciences, whereas Kim and Shao (2013) is geared towards official statistics. The *Handbook of Missing Data Methodology* (Molenberghs et al., 2015) presents a broad and up-to-date technical overview of the field of missing data. Raghunathan (2015) describes a variety of applications in social sciences and health using sequential regression multivariate imputation .

In addition to papers and books, high-quality software is now available to ease application of multiple imputation in practice. Rezvan et al. (2015) signal a wide adoption of multiple imputation, but warn that reporting is often substandard. Many more researchers have realized the full generality of the missing data problem. Effectively, missing data has now transformed into one of the great academic growth industries.

2.2   Concepts in incomplete data

2.2.1   Incomplete-data perspective

Many statistical techniques address some kind of incomplete-data problem. Suppose that we are interested in knowing the mean income Q in a given population. If we take a sample from the population, then the units not in the sample will have missing values because they will not be measured. It is not possible to calculate the population mean right away since the mean is undefined if one or more values are missing. The incomplete-data perspective is a conceptual framework for analyzing data as a missing data problem.

Estimating a mean from a population is a well known problem that can also be solved without a reference to missing data. It is nevertheless sometimes useful to think what we would have done had the data been complete, and what we could do to arrive at complete data. The incomplete-data perspective is general, and covers the sampling problem, the counterfactual model of causal inference, statistical modeling of the missing data, and statistical computation techniques. The books by Gelman et al. (2004, ch. 7) and Gelman and Meng (2004) provide in-depth discussions of the generality and richness of the incomplete data perspective. Little (2013) lists ten powerful ideas for the statistical scientist. His final advice reads as:

> My last simple idea is overarching: statistics is basically a missing data problem! Draw a picture of what's missing and find a good model to fill it in, along with a suitable (hopefully well calibrated) method to reflect uncertainty.

2.2.2   Causes of missing data

There is a broad distinction between two types of missing data: *intentional* and *unintentional* missing data. Intentional missing data are planned by the

Table 2.1: Examples of reasons for missingness for combinations of intentional/unintentional missing data with item/unit nonresponse.

	Intentional	Unintentional
Unit nonresponse	Sampling	Refusal
		Self-selection
Item nonresponse	Matrix sampling	Skip question
	Branching	Coding error

data collector. For example, the data of a unit can be missing because the unit was excluded from the sample. Another form of intentional missing data is the use of different versions of the same instrument for different subgroups, an approach known as matrix sampling. See Gonzalez and Eltinge (2007) or Graham (2012, Section 4) for an overview. Also, missing data that occur because of the routing in a questionnaire are intentional, as well as data (e.g., survival times) that are censored data at some time because the event (e.g., death) has not yet taken place. A related term in a multilevel context is systematically missing data. This term refers to variables that are missing for all individuals in a cluster because the variable was not measured in that cluster.(Resche-Rigon and White, 2018)

Though often foreseen, unintentional missing data are unplanned and not under the control of the data collector. Examples are: the respondent skipped an item, there was an error in the data transmission causing data to be missing, some of the objects dropped out before the study could be completed resulting in partially complete data, and the respondent was sampled but refused to cooperate. A related term in a multilevel context is sporadically missing data. This terms is used for variables with missing values for some but not all individuals in a cluster.

Another important distinction is *item nonresponse* versus *unit nonresponse*. Item nonresponse refers to the situation in which the respondent skipped one or more items in the survey. Unit nonresponse occurs if the respondent refused to participate, so all outcome data are missing for this respondent. Historically, the methods for item and unit nonresponse have been rather different, with unit nonresponse primarily addressed by weighting methods, and item nonresponse primarily addressed by edit and imputation techniques.

Table 2.1 cross-classifies both distinctions, and provides some typical examples in each of the four cells. The distinction between intentional/unintentional missing data is the more important one. The item/unit nonresponse distinction says *how much* information is missing, while the distinction between intentional and unintentional missing data says *why* some information is missing. Knowing the reasons why data are incomplete is a first step toward the solution.

2.2.3 Notation

The notation used in this book will be close to that of Rubin (1987a) and Schafer (1997), but there are some exceptions. The symbol m is used to indicate the number of multiple imputations. Compared to Rubin (1987a) the subscript m is dropped from most of the symbols. In Rubin (1987a), Y and R represent the data of the population, whereas in this book Y refers to data of the sample, similar to Schafer (1997). Rubin (1987a) uses X to represent the completely observed covariates in the population. Here we assume that the covariates are possibly part of Y, so there is not always a symbolic distinction between complete covariates and incomplete data. The symbol X is used to indicate the set of predictors in various types of models.

Let Y denote the $n \times p$ matrix containing the data values on p variables for all n units in the sample. We define the *response indicator* R as an $n \times p$ 0–1 matrix. The elements of Y and R are denoted by y_{ij} and r_{ij}, respectively, where $i = 1, \ldots, n$ and $j = 1, \ldots, p$. If y_{ij} is observed, then $r_{ij} = 1$, and if y_{ij} is missing, then $r_{ij} = 0$.

This book is restricted to the case where R is completely known, i.e., we know where the missing data are. This covers many applications of practical interest, but not all. For example, some questionnaires present a list of diseases and ask the respondent to place a "tick" at each disease that applies. If there is a "yes" we know that the field is not missing. However, if the field is not ticked, it could be because the person didn't have the disease (a genuine "no") or because the respondent skipped the question (a missing value). There is no way to tell the difference from the data, so these are *unknown unknowns*. In order to make progress in cases like these, we need additional assumptions about the response behavior.

The observed data are collectively denoted by Y_{obs}. The missing data are collectively denoted as Y_{mis}, and contain all elements y_{ij} where $r_{ij} = 0$. When taken together $Y = (Y_{\mathrm{obs}}, Y_{\mathrm{mis}})$ contain the hypothetically complete data. The part Y_{mis} has real values, but the values themselves are masked from us, where R indicates which values are masked. In their book, Little and Rubin (2002, p. 8) make the following key assumption:

> Missingness indicators hide the true values that are meaningful for analysis.

While this statement may seem obvious and uncomplicated, there are practical situations where it may not hold. In a trial where we are interested in both survival and quality of life, we may have missing values in either outcome. If we know that a person is alive, then an unknown quality of life outcome is simply missing because the quality of life score is defined for that person, but for some reason we haven't been able to see it. But if the person has died, quality of life becomes undefined, and that's the reason why we don't see it. It wouldn't make much sense to try to impute something that is undefined. A more sensible option is to stratify the analysis according to whether the concept is defined or not. The situation becomes more complex if we do not know the person's survival status. See Rubin (2000) for an analysis. In order to evade such complexities, we assume that Y contains values that are all defined, and that R indicates what we actually see.

If $Y = Y_{\text{obs}}$ (i.e., if the sample data are completely observed) and if we know the mechanism of how the sample was created, then it is possible to make a valid estimate of the population quantities of interest. For a simple random sample, we could just take the sample mean $\hat{Q}$ as an unbiased estimate of the population mean Q. We will assume throughout this book that we know how to do the correct statistical analysis on the complete data Y. If we cannot do this, then there is little hope that we can solve the more complex problem of analyzing Y_{obs}. This book addresses the problem of what to do if Y is observed incompletely. Incompleteness can incorporate intentional missing data, but also unintentional forms like refusals, self-selection, skipped questions, missed visits and so on.

Note that every unit in the sample has a row in Y. If no data have been obtained for a unit i (presumably because of unit nonresponse), the ith record will contain only the sample number and perhaps administrative data from the sampling frame. The remainder of the record will be missing.

A variable without any observed values is called a latent variable. Latent variables are often used to define concepts that are difficult to measure. Latent variables are theoretical constructs and not part of the manifest data, so they are typically not imputed. Mislevy (1991) showed how latent variable can be imputed, and provided several illustrative applications.

2.2.4   MCAR, MAR and MNAR again

Section 1.2 introduced MCAR, MAR and MNAR. This section provides more precise definitions.

The matrix R stores the locations of the missing data in Y. The distribution of R may depend on $Y = (Y_{\text{obs}}, Y_{\text{mis}})$, either by design or by happenstance, and this relation is described by the *missing data model*. Let ψ contain the parameters of the missing data model, then the general expression of the missing data model is $\Pr(R|Y_{\text{obs}}, Y_{\text{mis}}, \psi)$.

The data are said to be MCAR if

$$\Pr(R = 0|Y_{\text{obs}}, Y_{\text{mis}}, \psi) = \Pr(R = 0|\psi) \qquad (2.1)$$

so the probability of being missing depends only on some parameters ψ, the overall probability of being missing. The data are said to be MAR if

$$\Pr(R = 0|Y_{\text{obs}}, Y_{\text{mis}}, \psi) = \Pr(R = 0|Y_{\text{obs}}, \psi) \qquad (2.2)$$

so the missingness probability may depend on observed information, including any design factors. Finally, the data are MNAR if

$$\Pr(R = 0|Y_{\text{obs}}, Y_{\text{mis}}, \psi) \qquad (2.3)$$

does not simplify, so here the probability to be missing also depends on unobserved information, including Y_{mis} itself.

As explained in Chapter 1, simple techniques usually only work under MCAR, but this assumption is very restrictive and often unrealistic. Multiple imputation can handle both MAR and MNAR.

Several tests have been proposed to test MCAR versus MAR. These tests are not widely used, and their practical value is unclear. See Enders (2010, pp. 17–21) for an evaluation of two procedures. It is not possible to test MAR versus MNAR since the information that is needed for such a test is missing.

Numerical illustration. We simulate three archetypes of MCAR, MAR and MNAR. The data $Y = (Y_1, Y_2)$ are drawn from a standard bivariate normal distribution with a correlation between Y_1 and Y_2 equal to 0.5. Missing data are created in Y_2 using the missing data model

$$\Pr(R_2 = 0) = \psi_0 + \frac{e^{Y_1}}{1 + e^{Y_1}}\psi_1 + \frac{e^{Y_2}}{1 + e^{Y_2}}\psi_2 \qquad (2.4)$$

with different parameters settings for $\psi = (\psi_0, \psi_1, \psi_2)$. For MCAR we set $\psi_{\text{MCAR}} = (0.5, 0, 0)$, for MAR we set $\psi_{\text{MAR}} = (0, 1, 0)$ and for MNAR we set $\psi_{\text{MNAR}} = (0, 0, 1)$. Thus, we obtain the following models:

$$\text{MCAR} \quad : \quad \Pr(R_2 = 0) = 0.5 \qquad (2.5)$$
$$\text{MAR} \quad : \quad \text{logit}(\Pr(R_2 = 0)) = Y_1 \qquad (2.6)$$
$$\text{MNAR} \quad : \quad \text{logit}(\Pr(R_2 = 0)) = Y_2 \qquad (2.7)$$

where $\text{logit}(p) = \log(p/(1-p))$ for any $0 < p < 1$ is the logit function. In practice, it is more convenient to work with the inverse logit (or logistic) function inverse $\text{logit}^{-1}(x) = \exp(x)/(1 + \exp(x))$, which transforms a continuous x to the interval $\langle 0, 1 \rangle$. In R, it is straightforward to draw random values under these models as

```
logistic <- function(x) exp(x) / (1 + exp(x))
set.seed(80122)
n <- 300
y <- MASS::mvrnorm(n = n, mu = c(0, 0),
                   Sigma = matrix(c(1, 0.5, 0.5, 1), nrow = 2))
r2.mcar <- 1 - rbinom(n, 1, 0.5)
r2.mar  <- 1 - rbinom(n, 1, logistic(y[, 1]))
r2.mnar <- 1 - rbinom(n, 1, logistic(y[, 2]))
```

Figure 2.2 displays the distribution of Y_{obs} and Y_{mis} under the three missing data models. As expected, these are similar under MCAR, but become progressively more distinct as we move to the MNAR model.

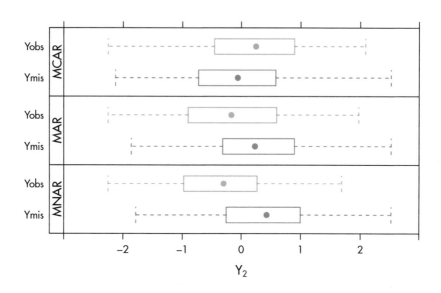

Figure 2.2: Distribution of Y_{obs} and Y_{mis} under three missing data models.

2.2.5   Ignorable and nonignorable♠

The example in the preceding section specified parameters ψ for three missing data models. The ψ-parameters have no intrinsic scientific value and are generally unknown. It would simplify the analysis if we could just ignore these parameters. The practical importance of the distinction between MCAR, MAR and MNAR is that it clarifies the conditions under which we can accurately estimate the scientifically interesting parameters without the need to know ψ.

The actually observed data consist of Y_{obs} and R. The joint density function $f(Y_{\mathrm{obs}}, R|\theta, \psi)$ of Y_{obs} and R together depends on parameters θ for the full data Y that are of scientific interest, and parameters ψ for the response indicator R that are seldom of interest. The joint density is proportional to the likelihood of θ and ψ, i.e.,

$$l(\theta, \psi|Y_{\mathrm{obs}}, R) \propto f(Y_{\mathrm{obs}}, R|\theta, \psi) \tag{2.8}$$

The question is: When can we determine θ without knowing ψ, or equivalently, the mechanism that created the missing data? The answer is given in Little and Rubin (2002, p. 119):

> The missing data mechanism is ignorable for likelihood inference if:
>
> 1. MAR: the missing data are missing at random; and

2. Distinctness: the parameters θ and ψ are distinct, in the sense that the joint parameter space of (ψ, θ) is the product of the parameter space of θ and the parameter space of ψ.

For valid Bayesian inference, the latter condition is slightly stricter: θ and ψ should be a priori independent: $p(\theta, \psi) = p(\theta)p(\psi)$ (Little and Rubin, 2002, p. 120). The MAR requirement is generally considered to be the more important condition. Schafer (1997, p. 11) says that in many situations the condition on the parameters is "intuitively reasonable, as knowing θ will provide little information about ψ and vice-versa." We should perhaps be careful in situations where the scientific interest focuses on the missing data process itself. For all practical purposes, the missing data model is said to be "ignorable" if MAR holds.

Note that the label "ignorable" does not mean that we can be entirely careless about the missing data. For inferences to be valid, we need to condition on those factors that influence the missing data rate. For example, in the MAR example of Section 2.2.4 the missingness in Y_2 depends on Y_1. A valid estimate of the mean of Y_2 cannot be made without Y_1, so we should include Y_1 somehow into the calculations for the mean of Y_2.

2.2.6 Implications of ignorability

The concept of ignorability plays an important role in the construction of imputation models. In imputation, we want to draw synthetic observations from the posterior distribution of the missing data, given the observed data and given the process that generated the missing data. The distribution is denoted as $P(Y_{\text{mis}}|Y_{\text{obs}}, R)$. If the nonresponse is ignorable, then this distribution does not depend on R (Rubin, 1987a, Result 2.3), i.e.,

$$P(Y_{\text{mis}}|Y_{\text{obs}}, R) = P(Y_{\text{mis}}|Y_{\text{obs}}) \qquad (2.9)$$

The implication is that

$$P(Y|Y_{\text{obs}}, R = 1) = P(Y|Y_{\text{obs}}, R = 0) \qquad (2.10)$$

so the distribution of the data Y is the same in the response and nonresponse groups. Thus, if the missing data model is ignorable we can model the posterior distribution $P(Y|Y_{\text{obs}}, R = 1)$ from the observed data, and use this model to create imputations for the missing data. Vice versa, techniques that (implicitly) assume equivalent distributions assume ignorability and thus MAR. On the other hand, if the nonresponse is nonignorable, we find

$$P(Y|Y_{\text{obs}}, R = 1) \neq P(Y|Y_{\text{obs}}, R = 0) \qquad (2.11)$$

so then we should incorporate R into the model to create imputations.

The assumption of ignorability is often sensible in practice, and generally provides a natural starting point. If, on the other hand, the assumption is not reasonable (e.g., when data are censored), we may specify $P(Y|Y_{\text{obs}}, R = 0)$ different from $P(Y|Y_{\text{obs}}, R = 1)$. The specification of $P(Y|Y_{\text{obs}}, R = 0)$ needs assumptions external to the data since, by definition, the information needed to estimate any regression weights for R is missing.

Example. Suppose that a growth study measures body weight in kg (Y_2) and gender (Y_1: 1 = boy, 0 = girl) of 15-year-old children, and that some of the body weights are missing. We can model the weight distribution for boys and girls separately for those with observed weights, i.e., $P(Y_2|Y_1 = 1, R_2 = 1)$ and $P(Y_2|Y_1 = 0, R_2 = 1)$. If we assume that the response mechanism is ignorable, then imputations for a boy's weight can be drawn from $P(Y_2|Y_1 = 1, R_2 = 1)$ since it will equal $P(Y_2|Y_1 = 1, R_2 = 0)$. The same can be done for the girls. This procedure leads to correct inferences on the combined sample of boys and girls, even if boys have substantially more missing values, or if the body weights of the boys and girls are very different.

The procedure outlined above is not appropriate if, within the boys or the girls, the occurrence of the missing data is related to body weight. For example, some of the heavier children may not want to be weighed, resulting in more missing values for the obese. It will be clear that assuming $P(Y_2|Y_1, R_2 = 0) = P(Y_2|Y_1, R_2 = 1)$ will underestimate the prevalence of overweight and obesity. In this case, it may be more realistic to specify $P(Y_2|Y_1, R_2 = 0)$ such that imputation accounts for the excess body weights in the children that were not weighed. There are many ways to do this. In all these cases the response mechanism will be nonignorable.

The assumption of ignorability is essentially the belief on the part of the user that the available data are sufficient to correct for the effects of the missing data. The assumption cannot be tested on the data itself, but it can be checked against suitable external validation data.

There are two main strategies that we may pursue if the response mechanism is not ignorable. The first is to expand the data, and assume ignorability on the expanded data (Collins et al., 2001). See also Section 6.2 for more details. In the above example, overweight children may simply not want anybody to know their weight, but perhaps have no objection if their waist circumference Y_3 is measured. As Y_3 predicts Y_2, R_2 or both, the ignorability assumption $P(Y_2|Y_1, Y_3, R_2 = 0) = P(Y_2|Y_1, Y_3, R_2 = 1)$ is less stringent, and hence more realistic.

The second strategy is to formulate the model for $P(Y_2|Y_1, R_2 = 0)$ different from $P(Y_2|Y_1, R_2 = 1)$, describing which body weights would have been observed if they had been measured. Such a model could simply add some extra kilos to the imputed values, but of course we need to be able to justify our choice in light of what we know about the data. See Section 3.8.1 for a more detailed discussion of the idea. In general, the formulation of nonignorable models should be driven by knowledge about the process that created the missing data. Any such methods need to be explained and justified as part of the statistical analysis.

2.3 Why and when multiple imputation works

2.3.1 Goal of multiple imputation

A *scientific estimand* Q is a quantity of scientific interest that we can calculate if we would observe the entire population. For example, we could be interested in the mean income of the population. In general, Q can be expressed as a known function of the population data. If we are interested in more than one quantity, Q will be a vector. Note that Q is a property of the population, so it does not depend on any design characteristics. Examples of scientific estimands include the population mean, the population (co)variance or correlation, and population factor loadings and regression coefficients, as well as these quantities calculated within known strata of the population. Examples of quantities that are not scientific estimands are sample means, standard errors and test statistics.

We can only calculate Q if the population data are fully known, but this is almost never the case. The goal of multiple imputation is to find an *estimate* $\hat{Q}$ that is *unbiased* and *confidence valid* (Rubin, 1996). We explain these concepts below.

Unbiasedness means that the average $\hat{Q}$ over all possible samples Y from the population is equal to Q. The formula is

$$E(\hat{Q}|Y) = Q \qquad (2.12)$$

The explanation of confidence validity requires some additional symbols. Let U be the estimated variance-covariance matrix of $\hat{Q}$. This estimate is *confidence valid* if the average of U over all possible samples is equal or larger than the variance of $\hat{Q}$. The formula is

$$E(U|Y) \geq V(\hat{Q}|Y) \qquad (2.13)$$

where the function $V(\hat{Q}|Y)$ denotes the variance caused by the sampling process. A statistical test with a stated nominal rejection rate of 5% should reject the null hypothesis in at most 5% of the cases when in fact the null hypothesis is true. A procedure is said to be confidence valid if this holds.

In summary, the goal of multiple imputation is to obtain estimates of the scientific estimand in the population. This estimate should on average be equal to the value of the population parameter. Moreover, the associated confidence intervals and hypothesis tests should achieve at least the stated nominal value.

2.3.2 Three sources of variation♠

The actual value of Q is unknown if some of the population data are unknown. Suppose we make an estimate $\hat{Q}$ of Q. The amount of uncertainty in $\hat{Q}$ about the true population value Q depends on what we know about

Y_{mis}. If we would be able to re-create Y_{mis} perfectly, then we can calculate Q with certainty. However, such perfect re-creation is almost never achievable. In other cases, we need to summarize the distribution of Q under varying Y_{mis}.

The possible values of Q given our knowledge of the data Y_{obs} are captured by the posterior distribution $P(Q|Y_{\text{obs}})$. In itself, $P(Q|Y_{\text{obs}})$ is often intractable, but it can be decomposed into two parts that are easier to solve as follows:

$$P(Q|Y_{\text{obs}}) = \int P(Q|Y_{\text{obs}}, Y_{\text{mis}})P(Y_{\text{mis}}|Y_{\text{obs}})dY_{\text{mis}} \qquad (2.14)$$

Here, $P(Q|Y_{\text{obs}})$ is the posterior distribution of Q given the observed data Y_{obs}. This is the distribution that we would like to know. $P(Q|Y_{\text{obs}}, Y_{\text{mis}})$ is the posterior distribution of Q in the hypothetically complete data, and $P(Y_{\text{mis}}|Y_{\text{obs}})$ is the posterior distribution of the missing data given the observed data.

The interpretation of Equation 2.14 is most conveniently done from right to left. Suppose that we use $P(Y_{\text{mis}}|Y_{\text{obs}})$ to draw imputations for Y_{mis}, denoted as $\dot{Y}_{\text{mis}}$. We can then use $P(Q|Y_{\text{obs}}, \dot{Y}_{\text{mis}})$ to calculate the quantity of interest Q from the imputed data $(Y_{\text{obs}}, \dot{Y}_{\text{mis}})$. We repeat these two steps with new draws $\dot{Y}_{\text{mis}}$, and so on. Equation 2.14 says that the actual posterior distribution of Q is equal to the average over the repeated draws of Q. This result is important since it expresses $P(Q|Y_{\text{obs}})$, which is generally difficult, as a combination of two simpler posteriors from which draws can be made.

It can be shown that the posterior mean of $P(Q|Y_{\text{obs}})$ is equal to

$$E(Q|Y_{\text{obs}}) = E(E[Q|Y_{\text{obs}}, Y_{\text{mis}}]|Y_{\text{obs}}) \qquad (2.15)$$

the average of the posterior means of Q over the repeatedly imputed data. This equation suggests the following procedure for combining the results of repeated imputations. Suppose that $\hat{Q}_l$ is the estimate of the ℓth repeated imputation, then the combined estimate is equal to

$$\bar{Q} = \frac{1}{m} \sum_{\ell=1}^{m} \hat{Q}_\ell \qquad (2.16)$$

where $\hat{Q}_\ell$ contains k parameters and is represented as a $k \times 1$ column vector.

The posterior variance of $P(Q|Y_{\text{obs}})$ is the sum of two variance components:

$$V(Q|Y_{\text{obs}}) = E[V(Q|Y_{\text{obs}}, Y_{\text{mis}})|Y_{\text{obs}}] + V[E(Q|Y_{\text{obs}}, Y_{\text{mis}})|Y_{\text{obs}}] \qquad (2.17)$$

This equation is well known in statistics, but can be difficult to grasp at first. The first component is the average of the repeated complete-data posterior variances of Q. This is called the within-variance. The second component is the variance between the complete-data posterior means of Q. This is called the between variance. Let $\bar{U}_\infty$ and B_∞ denote the estimated within and between components for an infinitely large number of imputations $m = \infty$. Then $T_\infty = \bar{U}_\infty + B_\infty$ is the posterior variance of Q.

Equation 2.17 suggests the following procedure to estimate T_∞ for finite m. We calculate the average of the complete-data variances as

$$\bar{U} = \frac{1}{m} \sum_{\ell=1}^{m} \bar{U}_\ell \qquad (2.18)$$

where the term $\bar{U}_\ell$ is the variance-covariance matrix of $\hat{Q}_\ell$ obtained for the ℓth imputation. The standard unbiased estimate of the variance between the m complete-data estimates is given by

$$B = \frac{1}{m-1} \sum_{\ell=1}^{m} (\hat{Q}_\ell - \bar{Q})(\hat{Q}_\ell - \bar{Q})' \qquad (2.19)$$

where $\bar{Q}$ is calculated by Equation 2.16.

It is tempting to conclude that the total variance T is equal to the sum of $\bar{U}$ and B, but that would be incorrect. We need to incorporate the fact that $\bar{Q}$ itself is estimated using finite m, and thus only approximates $\bar{Q}_\infty$. Rubin (1987a, eq. 3.3.5) shows that the contribution to the variance of this factor is systematic and equal to B_∞/m. Since B approximates B_∞, we may write

$$
\begin{aligned}
T &= \bar{U} + B + B/m \\
&= \bar{U} + \left(1 + \frac{1}{m}\right) B \qquad (2.20)
\end{aligned}
$$

for the total variance of $\bar{Q}$, and hence of $(Q-\bar{Q})$ if $\bar{Q}$ is unbiased. The procedure to combine the repeated-imputation results by Equations 2.16 and 2.20 is referred to as Rubin's rules.

In summary, the total variance T stems from three sources:

1. $\bar{U}$, the variance caused by the fact that we are taking a sample rather than observing the entire population. This is the conventional statistical measure of variability;

2. B, the extra variance caused by the fact that there are missing values in the sample;

3. B/m, the extra simulation variance caused by the fact that $\bar{Q}$ itself is estimated for finite m.

The addition of the latter term is critical to make multiple imputation work at low values of m. Not including it would result in p-values that are too low, or confidence intervals that are too short. Traditional choices for m are $m = 3$, $m = 5$ and $m = 10$. The current advice is to set m higher, e.g., $m = 50$ (cf. Section 2.8). The larger m gets, the smaller the effect of simulation error on the total variance.

Table 2.2: Role of symbols at three analytic levels and the relations between them. The relation $\Longrightarrow$ means "is an estimate of." The relation $\doteq$ means "is asymptotically equal to."

Incomplete Sample Y_{obs}		Complete Sample $Y = (Y_{\text{obs}}, Y_{\text{mis}})$		Population
$\bar{Q}$	$\Longrightarrow$	$\hat{Q}$	$\Longrightarrow$	Q
$\bar{U}$	$\Longrightarrow$	$U \doteq V(\hat{Q})$		
$B \doteq V(\bar{Q})$				

Steel et al. (2010) investigated alternatives for obtaining estimates of T using mixtures of normals. Under multivariate normality and for low m, these methods yield slightly more efficient estimates of T. The behavior of these methods is not known when normality is violated. Since application of the procedure is more complex than Rubin's rules, it is used sparingly.

2.3.3   Proper imputation

In order to yield valid statistical inferences, the imputed values should possess certain characteristics. Procedures that yield such imputations are called *proper* (Rubin, 1987a, pp. 118–128). Section 2.3.1 described two conditions needed for a valid estimate of Q. These requirements apply simultaneously to both the sampling and the nonresponse model. An analogous set of requirements exists if we zoom in on procedures that deal exclusively with response model. The important theoretical result is: If the imputation method is proper and if the complete-data model is valid in the sense of Section 2.3.1, the whole procedure is valid (Rubin, 1987a, p. 119).

Recall from Section 2.3.1 that the goal of multiple imputation is to find an estimate $\hat{Q}$ of Q with correct statistical properties. At the level of the sample, there is uncertainty about Q. This uncertainty is captured by U, the estimated variance-covariance of $\hat{Q}$ in the sample. If we have no missing data in the sample, the pair $(\hat{Q}, U)$ contains everything we know about Q.

If we have incomplete data, we can distinguish three analytic levels: the population, the sample and the incomplete sample. The problem of estimating Q in the population by $\hat{Q}$ from the sample is a traditional statistical problem. The key idea of the solution is to accompany $\hat{Q}$ by an estimate of its variability under repeated sampling U according to the sampling model.

Now suppose that we want to go from the incomplete sample to the complete sample. At the sample level we can distinguish two estimands, instead of one: $\hat{Q}$ and U. Thus, the role of the single estimand Q at the population level is taken over by the estimand pair $(\hat{Q}, U)$ at the sample level. Table 2.2 provides an overview of the three different analytic levels involved, the quantities defined at each level and their relations. Note that $\hat{Q}$ is both an estimate (of Q) as well as an estimand (of $\bar{Q}$). Also, U has two roles.

Imputation is the act of converting an incomplete sample into a complete sample. Imputation of data should, at the very least, lead to adequate estimates of both $\hat{Q}$ and U. Three conditions define whether an imputation procedure is considered proper. We use the slightly simplified version given by Brand (1999, p. 89) combined with Rubin (1987a). An imputation procedure is said to be *confidence proper* for complete-data statistics $(\hat{Q}, U)$ if at large m all of the following conditions hold approximately:

$$E(\bar{Q}|Y) = \hat{Q} \qquad (2.21)$$

$$E(\bar{U}|Y) = U \qquad (2.22)$$

$$\left(1 + \frac{1}{m}\right) E(B|Y) \geq V(\bar{Q}) \qquad (2.23)$$

The hypothetically complete sample data Y is now held fixed, and the response indicator R varies according to a specified model.

The first requirement is that $\bar{Q}$ is an unbiased estimate of $\hat{Q}$. This means that, when averaged over the response indicators R sampled under the assumed response model, the multiple imputation estimate $\bar{Q}$ is equal to $\hat{Q}$, the estimate calculated from the hypothetically complete data in the realized sample.

The second requirement is that $\bar{U}$ is an unbiased estimate of U. This means that, when averaged over the response indicator R sampled under the assumed response model, the estimate $\bar{U}$ of the sampling variance of $\hat{Q}$ is equal to U, the sampling variance estimate calculated from the hypothetically complete data in the realized sample.

The third requirement is that B is a confidence valid estimate of the variance due to missing data. Equation 2.23 implies that the extra inferential uncertainty about $\hat{Q}$ due to missing data is correctly reflected. On average, the estimate B of the variance due to missing data should be equal to $V(\bar{Q})$, the variance observed in the multiple imputation estimator $\bar{Q}$ over different realizations of the response mechanism. This requirement is analogous to Equation 2.13 for confidence valid estimates of U.

If we replace $\geq$ in Equation 2.23 by $>$, then the procedure is said to be *proper*, a stricter version. In practice, being confidence proper is enough to obtain valid inferences.

Note a procedure may be proper for the estimand pair $(\hat{Q}, U)$, while being improper for another pair $(\hat{Q}', U')$. Also, a procedure may be proper with respect to one response mechanism $P(R)$, but improper for an alternative mechanism $P(R')$.

It is not always easy to check whether a certain procedure is proper. Section 2.5 describes simulation-based tools for checking the adequacy of imputations for valid statistical inference. Chapter 3 provides examples of proper and improper procedures.

2.3.4   Scope of the imputation model

Imputation models vary in their scope. Models with a narrow scope are proper with respect to specific estimand $(\hat{Q}, U)$ and particular response mechanism, e.g., a particular proportion of nonresponse. Models with a broad scope are proper with respect to a wide range of estimates $\hat{Q}$, e.g., subgroup means, correlations, ratios and so on, and under a large variety of response mechanisms.

The scope is related to the setting in which the data are collected. The following list distinguishes three typical situations:

- *Broad.* Create one set of imputations to be used for all projects and analyses. A broad scope is appropriate for publicly released data, cohort data and registers, where different people use the data for different purposes.

- *Intermediate.* Create one set of imputations per project and use this set for all analyses. An intermediate scope is appropriate for analyses that estimate relatively similar quantities. The imputer and analyst can be different persons.

- *Narrow.* A separate imputed dataset is created for each analysis. The imputer and analyst are typically the same person. A narrow scope is appropriate if the imputed data are used only to estimate the same quantity. Different analyses require different imputations.

In general, imputations created under a broad scope can be applied more widely, and preferable for that reason. On the other hand, if we have a strong scientific model for the data, or if the parameters of interest have high-stakes consequences then using a narrow scope is better because the imputation model can be informed by the complete-data model, thus making sure that all interactions, non-linearities and distributional details are adequately met. In practice the correct model is often unknown. Therefore the techniques discussed in this book will emphasize imputations for the broader scope. Whatever is chosen, it is the responsibility of the imputer to indicate the scope of the generated imputations.

2.3.5   Variance ratios♠

For scalar Q, the ratio

$$\lambda = \frac{B + B/m}{T} \tag{2.24}$$

can be interpreted as the proportion of the variation attributable to the missing data. It is equal to zero if the missing data do not add extra variation to the sampling variance, an exceptional situation that can occur only if we can perfectly re-create the missing data. The maximum value is equal to 1, which occurs only if all variation is caused by the missing data. This is equally unlikely to occur in practice since it means that there is no information at all. If λ is high, say $\lambda > 0.5$, the influence of the imputation model on the final result is larger than that of the complete-data model.

The ratio

$$r = \frac{B + B/m}{\bar{U}} \tag{2.25}$$

is called the *relative increase in variance due to nonresponse* (Rubin, 1987a, eq. 3.1.7). The quantity is related to λ by $r = \lambda/(1 - \lambda)$.

Another related measure is the *fraction of information about Q missing due to nonresponse* (Rubin, 1987a, eq. 3.1.10). This measure is defined by

$$\gamma = \frac{r + 2/(\nu + 3)}{1 + r} \tag{2.26}$$

This measure needs an estimate of the degrees of freedom ν, and will be discussed in Section 2.3.6. The interpretations of γ and λ are similar, but γ is adjusted for the finite number of imputations. Both statistics are related by

$$\gamma = \frac{\nu + 1}{\nu + 3}\lambda + \frac{2}{\nu + 3} \tag{2.27}$$

The literature often confuses γ and λ, and erroneously labels λ as the fraction of missing information. The values of λ and γ are almost identical for large ν, but they could notably differ for low ν.

If Q is a vector, it is sometimes useful to calculate a compromise λ over all elements in $\bar{Q}$ as

$$\bar{\lambda} = \left(1 + \frac{1}{m}\right) \operatorname{tr}(BT^{-1})/k \tag{2.28}$$

where k is the dimension of $\bar{Q}$, and where B and T are now $k \times k$ matrices. The compromise expression for r is equal to

$$\bar{r} = \left(1 + \frac{1}{m}\right) \operatorname{tr}(B\bar{U}^{-1})/k \tag{2.29}$$

the average relative increase in variance.

The quantities λ, r and γ as well as their multivariate analogues $\bar{\lambda}$ and $\bar{r}$ are indicators of the severity of the missing data problem. Fractions of missing information up to 0.2 can be interpreted as "modest," 0.3 as "moderately large" and 0.5 as "high" (Li et al., 1991b). High values indicate a difficult problem in which the final statistical inferences are highly dependent on the way in which the missing data were handled. Note that estimates of λ, r and γ may be quite variable for low m (cf. Section 2.8).

2.3.6 Degrees of freedom♠

The degrees of freedom is the number of observations after accounting for the number of parameters in de model. The calculation of the degrees of freedom cannot be the same as for the complete data because part of the data is missing. The "old" formula (Rubin, 1987a, eq. 3.1.6) for the degrees of freedom can be written concisely as

$$
\begin{aligned}
\nu_{\text{old}} &= (m-1)\left(1+\frac{1}{r^2}\right) \\
&= \frac{m-1}{\lambda^2}
\end{aligned}
\tag{2.30}
$$

with r and λ defined as in Section 2.3.5. The lowest possible value is $\nu_{\text{old}} = m-1$, which occurs if essentially all variation is attributable to the nonresponse. The highest value $\nu_{\text{old}} = \infty$ indicates that all variation is sampling variation, either because there were no missing data, or because we could re-create them perfectly.

Barnard and Rubin (1999) noted that Equation 2.30 can produce values that are larger than the sample size in the complete data, a situation that is "clearly inappropriate." They developed an adapted version for small samples that is free of the problem. Let ν_{com} be the degrees of freedom of $\bar{Q}$ in the hypothetically complete data. In models that fit k parameters on data with a sample size of n we may set $\nu_{\text{com}} = n - k$. The estimated observed data degrees of freedom that accounts for the missing information is

$$
\nu_{\text{obs}} = \frac{\nu_{\text{com}} + 1}{\nu_{\text{com}} + 3}\nu_{\text{com}}(1 - \lambda)
\tag{2.31}
$$

The adjusted degrees of freedom to be used for testing in multiple imputation can be written concisely as

$$
\nu = \frac{\nu_{\text{old}}\nu_{\text{obs}}}{\nu_{\text{old}} + \nu_{\text{obs}}}
\tag{2.32}
$$

The quantity ν is always less than or equal to ν_{com}. If $\nu_{\text{com}} = \infty$, then Equation 2.32 reduces to 2.30. If $\lambda = 0$ then $\nu = \nu_{\text{com}}$, and if $\lambda = 1$ we find $\nu = 0$. Distributions with zero degrees of freedom are nonsensical, so for $\nu < 1$ we should refrain from any testing due to lack of information.

Alternative corrections were proposed by Reiter (2007) and Lipsitz et al. (2002). Wagstaff and Harel (2011) compared the four methods, and concluded that the sample-sample methods by Barnard-Rubin and Reiter performed satisfactory.

2.3.7   Numerical example

Many quantities introduced in the previous sections can be obtained by the pool() function in mice. The following code imputes the nhanes dataset, fits a simple linear model and pools the results:

```
library(mice)
imp <- mice(nhanes, print = FALSE, m = 10, seed = 24415)
fit <- with(imp, lm(bmi ~ age))
est <- pool(fit)
est
```

```
Class: mipo    m = 10
              estimate  ubar      b      t dfcom    df   riv lambda
(Intercept)      30.50 3.408  1.454  5.01      23  12.4 0.469  0.319
age              -2.13 0.906  0.238  1.17      23  15.1 0.289  0.224
                  fmi
(Intercept)     0.408
age             0.310
```

The column est imate is the value of $\bar{Q}$ as defined in Equation 2.16. Columns ubar, b and t are the variance estimates from Equations 2.18, 2.19 and 2.20, respectively. Column df com is the degrees of freedom is the hypothetically complete data ν_{com}, and df is the degrees of freedom after the Barnard-Rubin correction ν. The last three columns are the relative increase in variance r, the proportion of variance to due nonresponse λ and the fraction of missing information γ per parameter.

2.4 Statistical intervals and tests

2.4.1 Scalar or multi-parameter inference?

The ultimate objective of multiple imputation is to provide valid statistical estimates from incomplete data. For scalar Q, it is straightforward to calculate confidence intervals and p-values from multiply imputed data, the primary difficulty being the derivation of the appropriate degrees of freedom for the t- and F-distributions . Section 2.4.2 provides the relevant statistical procedures.

If Q is a vector, we have two options for analysis. The first option is to calculate confidence intervals and p-values for the individual elements in Q, and do all statistical tests per element. Such repeated-scalar inference is appropriate if we interpret each element as a separate, though perhaps related, model parameter. In this case, the test uses the fraction of missing information particular to each parameter.

The alternative option is to perform one statistical test that involves the elements of Q at once. This is appropriate in the context of multi-parameter or simultaneous inference, where we evaluate combinations of model parameters. Practical applications of such tests include the comparison of nested models and the testing of model terms that involved multiple parameters like regression estimates for dummy codings created from the same variable.

All methods assume that, under repeated sampling and with complete data, the parameter estimates $\hat{Q}$ are normally distributed around the population value Q as

$$\hat{Q} \sim N(Q, U) \tag{2.33}$$

where U is the variance-covariance matrix of $(Q - \hat{Q})$ (Rubin, 1987a, p. 75). For scalar Q, the quantity U reduces to σ_m^2, the variance of the estimate $\hat{Q}$ over repeated samples. Observe that U is not the variance of the measurements.

Several approaches for multi-parameter inference are available: Wald test, likelihood ratio test and χ^2-test. These methods are more complex than single-parameter inference, and their treatment is therefore deferred to Section 5.2. The next section shows how confidence intervals and p-values for scalar parameters can be calculated from multiply imputed data.

2.4.2   Scalar inference

Single parameter inference applies if $k = 1$, or if $k > 1$ and the test is repeated for each of the k components. Since the total variance of T is not known a priori, $\bar{Q}$ follows a t-distribution rather than the normal. Univariate tests are based on the approximation

$$\frac{Q - \bar{Q}}{\sqrt{T}} \sim t_\nu \tag{2.34}$$

where t_ν is the Student's t-distribution with ν degrees of freedom, with ν defined by Equation 2.32.

The $100(1 - \alpha)\%$ confidence interval of a $\bar{Q}$ is calculated as

$$\bar{Q} \pm t_{\nu,1-\alpha/2}\sqrt{T} \tag{2.35}$$

where $t_{\nu,1-\alpha/2}$ is the quantile corresponding to probability $1 - \alpha/2$ of t_ν. For example, use $t_{10,0.975} = 2.23$ for the 95% confidence interval with $\nu = 10$.

Suppose we test the null hypothesis $Q = Q_0$ for some specified value Q_0. We can find the p-value of the test as the probability

$$P_s = \Pr\left[F_{1,\nu} > \frac{(Q_0 - \bar{Q})^2}{T}\right] \tag{2.36}$$

where $F_{1,\nu}$ is an F where $F_{1,\nu}$ is an F-distribution with 1 and ν degrees of freedom.

2.4.3   Numerical example

Wald tests and confidence intervals for individual elements of Q are standard output of most statistical procedures. The `mice` package provides such output by running the `summary()` function on the `mipo` object created by `pool()`:

```
summary(est, conf.int = TRUE)
```

```
             estimate std.error statistic   df         p.value
(Intercept)     30.50      2.24     13.63 12.4 0.000000000694
age             -2.13      1.08     -1.97 15.1 0.067575739870
             2.5 % 97.5 %
(Intercept) 25.65 35.362
age         -4.43  0.174
```

The `estimate` and `df` columns are identical to the previous display. In addition, we get the standard error of the estimate, the Wald statistics, its associated p-value, and the nominal 95th percent confidence interval per parameter. In this toy example `age` is not a statistically significant predictor of `bmi` at a type I error rate of 5 percent. We may change the nominal length of the confidence intervals by the `conf.level` argument. It is possible to obtain all output by `summary(est, "all", conf.int = TRUE)`.

2.5 How to evaluate imputation methods

2.5.1 Simulation designs and performance measures

The advantageous properties of multiple imputation are only guaranteed if the imputation method used to create the missing data is proper. Equations 2.21–2.23 describe the conditions needed for proper imputation.

Checking the validity of statistical procedures is often done by simulation. There are generally two mechanisms that influence the observed data, the sampling mechanism and the missing data mechanism. Simulation can address sampling mechanism separately, the missing data mechanism separately, and both mechanisms combined. This leads to three general simulation designs.

1. *Sampling mechanism only.* The basic simulation steps are: choose Q, take samples $Y^{(s)}$, fit the complete-data model, estimate $\hat{Q}^{(s)}$ and $U^{(s)}$ and calculate the outcomes aggregated over s.

2. *Sampling and missing data mechanisms combined.* The basic simulation steps are: choose Q, take samples $Y^{(s)}$, generate incomplete data $Y_{obs}^{(s,t)}$, impute, estimate $\hat{Q}^{(s,t)}$ and $T^{(s,t)}$ and calculate outcomes aggregated over s and t.

3. *Missing data mechanism only.* The basic simulation steps are: choose

$(\hat{Q}, U)$, generate incomplete data $Y_{obs}^{(t)}$, impute, estimate $(\bar{Q}, \bar{U})^{(t)}$ and $B^{(t)}$ and calculate outcomes aggregated over t.

A popular procedure for testing missing-data applications is design 2 with settings $s = 1, \ldots, 1000$ and $t = 1$. As this design does not separate the two mechanisms, any problems found may result from both the sampling and the missing-data mechanism. Design 1 does not address the missing data, and is primarily of interest to study whether any problems are attributable to the complete-data model. Design 3 addresses the missing-data mechanism only, and thus allows for a more detailed assessment of any problem caused by the imputation step. An advantage of this procedure is that no population model is needed. Brand et al. (2003) describe this procedure in more detail.

If we are primarily interested in determining the quality of imputation methods, we may simplify evaluation by defining the sample equal to the population, and set the within-variance $\bar{U} = 0$ in Equation 2.20. See Vink and Van Buuren (2014) for a short exploration of the idea.

2.5.2   Evaluation criteria

The goal of multiple imputation is to obtain statistically valid inferences from incomplete data. The quality of the imputation method should thus be evaluated with respect to this goal. There are several measures that may inform us about the statistical validity of a particular procedure. These are:

1. *Raw bias (RB) and percent bias (PB)*. The raw bias of the estimate $\bar{Q}$ is defined as the difference between the expected value of the estimate and truth: $\text{RB} = \text{E}(\bar{Q}) - Q$. RB should be close to zero. Bias can also be expressed as percent bias: $\text{PB} = 100 \times |(\text{E}(\bar{Q}) - Q)/Q|$. For acceptable performance we use an upper limit for PB of 5%. (Demirtas et al., 2008)

2. *Coverage rate (CR)*. The coverage rate (CR) is the proportion of confidence intervals that contain the true value. The actual rate should be equal to or exceed the nominal rate. If CR falls below the nominal rate, the method is too optimistic, leading to false positives. A CR below 90 percent for a nominal 95 percent interval indicates poor quality. A high CR (e.g., 0.99) may indicate that confidence interval is too wide, so the method is inefficient and leads to inferences that are too conservative. Inferences that are "too conservative" are generally regarded a lesser sin than "too optimistic".

3. *Average width (AW)*. The average width of the confidence interval is an indicator of statistical efficiency. The length should be as small as possible, but not so small that the CR will fall below the nominal level.

4. *Root mean squared error (RMSE)*. The RMSE $= \sqrt{(\text{E}(\bar{Q}) - Q)^2}$ is a compromise between bias and variance, and evaluates $\bar{Q}$ on both accuracy and precision.

If all is well, then RB should be close to zero, and the coverage should be near 0.95. Methods having no bias and proper coverage are called randomization-valid (Rubin, 1987a). If two methods are both randomization-valid, the method with the shorter confidence intervals is more efficient. While the RMSE is widely used, we will see in Section 2.6 that it is not a suitable metric to evaluate multiple imputation methods.

2.5.3 Example

This section demonstrates the measures defined in Section 2.5.2 can be calculated using simulation. The process starts by specifying a model that is of scientific interest and that fixes Q. Pseudo-observations according to the model are generated, and part of these observations is deleted, resulting in an incomplete dataset. The missing values are then filled using the new imputation procedure, and Rubin's rules are applied to calculate the estimates $\bar{Q}$ and T. The whole process is repeated a large number of times, say in 1000 runs, each starting from different random seeds.

For the sake of simplicity, suppose scientific interest focuses on determining β in the linear model $y_i = \alpha + x_i\beta + \epsilon_i$. Here $\epsilon_i \sim N(0, \sigma^2)$ are random errors uncorrelated with x. Suppose that the true values are $\alpha = 0$, $\beta = 1$ and $\sigma^2 = 1$. We have 50% random missing data in x, and compare two imputation methods: regression imputation (cf. Section 1.3.4) and stochastic regression imputation (cf. Section 1.3.5).

It is convenient to create a series of small R functions. The `create.data()` function randomly draws artificial data from the specified linear model.

```
create.data <- function(beta = 1, sigma2 = 1, n = 50,
                        run = 1) {
  set.seed(seed = run)
  x <- rnorm(n)
  y <- beta * x + rnorm(n, sd = sqrt(sigma2))
  cbind(x = x, y = y)
}
```

Next, we remove some data in order to make the data incomplete. Here we use a simple random missing data mechanism (MCAR) to generate approximately 50% missing values.

```
make.missing <- function(data, p = 0.5){
  rx <- rbinom(nrow(data), 1, p)
  data[rx == 0, "x"] <- NA
  data
}
```

We then define a small test function that calls `mice()` and applies Rubin's rules to the imputed data.

```
test.impute <- function(data, m = 5, method = "norm", ...) {
  imp <- mice(data, method = method, m = m, print = FALSE, ...)
  fit <- with(imp, lm(y ~ x))
  tab <- summary(pool(fit), "all", conf.int = TRUE)
  as.numeric(tab["x", c("estimate", "2.5 %", "97.5 %")])
}
```

The following function puts everything together:

```
simulate <- function(runs = 10) {
  res <- array(NA, dim = c(2, runs, 3))
  dimnames(res) <- list(c("norm.predict", "norm.nob"),
                    as.character(1:runs),
                    c("estimate", "2.5 %","97.5 %"))
  for(run in 1:runs) {
    data <- create.data(run = run)
    data <- make.missing(data)
    res[1, run, ] <- test.impute(data, method = "norm.predict",
                          m = 2)
    res[2, run, ] <- test.impute(data, method = "norm.nob")
  }
  res
}
```

Performing 1000 simulations is now done by calling `simulate()`, thus

```
res <- simulate(1000)
```

The means of the estimate, the lower and upper bounds of the confidence intervals per method, can be obtained by

```
apply(res, c(1, 3), mean, na.rm = TRUE)
```

```
              estimate 2.5 % 97.5 %
norm.predict     1.343 1.065   1.62
norm.nob         0.995 0.648   1.34
```

The function of the following code is to calculate the quality statistics as defined in Section 2.5.2.

```
true <- 1
RB <- rowMeans(res[,, "estimate"]) - true
PB <- 100 * abs((rowMeans(res[,, "estimate"]) - true)/ true)
CR <- rowMeans(res[,, "2.5 %"] < true & true < res[,, "97.5 %"])
AW <- rowMeans(res[,, "97.5 %"] - res[,, "2.5 %"])
```

```
RMSE <- sqrt(rowMeans((res[,, "estimate"] - true)^2))
data.frame(RB, PB, CR, AW, RMSE)

                RB   PB    CR    AW  RMSE
norm.predict  0.343 34.3 0.364 0.555 0.409
norm.nob     -0.005  0.5 0.925 0.693 0.201
```

The interpretation of the results is as follows. Regression imputation by method `norm.predict` produces severely biased estimates of β. The true β is 1, but the average estimate after regression imputation is 1.343. Moreover, the true value is located within the confidence interval in only 36% of the cases, far below the nominal value of 95%. Hence, regression imputation is not randomization-valid for β, even under MCAR. Because of this, the estimates for AW and RMSE are not relevant. This example shows that statistical inference on incomplete data that were imputed by regression imputation can produce the wrong answer.

The story for stochastic regression imputation is different. The `norm.nob` method is unbiased and has a coverage of 92.5%. The method is not randomization-valid, but it is near. The AW and RMSE serve as useful indicators, though both may be a little low as the confidence intervals appear somewhat short. Chapter 3 shows how to make it fully randomization-valid.

Of course, simulation studies do not guarantee fitness for a particular application. However, if simulation studies illustrate limitations in simple examples, we may expect these will also be present in applications.

2.6 Imputation is not prediction

In the world of simulation we have access to both the true and imputed values, so an obvious way to quantify the quality of a method is to see how well it can recreate the true data. The method that best recovers the true data "wins." An early paper developing this idea is Gleason and Staelin (1975), but the literature is full of examples. The approach is simple and appealing. But will it also select the best imputation method?

The answer is "No". Suppose that we would measure discrepancy by the RMSE of the imputed values:

$$\text{RMSE} = \sqrt{\frac{1}{n_{\text{mis}}} \sum_{i=1}^{n_{\text{mis}}} (y_i^{\text{mis}} - \dot{y}_i)^2} \qquad (2.37)$$

where y_i^{mis} represents the true (removed) data value for unit i and where $\dot{y}_i$ is imputed value for unit i. For multiply imputed data we calculate RMSE for each imputed dataset, and average these.

It is well known that the minimum RMSE is attained by predicting the missing $\dot{y}_i$ by the linear model with the regression weights set to their least squares estimates. According to this reasoning the "best" method replaces each missing value by its most likely value under the model. However, this will find the same values over and over, and is single imputation. This method ignores the inherent uncertainty of the missing values (and acts as if they were known after all), resulting in biased estimates and invalid statistical inferences. Hence, the method yielding the lowest RMSE is bad for imputation. More generally, measures based on similarity between the true and imputed values do not separate valid from invalid imputation methods.

Let us check this claim with a short simulation. The `rmse()` below calculates the RMSE from the true and multiply imputed data for missing data in variable x.

```r
rmse <- function(truedata, imp, v = "x") {
  mx <- is.na(mice::complete(imp, 0))[, v]
  mse <- rep(NA, imp$m)
  for (k in seq_len(imp$m)) {
    filled <- mice::complete(imp, k)[mx, v]
    true <- truedata[mx, v]
    mse[k] <- mean((filled - true)^2)
  }
  sqrt(mean(mse))
}
```

The `simulate2()` function creates the same missing data as before.

```r
simulate2 <- function(runs = 10) {
  res <- array(NA, dim = c(2, runs, 1))
  dimnames(res) <- list(c("norm.predict", "norm.nob"),
                        as.character(1:runs),
                        "RMSE")
  for(run in 1:runs) {
    truedata <- create.data(run = run)
    data <- make.missing(truedata)
    imp <- mice(data, method = "norm.predict", m = 1,
            print = FALSE)
    res[1, run, ] <- rmse(truedata, imp)
    imp <- mice(data, method = "norm.nob", print = FALSE)
    res[2, run, ] <- rmse(truedata, imp)
  }
  res
}
```

```
res2 <- simulate2(1000)
apply(res2, c(1, 3), mean, na.rm = TRUE)

              RMSE
norm.predict 0.725
norm.nob     1.025
```

The simulation confirms that regression imputation is better at recreating the missing data. Remember from Section 1.3.4 that regression imputation is fundamentally flawed. Its estimate of β is biased (even under MCAR) and the accompanying confidence interval is too short.

The example demonstrates that the RMSE is not informative for evaluating imputation methods. Assessing the discrepancy between true data and the imputed data may seem a simple and attractive way to select the best imputation method. However, it is not useful to evaluate methods solely based on their ability to recreate the true data. On the contrary, selecting such methods may be harmful as these might increase the rate of false positives. Imputation is not prediction.

2.7 When not to use multiple imputation

Should we always use multiple imputation for the missing data? We probably could, but there are good alternatives in some situations. Section 1.6 already discussed some approaches not covered in this book, each of which has its merits. This section revisits complete-case analysis. Apart from being simple to apply, it can be a viable alternative to multiple imputation in particular situations.

Suppose that the complete-data model is a regression with outcome Y and predictors X. If the missing data occur in Y only, complete-case analysis and multiple imputation are equivalent, so then complete-case analysis is preferred since it is easier, more efficient and more robust (Von Hippel, 2007). This applies to the regression weights. Quantities that depend on the correct marginal distribution of Y, such as the mean or R^2, require the stronger MCAR assumption. Multiple imputation gains an advantage over complete-case analysis if additional predictors for Y are available that are not part of X. The efficiency of complete-case analysis declines if X contains missing values, which may result in inflated type II error rates. Complete-case analysis can perform quite badly under MAR and some MNAR cases (Schafer and Graham, 2002), but there are two special cases where listwise deletion outperforms multiple imputation.

The first special case occurs if the probability to be missing does not depend on Y. Under the assumption that the complete-data model is correct, the regression coefficients are free of bias (Little, 1992; King et al., 2001). This holds for any type of regression analysis, and for missing data in both Y and X. Since the missing data rate may depend on X, complete-case analysis will in fact work in a relevant class of MNAR models. White and Carlin (2010) confirmed the superiority of complete-case analysis by simulation. The differences were often small, and multiple imputation gained the upper hand as more predictive variables were included. The property is useful though in practice.

The second special case holds only if the complete data model is logistic regression. Suppose that the missing data are confined to either a dichotomous Y or to X, but not to both. Assuming that the model is correctly specified, the regression coefficients (except the intercept) from the complete-case analysis are unbiased if the probability to be missing depends only on Y and not on X (Vach, 1994). This property provides the statistical basis of the estimation of the odds ratio from case-control studies in epidemiology. If missing data occur in both Y and X the property does not hold.

At a minimum, application of listwise deletion should be a conscious decision of the analyst, and should preferably be accompanied by an explicit statement that the missing data fit in one of the three categories described above.

Other alternatives to multiple imputation were briefly reviewed in Section 1.6, and may work well in particular applications. However, none of these is as general as multiple imputation.

2.8   How many imputations?

One of the distinct advantages of multiple imputation is that it can produce unbiased estimates with correct confidence intervals with a low number of imputed datasets, even as low as $m = 2$. Multiple imputation is able to work with low m since it enlarges the between-imputation variance B by a factor $1/m$ before calculating the total variance in $T = \bar{U} + (1 + m^{-1})B$.

The classic advice is to use a low number of imputation, somewhere between 3 and 5 for moderate amounts of missing information. Several authors investigated the influence of m on various aspects of the results. The picture emerging from this work is that it is often beneficial to set m higher, somewhere in the range of 20–100 imputations. This section reviews the relevant work in the area.

The advice for low m rests on the following argument. Multiple imputation is a simulation technique, and hence $\bar{Q}$ and its variance estimate T are subject to simulation error. Setting $m = \infty$ causes all simulation error to disappear, so $T_\infty < T_m$ if $m < \infty$. The question is when T_∞ is close enough to T_m.

(Rubin, 1987a, p. 114) showed that the two variances are related by

$$T_m = \left(1 + \frac{\gamma_0}{m}\right) T_\infty \tag{2.38}$$

where γ_0 is the (true) population fraction of missing information. This quantity is equal to the expected fraction of observations missing if Y is a single variable without covariates, and commonly less than this if there are covariates that predict Y. For example, for $\gamma_0 = 0.3$ (e.g., a single variable with 30% missing) and $m = 5$ we find that the calculated variance T_m is $1 + 0.3/5 = 1.06$ times (i.e., 6%) larger than the ideal variance T_∞. The corresponding confidence interval would thus be $\sqrt{1.06} = 1.03$ (i.e., 3%) longer than the ideal confidence interval based on $m = \infty$. Increasing m to 10 or 20 would bring the factor down 1.5% and 0.7%, respectively. The argument is that "the additional resources that would be required to create and store more than a few imputations would not be well spent" (Schafer, 1997, p. 107), and "in most situations there is simply little advantage to producing and analyzing more than a few imputed datasets" (Schafer and Olsen, 1998, p. 549).

Royston (2004) observed that the length of the confidence interval also depends on ν, and thus on m (cf. Equation 2.30). He suggested to base the criterion for m on the confidence coefficient $t_\nu \sqrt{T}$, and proposed that the coefficient of variation of $\ln(t_\nu \sqrt{T})$ should be smaller than 0.05. This effectively constrains the range of uncertainty about the confidence interval to roughly within 10%. This rule requires m to be "at least 20 and possibly more."

Graham et al. (2007) investigated the effect of m on the statistical power of a test for detecting a small effect size (< 0.1). Their advice is to set m high in applications where high statistical power is needed. For example, for $\gamma_0 = 0.3$ and $m = 5$ the statistical power obtained is 73.1% instead of the theoretical value of 78.4%. We need $m = 20$ to increase the power to 78.2%. In order to have an attained power within 1% of the theoretical power, then for fractions of missing information $\gamma = (0.1, 0.3, 0.5, 0.7, 0.9)$ we need to set $m = (20, 20, 40, 100, > 100)$, respectively.

Bodner (2008) explored the variability of three quantities under various m: the width of the 95% confidence interval, the p-value, and γ_0. Bodner selected m such that the width of the 95% confidence interval is within 10% of its true value 95% of the time. For $\gamma_0 = (0.05, 0.1, 0.2, 0.3, 0.5, 0.7, 0.9)$, he recommends $m = (3, 6, 12, 24, 59, 114, 258)$, respectively, using a linear rule. Since the true γ_0 is unknown, Bodner suggested the proportion of complete cases as a conservative estimate of γ_0. Von Hippel (2018) showed that a relation between m and γ_0 is better explained by a quadratic rule

$$m = 1 + \frac{1}{2} \left(\frac{\gamma_0}{\mathrm{SD}(\sqrt{U_\ell})\mathrm{E}(\sqrt{U_\ell})}\right)^2, \tag{2.39}$$

where $\mathrm{E}(\sqrt{U_\ell})$ and $\mathrm{SD}(\sqrt{U_\ell})$ are the mean and standard deviation of the standard errors calculated from the imputed datasets. The rule is used in a two-step procedure, where the first step estimates γ_0 and its 95% confidence interval. The upper limit of the confidence interval is then plugged into Equation 2.39. Compared to Bodner, the rule suggests somewhat lower m if $\gamma_0 < 0.5$ and substantially higher m if $\gamma_0 > 0.5$.

The starting point of White et al. (2011b) is that all essential quantities in the analysis should be reproducible within some limit, including confidence intervals, p-values and estimates of the fraction of missing information. They take a quote from Von Hippel (2009) as a rule of thumb: *the number of imputations should be similar to the percentage of cases that are incomplete.* This rule applies to fractions of missing information of up to 0.5. If $m \approx 100\lambda$, the following properties will hold for a parameter β:

1. The Monte Carlo error of $\hat{\beta}$ is approximately 10% of its standard error;

2. The Monte Carlo error of the test statistic $\hat{\beta}/\mathrm{se}(\hat{\beta})$ is approximately 0.1;

3. The Monte Carlo error of the p-value is approximately 0.01 when the true p-value is 0.05.

White et al. (2011b) suggest these criteria provide an adequate level of reproducibility in practice. The idea of reproducibility is sensible, the rule is simple to apply, so there is much to commend it. The rule has now become the de-facto standard, especially in medical applications. One potential difficulty might be that the percentage of complete cases is sensitive to the number of variables in the data. If we extend the active dataset by adding more variables, then the percentage of complete cases can only drop. An alternative would be to use the average missing data rate as a less conservative estimate.

Theoretically it is always better to use higher m, but this involves more computation and storage. Setting m very high (say $m = 200$) may be useful for low-level estimands that are very uncertain, and for which we want to approximate the full distribution, or for parameters that are notoriously different to estimates, like variance components. On the other hand, setting m high may not be worth the extra wait if the primary interest is on the point estimates (and not on standard errors, p-values, and so on). In that case using $m = 5 - 20$ will be enough under moderate missingness.

Imputing a dataset in practice often involves trial and error to adapt and refine the imputation model. Such initial explorations do not require large m. It is convenient to set $m = 5$ during model building, and increase m only after being satisfied with the model for the "final" round of imputation. So if calculation is not prohibitive, we may set m to the average percentage of missing data. The substantive conclusions are unlikely to change as a result of raising m beyond $m = 5$.

2.9 Exercises

1. *Nomogram.* Construct a graphic representation of Equation 2.27 that allows the user to convert λ and γ for different values of ν. What influence does ν have on the relation between λ and γ?

2. *Models.* Explain the difference between the response model and the imputation model.

3. *Listwise deletion.* In the `airquality` data, predict `Ozone` from `Wind` and `Temp`. Now randomly delete the half of the wind data above 10 mph, and randomly delete half of the temperature data above 80°F.

    (a) Are the data MCAR, MAR or MNAR?

    (b) Refit the model under listwise deletion. Do you notice a change in the estimates? What happens to the standard errors?

    (c) Would you conclude that listwise deletion provides valid results here?

    (d) If you add a quadratic term to the model, would that alter your conclusion?

4. *Number of imputations.* Consider the `nhanes` dataset in `mice`.

    (a) Use the functions `ccn()` to calculate the number of complete cases. What percentage of the cases is incomplete?

    (b) Impute the data with `mice` using the defaults with `seed=1`, predict `bmi` from `age`, `hyp` and `chl` by the normal linear regression model, and pool the results. What are the proportions of variance due to the missing data for each parameter? Which parameters appear to be most affected by the nonresponse?

    (c) Repeat the analysis for `seed=2` and `seed=3`. Do the conclusions remain the same?

    (d) Repeat the analysis with $m = 50$ with the same seeds. Would you prefer this analysis over those with $m = 5$? Explain why.

5. *Number of imputations (continued).* Continue with the data from the previous exercise.

    (a) Write an R function that automates the calculations of the previous exercise. Let `seed` run from 1 to 100 and let m take on values m = `c(3, 5, 10, 20, 30, 40, 50, 100, 200)`.

    (b) Plot the estimated proportions of explained variance due to missing data for the `age`-parameter against m. Based on this graph, how many imputations would you advise?

(c) Check White's conditions 1 and 2 (cf. Section 2.8). For which m do these conditions true?

(d) Does this also hold for categorical data? Use the nhanes2 to study this.

6. *Automated choice of m.* Write an R function that implements the methods discussed in Section 2.8.

Chapter 3

Univariate missing data

Statistics is a missing-data problem.
Roderick J.A. Little

Chapter 2 described the theory of multiple imputation. This chapter looks into ways of creating the actual imputations. In order to avoid unnecessary complexities at this point, the text is restricted to univariate missing data. The incomplete variable is called the *target* variable. Thus, in this chapter there is only one variable with missing values. The consequences of the missing data depend on the role of the target variables within the complete-data model that is applied to the imputed data.

There are many ways to create imputations, but only a few of those lead to valid statistical inferences. This chapter outlines ways to check the correctness of a procedure, and how this works out for selected procedures. Most of the methods are designed to work under the assumption that the relations within the missing parts are similar to those in the observed parts, or more technically, the assumption of ignorability. The chapter closes with a description of some alternatives of what we might do when that assumption is suspect.

3.1 How to generate multiple imputations

This section illustrates five ways to create imputations for a single incomplete continuous target variable. We use dataset number 88 in Hand et al. (1994), which is also part of the MASS library under the name `whiteside`. Mr. Whiteside of the UK Building Research Station recorded the weekly gas consumption (in 1000 cubic feet) and average external temperature (in °C) at his own house in south-east England for two heating seasons (1960 and 1961). The house thermostat was set at 20°C throughout.

Figure 3.1a plots the observed data. More gas is needed in colder weeks, so there is an obvious relation in the data. The dataset is complete, but for the sake of argument suppose that the gas consumption in row 47 of the data is missing. The temperature at this deleted observation is equal to 5°C. How would we create multiple imputations for the missing gas consumption?

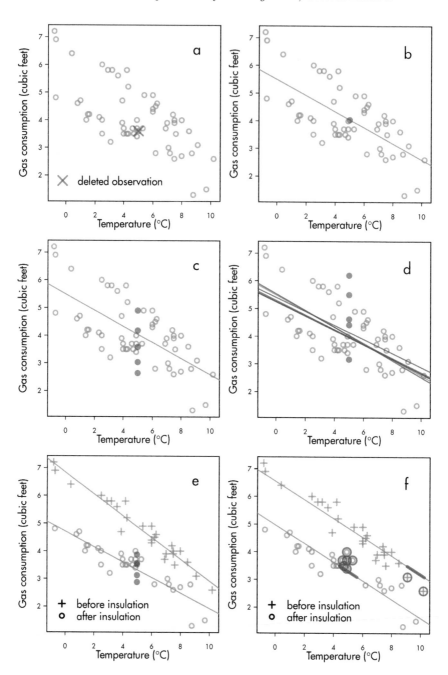

Figure 3.1: Five ways to impute missing gas consumption for a temperature of 5°C: (a) no imputation; (b) predict; (c) predict + noise; (d) predict + noise + parameter uncertainty; (e) two predictors; (f) drawing from observed data.

3.1.1 Predict method

A first possibility is to calculate the regression line, and take the imputation from the regression line. The estimated regression line is equal to $y = 5.49 - 0.29x$, so the value at $x = 5$ is $5.49 - 0.29 \times 5 = 4.04$. Figure 3.1b shows where the imputed value is. This is actually the "best" value in the sense that it is the most likely one under the regression model. However, even the best value may differ from the actual (unknown) value. In fact, we are uncertain about the true gas consumption. Predicted values, however, do not portray this uncertainty, and therefore cannot be used as multiple imputations.

3.1.2 Predict + noise method

We can improve upon the prediction method by adding an appropriate amount of random noise to the predicted value. Let us assume that the observed data are normally distributed around the regression line. The estimated standard deviation in the Whiteside data is equal to 0.86 cubic feet. The idea is now to draw a random value from a normal distribution with a mean of zero and a standard deviation of 0.86, and add this value to the predicted value. The underlying assumption is that the distribution of gas consumption of the incomplete observation is identical to that in the complete cases.

We can repeat the draws to get multiple synthetic values around the regression line. Figure 3.1c illustrates five such drawn values. On average, the synthetic values will be equal to the predicted value. The variability in the values reflects that fact that we cannot accurately predict gas consumption from temperature.

3.1.3 Predict + noise + parameter uncertainty

Adding noise is a major step forward, but not quite right. The method in the previous section requires that the intercept, the slope and the standard deviation of the residuals are known. However, the values of these parameters are typically unknown, and hence must be estimated from the data. If we had drawn a different sample from the same population, then our estimates for the intercept, slope and standard deviation would be different, perhaps slightly. The amount of extra variability is strongly related to the sample size, with smaller samples yielding more variable estimates.

The parameter uncertainty also needs to be included in the imputations. There are two main methods for doing so. Bayesian methods draw the parameters directly from their posterior distributions, whereas bootstrap methods resample the observed data and re-estimate the parameters from the resampled data.

Figure 3.1d shows five sampled regression lines calculated by Bayesian sampling. Imputed values are now defined as the predicted value of the sampled line added with noise, as in Section 3.1.2.

3.1.4   A second predictor

The dataset actually contains a second predictor that indicates whether the house was insulated or not. Incorporating this extra information reduces the uncertainty of the imputed values.

Figure 3.1e shows the same data, but now flagged according to insulation status. Two regression lines are shown, one for the insulated houses and the other for the non-insulated houses. It is clear that less gas is needed after insulation. Suppose we know that the external temperature is 5°C *and* that the house was insulated. How do we create multiple imputation given these two predictors?

We apply the same method as in Section 3.1.3, but now using the regression line for the insulated houses. Figure 3.1e shows the five values drawn for this method. As expected, the distribution of the imputed gas consumption has shifted downward. Moreover, its variability is lower, reflecting that fact that gas consumption can be predicted more accurately as insulation status is also known.

3.1.5   Drawing from the observed data

Figure 3.1f illustrates an alternative method to create imputations. As before, we calculate the predicted value at 5°C for an insulated house, but now select a small number of candidate donors from the observed data. The selection is done such that the predicted values are close. We then randomly select one donor from the candidates, and use the *observed* gas consumption that belongs to that donor as the synthetic value. The figure illustrates the candidate donors, not the imputations.

This method is known as *predictive mean matching*, and always finds values that have been actually observed in the data. The underlying assumption is that within the group of candidate donors gas consumption has the same distribution in donors and receivers. The variability between the imputations over repeated draws is again a reflection of the uncertainty of the actual value.

3.1.6   Conclusion

In summary, prediction methods are not suitable to create multiple imputations. Both the inherent prediction error and the parameter uncertainty should be incorporated into the imputations. Adding a relevant extra predictor reduces the amount of uncertainty, and leads to more efficient estimates later on. The text also highlights an alternative that draws imputations from the observed data. The imputation methods discussed in this chapter are all variations on this basic idea.

3.2 Imputation under the normal linear normal

3.2.1 Overview

For univariate Y we write lower-case y for Y. Any predictors in the imputation model are collected in X. Symbol X_{obs} indicates the subset of n_1 rows of X for which y is observed, and X_{mis} is the complementing subset of n_0 rows of X for which y is missing. The vector containing the n_1 observed data in y is denoted by y_{obs}, and the vector of n_0 imputed values in y is indicated by $\dot{y}$. This section reviews four different ways of creating imputations under the normal linear model. The four methods are:

1. *Predict.* $\dot{y} = \hat{\beta}_0 + X_{\text{mis}}\hat{\beta}_1$, where $\hat{\beta}_0$ and $\hat{\beta}_1$ are least squares estimates calculated from the observed data. Section 1.3.4 named this regression imputation. In `mice` this method is available as method `norm.predict`.

2. *Predict + noise.* $\dot{y} = \hat{\beta}_0 + X_{\text{mis}}\hat{\beta}_1 + \dot{\epsilon}$, where $\dot{\epsilon}$ is randomly drawn from the normal distribution as $\dot{\epsilon} \sim N(0, \hat{\sigma}^2)$. Section 1.3.5 named this stochastic regression imputation. In `mice` this method is available as method `norm.nob`.

3. *Bayesian multiple imputation.* $\dot{y} = \dot{\beta}_0 + X_{\text{mis}}\dot{\beta}_1 + \dot{\epsilon}$, where $\dot{\epsilon} \sim N(0, \dot{\sigma}^2)$ and $\dot{\beta}_0$, $\dot{\beta}_1$ and $\dot{\sigma}$ are random draws from their posterior distribution, given the data. Section 3.1.3 named this "predict + noise + parameters uncertainty." The method is available as method `norm`.

4. *Bootstrap multiple imputation.* $\dot{y} = \dot{\beta}_0 + X_{\text{mis}}\dot{\beta}_1 + \dot{\epsilon}$, where $\dot{\epsilon} \sim N(0, \dot{\sigma}^2)$, and where $\dot{\beta}_0$, $\dot{\beta}_1$ and $\dot{\sigma}$ are the least squares estimates calculated from a bootstrap sample taken from the observed data. This is an alternative way to implement "predict + noise + parameters uncertainty." The method is available as method `norm.boot`.

3.2.2 Algorithms♠

The calculations of the first two methods are straightforward and do not need further explanation. This section describes the algorithms used to introduce sampling variability into the parameters estimates of the imputation model.

The Bayesian sampling draws $\dot{\beta}_0$, $\dot{\beta}_1$ and $\dot{\sigma}$ from their respective posterior distributions. Box and Tiao (1973, Section 2.7) explains the Bayesian theory behind the normal linear model. We use the method that draws imputations under the normal linear model using the standard noninformative priors for each of the parameters. Given these priors, the required inputs are:

Algorithm 3.1: Bayesian imputation under the normal linear model.

1. Calculate the cross-product matrix $S = X'_{\text{obs}}X_{\text{obs}}$.

2. Calculate $V = (S + \text{diag}(S)\kappa)^{-1}$, with some small κ.

3. Calculate regression weights $\hat{\beta} = VX'_{\text{obs}}y_{\text{obs}}$.

4. Draw a random variable $\dot{g} \sim \chi^2_\nu$ with $\nu = n_1 - q$.

5. Calculate $\dot{\sigma}^2 = (y_{\text{obs}} - X_{\text{obs}}\hat{\beta})'(y_{\text{obs}} - X_{\text{obs}}\hat{\beta})/\dot{g}$.

6. Draw q independent $N(0,1)$ variates in vector $\dot{z}_1$.

7. Calculate $V^{1/2}$ by Cholesky decomposition.

8. Calculate $\dot{\beta} = \hat{\beta} + \dot{\sigma}\dot{z}_1 V^{1/2}$.

9. Draw n_0 independent $N(0,1)$ variates in vector $\dot{z}_2$.

10. Calculate the n_0 values $\dot{y} = X_{\text{mis}}\dot{\beta} + \dot{z}_2\dot{\sigma}$.

- y_{obs}, the $n_1 \times 1$ vector of observed data in the incomplete (or target) variable y;

- X_{obs}, the $n_1 \times q$ matrix of predictors of rows with observed data in y;

- X_{mis}, the $n_0 \times q$ matrix of predictors of rows with missing data in y.

The algorithm assumes that both X_{obs} and X_{mis} contain no missing values. Chapter 4 deals with the case where X_{obs} and X_{mis} also could be incomplete.

Algorithm 3.1 is adapted from Rubin (1987a, p. 167), and is implemented as the method `norm` (or, equivalently, as the function `mice.impute.norm()`) in the `mice` package. Any drawn values are identified with a dot above the symbol, so $\dot{\beta}$ is a value of β drawn from the posterior distribution. The algorithm uses a ridge parameter κ to evade problems with singular matrices. This number should be set to a positive number close to zero, e.g., $\kappa = 0.0001$. For some data, larger κ may be needed. High values of κ, e.g., $\kappa = 0.1$, may introduce a systematic bias toward the null, and should thus be avoided.

The bootstrap is a general method for estimating sampling variability through resampling the data (Efron and Tibshirani, 1993). Algorithm 3.2 calculates univariate imputations by drawing a bootstrap sample from the complete part of the data, and subsequently takes the least squares estimates given the bootstrap sample as a "draw" that incorporates sampling variability into the parameters (Heitjan and Little, 1991). Compared to the Bayesian method, the bootstrap method avoids the Choleski decomposition and does not need to draw from the χ^2-distribution.

Algorithm 3.2: Imputation under the normal linear model with bootstrap.♠

1. Draw a bootstrap sample $(\dot{y}_{obs}, \dot{X}_{obs})$ of size n_1 from (y_{obs}, X_{obs}).

2. Calculate the cross-product matrix $\dot{S} = \dot{X}'_{obs}\dot{X}_{obs}$.

3. Calculate $\dot{V} = (\dot{S} + \text{diag}(\dot{S})\kappa)^{-1}$, with some small κ.

4. Calculate regression weights $\dot{\beta} = \dot{V}\dot{X}'_{obs}\dot{y}_{obs}$.

5. Calculate $\dot{\sigma}^2 = (\dot{y}_{obs} - \dot{X}_{obs}\dot{\beta})'(\dot{y}_{obs} - \dot{X}_{obs}\dot{\beta})/(n_1 - q - 1)$.

6. Draw n_0 independent $N(0,1)$ variates in vector $\dot{z}_2$.

7. Calculate the n_0 values $\dot{y} = X_{mis}\dot{\beta} + \dot{z}_2\dot{\sigma}$.

Table 3.1: Properties of β_1 under imputation of missing y by five methods for the normal linear model ($n_{sim} = 10000$).

Method	Bias	% Bias	Coverage	CI Width	RMSE
norm.predict	0.0000	0.0	0.652	0.114	0.063
norm.nob	-0.0001	0.0	0.908	0.226	0.064
norm	-0.0001	0.0	0.951	0.314	0.066
norm.boot	-0.0001	0.0	0.941	0.299	0.066
Listwise deletion	0.0001	0.0	0.946	0.251	0.063

3.2.3 Performance

Which of these four imputation methods of Section 3.2 is best? In order to find out let us conduct a small simulation experiment where we calculate the performance statistics introduced in Section 2.5.3. We keep close to the original data by assuming that $\beta_0 = 5.49$, $\beta_1 = -0.29$ and $\sigma = 0.86$ are the population values. These values are used to generate artificial data with known properties.

Table 3.1 summarizes the results for the situation where we have 50% completely random missing in y and $m = 5$. All methods are unbiased for β_1. The confidence interval of method norm.predict is much too short, leading to substantial undercoverage and p-values that are "too significant." This result confirms the problems already noted in Section 2.6. The norm.nob method performs better, but the coverage of 0.908 is still too low. Methods norm and norm.boot and complete-case analysis are correct. Complete-case analysis is a correct analysis here (Little and Rubin, 2002), and in fact the most efficient choice for this problem as it yields the shortest confidence interval (cf. Section 2.7). This result does not hold more generally. In realistic situations involving more covariates multiple imputation will rapidly catch up and pass

Table 3.2: Properties of β_1 under imputation of missing x by five methods for the normal linear model ($n_{\text{sim}} = 10000$).

Method	Bias	% Bias	Coverage	CI Width	RMSE
norm.predict	-0.1007	34.7	0.359	0.160	0.118
norm.nob	0.0006	0.2	0.924	0.202	0.056
norm	0.0075	2.6	0.955	0.254	0.058
norm.boot	-0.0014	0.5	0.946	0.238	0.058
Listwise deletion	-0.0001	0.0	0.946	0.251	0.063

complete-case analysis. Note that the RMSE values are uninformative for separating correct and incorrect methods, and are in fact misleading.

While method norm.predict is simple and fast, the variance estimate is too low. Several methods have been proposed to correct the estimate (Lee et al., 1994; Fay, 1996; Rao, 1996; Schafer and Schenker, 2000). Though such methods require special adaptation of formulas to calculate the variance, they may be useful when the missing data are restricted to the outcome.

It is straightforward to adapt the simulations to other, perhaps more interesting situations. Investigating the effect of missing data in the explanatory x instead of the outcome variable requires only a small change in the function to create the missing data. Table 3.2 displays the results. Method norm.predict is now severely biased, whereas the other methods remain unbiased. The confidence interval of norm.nob is still too short, but less than in Table 3.1. Methods norm, norm.boot and listwise deletion are correct, in the sense that these are unbiased and have appropriate coverage. Again, under the simulation conditions, listwise deletion is the optimal analysis. Note that norm is slightly biased, whereas method norm.boot slightly underestimates the variance. Both tendencies are small in magnitude. The RMSE values are uninformative, and are only shown to illustrate that point.

We could increase the number of explanatory variables and the number of imputations m to see how much the average confidence interval width would shrink. It is also easy to apply more interesting missing data mechanisms, such as those discussed in Section 3.2.4. Data can be generated from skewed distributions, the sample size n can be varied and so on. Extensive simulation work is available (Rubin and Schenker, 1986b; Rubin, 1987a).

3.2.4 Generating MAR missing data

Just making random missing data is not always interesting. We obtain more informative simulations if the missingness probability is a function of the observed, and possibly of the unobserved, information. This section considers some methods for creating missing data.

Let us first consider three methods to create missing data in artificial data. The data are generated as 1000 draws from the bivariate normal distribution $P(Y_1, Y_2)$ with means $\mu_1 = \mu_2 = 5$, variances $\sigma_1^2 = \sigma_2^2 = 1$, and covariance

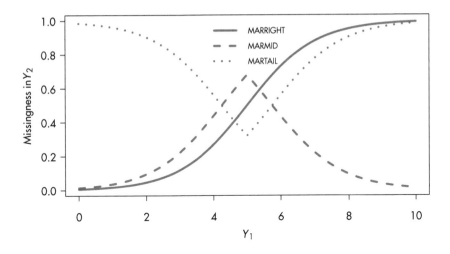

Figure 3.2: Probability that Y_2 is missing as a function of the values of Y_1 under three models for the missing data.

$\sigma_{12} = 0.6$. We assume that all values generated are positive. Missing data in Y_2 can be created in many ways. Let R_2 be the response indicator for Y_2. We study three examples, each of which affects the distribution in different ways:

$$
\begin{array}{lll}
\text{MARRIGHT} & : & \text{logit}(\Pr(R_2 = 0)) = -5 + Y_1 & (3.1) \\
\text{MARMID} & : & \text{logit}(\Pr(R_2 = 0)) = 0.75 - |Y_1 - 5| & (3.2) \\
\text{MARTAIL} & : & \text{logit}(\Pr(R_2 = 0)) = -0.75 + |Y_1 - 5| & (3.3)
\end{array}
$$

where $\text{logit}(p) = \log(p) - \log(1 - p)$ with $0 \le p \le 1$ is the logit function. Its inverse $\text{logit}^{-1}(x) = \exp(x)/(1 + \exp(x))$ is known as the logistic function.

Generating missing data under these models in R can be done in three steps: calculate the missingness probability of each data point, make a random draw from the binomial distribution and set the corresponding observations to NA. The following script creates missing data according to MARRIGHT:

```
set.seed(1)
n <- 10000
sigma <- matrix(c(1, 0.6, 0.6, 1), nrow = 2)
cmp <- MASS::mvrnorm(n = n, mu = c(5, 5), Sigma = sigma)
p2.marright <- 1 - plogis(-5 + cmp[, 1])
r2.marright <- rbinom(n, 1, p2.marright)
yobs <- cmp
yobs[r2.marright == 0, 2] <- NA
```

Figure 3.2 displays the probability of being missing under the three MAR

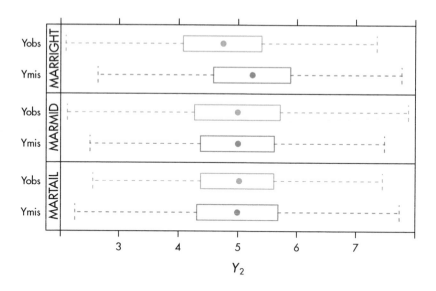

Figure 3.3: Box plot of Y_2 separated for the observed and missing parts under three models for the missing data based on $n = 10000$.

mechanisms. All mechanisms yield approximately 50% of missing data, but do so in very different ways. Figure 3.3 displays the distributions of Y_2 under the three models. MARRIGHT deletes more high values, so the distribution of the observed data shifts to the left. MARMID deletes more data in the center, so the variance of the observed data grows, but the mean is not affected. MARTAIL shows the reverse effect. The variance of observed data reduces because the missing data occur in the tails.

These mechanisms are more extreme than we are likely to see in practice. Not only is there a strong relation between Y_1 and R_2, but the percentage of missing data is also quite high (50%). On the other hand, if methods perform well under these data deletion schemes, they will also do so in less extreme situations that are more likely to be encountered in practice.

3.2.5   MAR missing data generation in multivariate data

Creating missing data from complete data is easy to do for simple scenarios with one missing value per row. Things become more complicated for multiple missing values per unit, as we need to be careful not to delete any values that are needed to make the problem MAR.

Brand (1999, pp. 110–113) developed the following method for generating non-monotone multivariate missing data in p variables $Y_1, \ldots, Y_p$. We assume that $Y = (Y_1, \ldots, Y_p)$ is initially completely known. The method requires specification of

- α, the desired proportion of incomplete cases,

- R_{pat}, a binary $n_{\text{pat}} \times p$ matrix defining n_{pat} allowed patterns of missing data, where all response patterns except $(0, 0, \ldots, 0)$ and $(1, 1, \ldots, 1)$ may occur,

- $f = (f_{(1)}, \ldots, f_{(n_{\text{pat}})})$, a vector containing the relative frequencies of each pattern, scaled such that $\sum_s^{n_{\text{pat}}} f_{(s)} = 1$,

- $P(R|Y) = (P(R_{(1)}|Y(1)), \ldots, P(R_{(n_{\text{pat}})}|Y_{(n_{\text{pat}})}))$, a set of n_{pat} response probability models, one for each pattern.

The general procedure is as follows: Each case is allocated to one of n_{pat} candidate blocks using a random draw from the multinomial distribution with probabilities $f_{(1)}, \ldots, f_{(n_{\text{pat}})}$. Within the sth candidate block, a subgroup of $\alpha n f_{(s)}$ cases is made incomplete according to pattern $R_{(s)}$ using the missing data model $P(R_{(s)}|Y_{(s)})$, where $s = 1, \ldots, n_{\text{pat}}$. The procedure results in approximately α incomplete cases, that are distributed over the allowed response patterns. If the missing data are to be MAR, then the missing variables in the sth pattern should not influence the missingness probability defined by the missing data model $P(R_{(s)}|Y_{(s)}$ for block s.

The `ampute()` function in `mice` implements the method. For example, we can create 50% missing data in both Y_1 and Y_2 according to a MARRIGHT scenario by

```
amp <- ampute(cmp, type = "RIGHT")
apply(amp$amp, 2, mean, na.rm = TRUE)

  V1   V2
4.91 4.91
```

As expected, the means in the amputed data are lower than in the complete data. It is possible to inspect the distributions of the observed data more closely by `md.pattern(amp$amp)`, `bwplot(amp)` and `xyplot(amp)`xyplot. Many options are available that allows the user to tailor the missing data patterns to the data at hand. See Schouten et al. (2018) for details.

3.2.6 Conclusion

Tables 3.1 and 3.2 show that methods `norm.predict` (regression imputation) and `norm.nob` (stochastic regression imputation) fail in terms of understating the uncertainty in the imputations. If the missing data occur in y only, then it is possible to correct the variance formulas of method `norm.predict`. However, if the missing data occur in X, `norm.predict` is severely biased, so then variance correction is not useful. Methods `norm` and `norm.boot` account for the uncertainty of the imputation model provide statistically correct inferences. For missing y, the efficiency of these methods is less than theoretically possible, presumably due to simulation error.

It is always better to include parameter uncertainty, either by the Bayesian or the bootstrap method. The effect of doing so will diminish with increasing sample size (Exercise 2), so for estimates based on a large sample one may opt for the simpler `norm.nob` method if speed of calculation is at premium. Note that in subgroup analyses, the large-sample requirement applies to the subgroup size, and not to the total sample size.

3.3   Imputation under non-normal distributions

3.3.1   Overview

The imputation methods discussed in Section 3.2 produce imputations drawn from a normal distribution. In practice the data could be skewed, long tailed, non-negative, bimodal or rounded, to name some deviations from normality. This creates an obvious mismatch between observed and imputed data which could adversely affect the estimates of interest.

The effect of non-normality is generally small for measures that rely on the center of the distribution, like means or regression weights, but it could be substantial for estimates like a variance or a percentile. In general, normal imputations appear to be robust against violations of normality. Demirtas et al. (2008) found that flatness of the density, heavy tails, non-zero peakedness, skewness and multimodality do not appear to hamper the good performance of multiple imputation for the mean structure in samples $n > 400$, even for high percentages (75%) of missing data in one variable. The variance parameter is more critical though, and could be off-target in smaller samples.

One approach is to transform the data toward normality before imputation, and back-transform them after imputation. A beneficial side effect of transformation is that the relation between x and y may become closer to a linear relation. Sometimes applying a simple function to the data, like the logarithmic or inverse transform, is all that is needed. More generally, the transformation could be made to depend on known covariates like age and sex, for example as done in the LMS model (Cole and Green, 1992) or the GAMLSS model (Rigby and Stasinopoulos, 2005).

Von Hippel (2013) warns that application of tricks to make the distribution of skewed variables closer to normality (e.g., censoring, truncation, transformation) may make matters worse. Censoring (rounding a disallowed value to the nearest allowed value) and truncation (redrawing a disallowed value until it is within the allowed range) can change both the mean and variability in the data. Transformations may fail to achieve near-normality, and even if that succeeds, bivariate relations may be affected when imputed by a method that assumes normality. The examples of Von Hippel are somewhat extreme, but they do highlight the point that simple fixes to achieve normality are limited by what they can do.

There are two possible strategies to progress. The first is to use predictive mean matching. Section 3.4 will describe this approach in more detail. The other strategy is to model the non-normal data, and to directly draw imputations from those models. Liu (1995) proposed methods for drawing imputations under the *t*-distribution instead of the normal. He and Raghunathan (2006) created imputations by drawing from Tukey's *gh*-distribution, which can take many shapes. Demirtas and Hedeker (2008a) investigated the behavior of methods for drawing imputation from the Beta and Weibull distributions. Likewise, Demirtas and Hedeker (2008b) took draws from Fleishman polynomials, which allows for combinations of left and right skewness with platykurtic and leptokurtic distributions.

The GAMLSS method (Rigby and Stasinopoulos, 2005; Stasinopoulos et al., 2017) extends both the generalized linear model and the generalized additive model. A unique feature of GAMLSS is its ability to specify a (possibly nonlinear) model for each of the parameters of the distribution, thus giving rise to an extremely flexible toolbox that can be used to model almost any distribution. The gamlss package contains over 60 built-in distributions. Each distribution comes with a function to draw random variates, so once the gamlss model is fitted, it can also be used to draw imputations. The first edition of this book showed how to construct a new univariate imputation function that mice could call. This is not needed any more. De Jong (2012) and De Jong et al. (2016) developed a series of imputation methods based on GAMLSS, so it is now easy to perform multiple imputation under variety of distributions. The ImputeRobust package (Salfran and Spiess, 2017), implements various mice methods for continuous data: gamlss (normal), gamlssJSU (Johnson's SU), gamlssTF (*t*-distribution) and gamlssGA (gamma distribution). The following section demonstrates the use of the package.

3.3.2 Imputation from the *t*-distribution

We illustrate imputation from the *t*-distribution. The *t*-distribution is favored for more robust statistical modeling in a variety of settings (Lange et al., 1989). Van Buuren and Fredriks (2001) observed unexplained kurtosis in the distribution of head circumference in children. Rigby and Stasinopoulos (2006) fitted a *t*-distribution to these data, and observed a substantial improvement of the fit.

Figure 3.4 plots the data for Dutch boys aged 1–2 years. Due to the presence of several outliers, the *t*-distribution with 6.7 degrees of freedom fits the data substantially better than the normal distribution (Akaike Information Criterion (AIC): 2974.5 (normal model) versus 2904.3 (*t*-distribution). If the

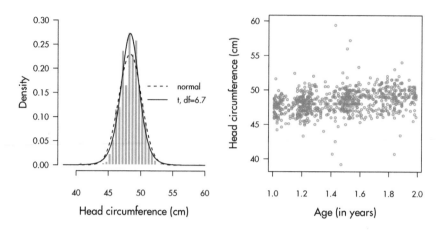

Figure 3.4: Measured head circumference of 755 Dutch boys aged 1–2 years (Fredriks et al., 2000a).

outliers are genuine data, then the *t*-distribution should provide imputations that are more realistic than the normal.

We create a synthetic dataset by imputing head circumference of the same 755 boys. Imputation is easily done with the following steps: append the data with a duplicate, create missing data in `hc` and run `mice()` calling the `gamlssTF` method as follows:

```
library(ImputeRobust)
library(gamlss)
data(db)
data <- subset(db, age > 1 & age < 2, c("age", "head"))
names(data) <- c("age", "hc")
synthetic <- rep(c(FALSE, TRUE), each = nrow(data))
data2 <- rbind(data, data)
row.names(data2) <- 1:nrow(data2)
data2[synthetic, "hc"] <- NA
imp <- mice(data2, m = 1, meth = "gamlssTF", seed = 88009,
            print = FALSE)
syn <- subset(mice::complete(imp), synthetic)
```

Figure 3.5 is the equivalent of Figure 3.4, but now calculated from the synthetic data. Both configurations are similar. As expected, some outliers also occur in the imputed data, but these are a little less extreme than in the observed data due to the smoothing by the *t*-distribution. The estimated degrees of freedom varies over replications, and appears to be somewhat larger than the value of 6.7 estimated from the observed data. For this replication, it is larger (11.5). The distribution of the imputed data is better behaved

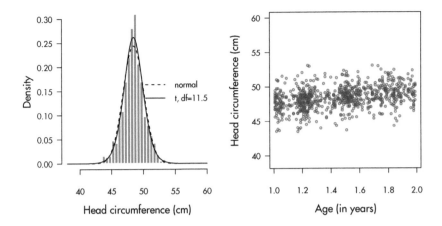

Figure 3.5: Fully synthetic data of head circumference of 755 Dutch boys aged 1–2 years using a *t*-distribution.

compared to the observed data. The typical rounding patterns seen in the real measurements are not present in the imputed data. Though these are small differences, they may be of relevance in particular analyses.

3.4 Predictive mean matching

3.4.1 Overview

Predictive mean matching calculates the predicted value of target variable Y according to the specified imputation model. For each missing entry, the method forms a small set of candidate donors (typically with 3, 5 or 10 members) from all complete cases that have predicted values closest to the predicted value for the missing entry. One donor is randomly drawn from the candidates, and the observed value of the donor is taken to replace the missing value. The assumption is the distribution of the missing cell is the same as the observed data of the candidate donors.

Predictive mean matching is an easy-to-use and versatile method. It is fairly robust to transformations of the target variable, so imputing $\log(Y)$ often yields results similar to imputing $\exp(Y)$. The method also allows for discrete target variables. Imputations are based on values observed elsewhere, so they are realistic. Imputations outside the observed data range will not occur, thus evading problems with meaningless imputations (e.g., negative body height). The model is implicit (Little and Rubin, 2002), which means that there is no need to define an explicit model for the distribution of the

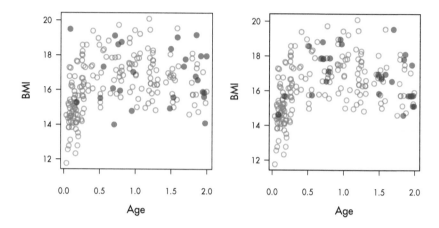

Figure 3.6: Robustness of predictive mean matching (right) relative to imputation under the linear normal model (left).

missing values. Because of this, predictive mean matching is less vulnerable to model misspecification than the methods discussed in Sections 3.2 and 3.3.

Figure 3.6 illustrates the robustness of predictive mean matching relative to the normal model. The figure displays the body mass index (BMI) of children aged 0–2 years. BMI rapidly increases during the first half year of life, has a peak around 1 year and then slowly drops at ages when the children start to walk. The imputation model is, however, incorrectly specified, being linear in age. Imputations created under the normal model display in an incorrect slowly rising pattern, and contain several implausible values. In contrast, the imputations created by predictive mean matching follow the data quite nicely, even though the predictive mean itself is clearly off-target for some of the ages. This example shows that predictive mean matching is robust against misspecification, where the normal model is not.

Predictive mean matching is an example of a hot deck method, where values are imputed using values from the complete cases matched with respect to some metric. The expression "hot deck" literally refers to a pack of computer control cards containing the data of the cases that are in some sense close. Reviews of hot deck methods can be found in Ford (1983), Brick and Kalton (1996), Koller-Meinfelder (2009), Andridge and Little (2010) and De Waal et al. (2011, pp. 249–255, 349–355).

Figure 3.7 is an illustration of the method using the `whiteside` data. The predictor is equal to 5°C and the bandwidth is 1.2. The thick blue line indicates the area of the target variable where matches should be sought. The blue part of the figure are considered fixed. The red line correspond to one random draw of the line parameters to incorporate sampling uncertainty. The two light-red bands indicate the area where matches are permitted. In this

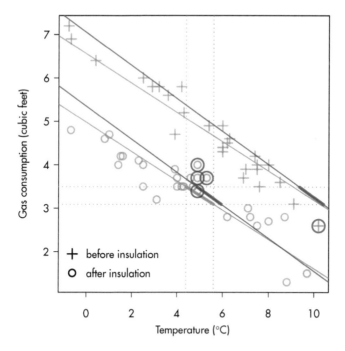

particular instance, five candidate donors are found, four from the subgroup "after insulation" and one from the subgroup "before insulation." The last step is to make a random draw among these five candidates. The red parts in the figure will vary between different imputed datasets, and thus the set of candidates will also vary over the imputed datasets.

The data point at coordinate (10.2, 2.6) is one of the candidate donors. This point differs from the incomplete unit in both temperature and insulation status, yet it is selected as a candidate donor. The advantage of including the point is that closer matches in terms of the predicted values are possible. Under the assumption that the distribution of the target in different bands is similar, including points from different bands is likely to be beneficial.

3.4.2 Computational details♠

Various metrics are possible to define the distance between the cases. The predictive mean matching metric was proposed by Rubin (1986) and Little (1988). This metric is particularly useful for missing data applications because it is optimized for each target variable separately. The predicted value only needs to be a convenient one-number summary of the important information that relates the covariates to the target. Calculation is straightforward, and it is easy to include nominal and ordinal variables.

Once the metric has been defined, there are various ways to select the donor. Let $\hat{y}_i$ denote the predicted value of the rows with an observed y_i where $i = 1, \ldots, n_1$. Likewise, let $\hat{y}_j$ denote the predicted value of the rows with missing y_j where $j = 1, \ldots, n_0$. Andridge and Little (2010) distinguish four methods:

1. Choose a threshold η, and take all i for which $|\hat{y}_i - \hat{y}_j| < \eta$ as candidate donors for imputing j. Randomly sample one donor from the candidates, and take its y_i as replacement value.

2. Take the closest candidate, i.e., the case i for which $|\hat{y}_i - \hat{y}_j|$ is minimal as the donor. This is known as "nearest neighbor hot deck," "deterministic hot deck" or "closest predictor."

3. Find the d candidates for which $|\hat{y}_i - \hat{y}_j|$ is minimal, and sample one of them. Usual values for d are 3, 5 and 10. There is also an adaptive method to specify the number of donors (Schenker and Taylor, 1996).

4. Sample one donor with a probability that depends on $|\hat{y}_i - \hat{y}_j|$ (Siddique and Belin, 2008).

In addition, it is useful to distinguish four types of matching:

1. *Type 0*: $\hat{y} = X_{\text{obs}}\hat{\beta}$ is matched to $\hat{y}_j = X_{\text{mis}}\hat{\beta}$;

2. *Type 1*: $\hat{y} = X_{\text{obs}}\hat{\beta}$ is matched to $\dot{y}_j = X_{\text{mis}}\dot{\beta}$;

3. *Type 2*: $\dot{y} = X_{\text{obs}}\dot{\beta}$ is matched to $\dot{y}_j = X_{\text{mis}}\dot{\beta}$;

4. *Type 3*: $\dot{y} = X_{\text{obs}}\dot{\beta}$ is matched to $\ddot{y}_j = X_{\text{mis}}\ddot{\beta}$.

Here $\hat{\beta}$ is the estimate of β, while $\dot{\beta}$ is a value randomly drawn from the posterior distribution of β. Type 0 matching ignores the sampling variability in $\hat{\beta}$, leading to improper imputations. Type 2 matching appears to solve this. However, it is insensitive to the process of taking random draws of β if there are only a few variables. In the extreme case, with a single X, the set of candidate donors based on $|\dot{y}_i - \dot{y}_j|$ remains unchanged under different values of $\dot{\beta}$, so the same donor(s) get selected too often. Type 1 matching is a small but nifty adaptation of the matching distance that seems to alleviate the problem. The difference with Type 0 and Type 2 matching is that in Type 1 matching only $X_{\text{mis}}\dot{\beta}$ varies stochastically and does not cancel out any more. As a result $\dot{\eta}$ incorporates between-imputation variation. Type 3 matching creates two draws for β, one for the donor set and one for the recipient set. In retrospect, it is interesting to note that Type 1 matching was already described by Little (1988, eq. 4). It disappeared from the literature, only to reappear two decades

1. Calculate $\dot{\beta}$ and $\hat{\beta}+$ by Steps 1-8 of Algorithm 3.1.

2. Calculate $\dot{\eta}(i,j) = |X_{\text{obs},[i]}\hat{\beta} - X_{\text{mis},[j]}\dot{\beta}|$ with $i = 1, \ldots, n_1$ and $j = 1, \ldots, n_0$.

3. Construct n_0 sets Z_j, each containing d candidate donors, from Y_{obs} such that $\sum_d \dot{\eta}(i,j)$ is minimum for all $j = 1, \ldots, n_0$. Break ties randomly.

4. Draw one donor i_j from Z_j randomly for $j = 1, \ldots, n_0$.

5. Calculate imputations $\dot{y}_j = y_{i_j}$ for $j = 1, \ldots, n_0$.

later in the works of Koller-Meinfelder (2009, p. 43) and White et al. (2011b, p. 383).

Algorithm 3.3 provides the steps used in predictive mean matching using Bayesian parameter draws for β. It is possible to create the bootstrap version of this algorithm that will also evade the need to draw β along the same lines as Algorithm 3.2. Given that the number of candidate donors and the model for the mean is provided by the user, the algorithm does not need an explicit specification of the distribution.

Morris et al. (2014) suggested a variation called *local residuals draws*. Rather than taking the observed value of the donor, this method borrows the residual from the donor, and adds that to the predicted value from the target case. Thus, imputations are not equal to observed values, and can extend beyond the range of the observed data. This may address concerns about variability of imputations.

3.4.3   Number of donors

There are different strategies for defining the set and number of candidate donors. Setting $d = 1$ is generally considered to be too low, as it may reselect the same donor over and over again. Predictive mean matching performs very badly when d is small and there are lots of ties for the predictors among the individuals to be imputed. The reason is that the tied individuals all get the same imputed value in each imputed dataset when $d = 1$ (Ian White, personal communication). Setting d to a high value (say $n/10$) alleviates the duplication problem, but may introduce bias since the likelihood of bad matches increases. Schenker and Taylor (1996) evaluated $d = 3$, $d = 10$ and an adaptive scheme. The adaptive method was slightly better than using a fixed number of candidates, but the differences were small. Morris et al. (2014) compared various settings for d, and found that $d = 5$ and $d = 10$ generally provided the best results. Kleinke (2017) found that $d = 5$ may be too high for sample size lower than $n = 100$, and suggested setting $d = 1$ for better point estimates for small samples. Gaffert et al. (2016) explored scenarios in which candidate donors have different probabilities to be drawn, where the probability depends on the distance between the donor and recipient cases. As all observed cases can be donors in this scenario, there is no need to specify d. Instead a closeness parameter needs to be specified, and this was made adaptive to the data. An advantage of using all donors is that the variance of the imputations can be corrected by the Parzen correction, which alleviates concerns about insufficient variability of the imputes. Their simulations showed that with a small sample ($n = 10$), the adaptive method is clearly superior to methods with a fixed donor pool. The method is available in `mice` as the `midastouch` method. There is also a separate `midastouch` package in R. Related work can be found in Tutz and Ramzan (2015).

The default in `mice` is $d = 5$, and represents a compromise. The above results suggest that an adaptive method for setting d could improve small sample behavior. Meanwhile, the number of donors can be changed through the `donors` argument.

Table 3.3 repeats the simulation experiment done in Tables 3.1 and 3.2 for predictive mean matching for three different choices of the number d of candidate donors. Results are given for $n = 50$ and $n = 1000$. For $n = 50$ we find that β_1 is increasingly biased towards the null for larger d. Because of the bias, the coverage is lower than nominal. For missing x the bias is much smaller. Setting d to a lower value, as recommended by Kleinke (2017), improves point estimates, but the magnitude of the effect depends on whether the missing values occur in x or y. For the sample size $n = 1000$ predictive mean matching appears well calibrated for $d = 5$ for missing data in y, and has slight undercoverage for missing data in x. Note that Table 3.3 in the first edition of this book presented incorrect information because it had erroneously imputed the data by `norm` instead of `pmm`.

3.4.4   Pitfalls

The obvious danger of predictive mean matching is the duplication of the same donor value many times. This problem is more likely to occur if the sample is small, or if there are many more missing data than observed data in a particular region of the predicted value. Such unbalanced regions are more likely if the proportion of incomplete cases is high, or if the imputation model contains variables that are very strongly related to the missingness. For small samples the donor pool size can be reduced, but be aware that this may not work if there are only a few predictors.

The traditional method does not work for a small number of predictors. Heitjan and Little (1991) report that for just two predictors the results were "disastrous." The cause of the problem appears to be related to their use

Table 3.3: Properties of β_1 under multiple imputation by predictive mean matching and $m = 5$, 50% MCAR missing data and $n_{\text{sim}} = 1000$.

Method		Bias	% Bias	Coverage	CI Width	RMSE
Missing y, $n = 50$	d					
pmm	1	0.016	5.4	0.884	0.252	0.071
pmm	3	0.028	9.7	0.890	0.242	0.070
pmm	5	0.039	13.6	0.876	0.241	0.075
pmm	10	0.065	22.4	0.806	0.245	0.089
Missing x						
pmm	1	-0.002	0.8	0.916	0.223	0.063
pmm	3	0.002	0.9	0.931	0.228	0.061
pmm	5	0.008	2.8	0.938	0.237	0.062
pmm	10	0.028	9.6	0.946	0.261	0.067
Listwise deletion		0.000	0.0	0.946	0.251	0.063
Missing y, $n = 50$	κ					
midastouch	auto	0.013	4.5	0.920	0.265	0.066
midastouch	2	0.032	11.1	0.917	0.273	0.068
midastouch	3	0.018	6.2	0.927	0.261	0.064
midastouch	4	0.012	4.1	0.926	0.260	0.064
Missing x						
midastouch	auto	-0.003	0.9	0.932	0.241	0.060
midastouch	2	0.013	4.4	0.959	0.264	0.059
midastouch	3	0.000	0.2	0.947	0.245	0.058
midastouch	4	-0.004	1.4	0.940	0.237	0.058
Listwise deletion		0.000	0.0	0.946	0.251	0.063
Missing y, $n = 1000$	d					
pmm	1	0.001	0.2	0.929	0.056	0.014
pmm	3	0.001	0.4	0.950	0.056	0.013
pmm	5	0.002	0.6	0.951	0.055	0.013
pmm	10	0.003	1.2	0.932	0.054	0.013
Missing x						
pmm	1	0.000	0.2	0.926	0.041	0.011
pmm	3	0.000	0.1	0.933	0.041	0.011
pmm	5	0.000	0.1	0.937	0.042	0.011
pmm	10	0.000	0.1	0.928	0.042	0.011
Listwise deletion		0.000	0.1	0.955	0.050	0.012

of Type 0 matching. The default in `mice` is Type 1 matching, which works better for small number of predictors. The setting can be changed to Type 0 or Type 2 matching through the `matchtype` argument.

Predictive mean matching is no substitute for sloppy modeling. If the imputation model is misspecified, performance can become poor if there are strong relations in the data that are not modeled (Morris et al., 2014). The default imputation model in `mice` consists of a linear main effect model conditional on all other variables, but this may be inadequate in the presence of strong nonlinear relations. More generally, any terms appearing in the complete-data model need to be accounted for in the imputation model. Morris et al. (2014) advise to spend efforts on specifying the imputation model correctly, rather than expecting predictive mean matching to do the work.

3.4.5   Conclusion

Predictive mean matching with $d = 5$ is the default in `mice()` for continuous data. The method is robust against misspecification of the imputation model, yet performs as well as theoretically superior methods. In the context of missing covariate data, Marshall et al. (2010a) concluded that predictive mean matching "produced the least biased estimates and better model performance measures." Another simulation study that addressed skewed data concluded that predictive mean matching "may be the preferred approach provided that less than 50% of the cases have missing data and the missing data are not MNAR" (Marshall et al., 2010b). Kleinke (2017) found that the method works well across a wide variety of scenarios, but warned the default cannot address severe skewness or small samples.

The method works best with large samples, and provides imputations that possess many characteristics of the complete data. Predictive mean matching cannot be used to extrapolate beyond the range of the data, or to interpolate within the range of the data if the data at the interior are sparse. Also, it may not perform well with small datasets. Bearing these points in mind, predictive mean matching is a great all-around method with exceptional properties.

3.5   Classification and regression trees

3.5.1   Overview

Classification and regression trees (CART) (Breiman et al., 1984) are a popular class of machine learning algorithms. CART models seek predictors and cut points in the predictors that are used to split the sample. The cut points divide the sample into more homogeneous subsamples. The splitting process is repeated on both subsamples, so that a series of splits defines a

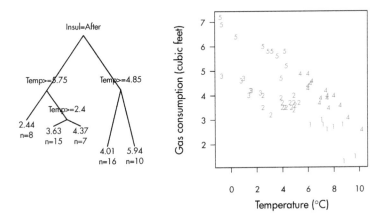

binary tree. The target variable can be discrete (classification tree) or continuous (regression tree).

Figure 3.8 illustrates a simple CART solution for the whiteside data. The left-hand side contains the optimal binary tree for predicting gas consumption from temperature and insulation status. The right-hand side shows the scatterplot in which the five groups are labeled by their terminal nodes.

CART methods have properties that make them attractive for imputation: they are robust against outliers, can deal with multicollinearity and skewed distributions, and are flexible enough to fit interactions and nonlinear relations. Furthermore, many aspects of model fitting have been automated, so there is "little tuning needed by the imputer" (Burgette and Reiter, 2010).

The idea of using CART methods for imputation has been suggested by a wide variety of authors in a variety of ways. See Saar-Tsechansky and Provost (2007) for an introductory overview. Some investigators (He, 2006; Vateekul and Sarinnapakorn, 2009) simply fill in the mean or mode. The majority of tree-based imputation methods use some form of single imputation based on prediction (Bárcena and Tusell, 2000; Conversano and Cappelli, 2003; Siciliano et al., 2006; Creel and Krotki, 2006; Ishwaran et al., 2008; Conversano and Siciliano, 2009). Multiple imputation methods have been developed by Harrell (2001), who combined it with optimal scaling of the input variables, by Reiter (2005b) and by Burgette and Reiter (2010). Wallace et al. (2010) present a multiple imputation method that averages the imputations to produce a single tree and that does not pool the variances. Parker (2010) investigates multiple imputation methods for various unsupervised and supervised learning algorithms.

The missForest method (Stekhoven and Bühlmann, 2011) successfully

---

Algorithm 3.4: Imputation under a tree model using the bootstrap.♠

1. Draw a bootstrap sample $(\dot{y}_{\text{obs}}, \dot{X}_{\text{obs}})$ of size n_1 from $(y_{\text{obs}}, X_{\text{obs}})$.

2. Fit $\dot{y}_{\text{obs}}$ by $\dot{X}_{\text{obs}}$ by a tree model $f(X)$.

3. Predict the n_0 terminal nodes g_j from $f(X_{\text{mis}})$.

4. Construct n_0 sets Z_j of all cases at node g_j, each containing d_j candidate donors.

5. Draw one donor i_j from Z_j randomly for $j = 1, \ldots, n_0$.

6. Calculate imputations $\dot{y}_j = y_{i_j}$ for $j = 1, \ldots, n_0$.

---

used regression and classification trees to predict the outcomes in mixed continuous/categorical data. `MissForest` is popular, presumably because it produces a *single* complete dataset, which at the same time is the reason why it fails as a scientific method. The `missForest` method does not account for the uncertainty caused by the missing data, treats the imputed data as if they were real (which they are not), and thus invents information. As a consequence, p-values calculated after application of `missForest` will be more significant than they actually are, confidence intervals will be shorter than they actually are, and relations between variables will be stronger than they actually are. These problems worsen as more missing values are imputed. Unfortunately, comparisons studies that evaluate only accuracy, such as Waljee et al. (2013), will fail to detect these problems.

As a alternative, multiple imputations can be created using the tree in Figure 3.8. For a given temperature and insulation status, traverse the tree and find the appropriate terminal node. Form the donor group of all observed cases at the terminal node, randomly draw a case from the donor group, and take its reported gas consumption as the imputed value. The idea is identical to predictive mean matching (cf. Section 3.4), where the "predictive mean" is now calculated by a tree model instead of a regression model. As before, the parameter uncertainty can be incorporated by fitting the tree on a bootstrapped sample.

Algorithm 3.4 describes the major steps of an algorithm for creating imputations using a classification or regression tree. There is considerable freedom at step 2, where the tree model is fitted to the training data $(\dot{y}_{\text{obs}}, \dot{X}_{\text{obs}})$. It may be useful to fit the tree such that the number of cases at each node is equal to some pre-set number, say 5 or 10. The composition of the donor groups will vary over different bootstrap replications, thus incorporating sampling uncertainty about the tree.

Multiple imputation methodology using trees has been developed by Burgette and Reiter (2010), Shah et al. (2014) and Doove et al. (2014). The main motivation given in these papers was to improve our ability to account for interactions and other non-linearities, but these are generic methods that apply to both continuous and categorical outcomes and predictors. Burgette and Reiter (2010) used the `tree` package, and showed that the CART results for recovering interactions were uniformly better than standard techniques. Shah et al. (2014) applied random forest techniques to both continuous and categorical outcomes, which produced more efficient estimates than standard procedures. The techniques are available as methods `rfcat` and `rfcont` in the `CALIBERrfimpute` package. Doove et al. (2014) independently developed a similar set of routines building on the `rpart` (Therneau et al., 2017) and `randomForest` (Liaw and Wiener, 2002) packages. Methods `cart` and `rf` are part of `mice`.

A recent development is the growing interest from the machine learning community for the idea of multiple imputation. The problem of imputing missing values has now been discovered by many, but unfortunately nearly all new algorithms produce single imputations. An exception is the paper by Sovilj et al. (2016), who propose the *extreme learning machine* using conditional Gaussian mixture models to generate multiple imputations. It is a matter of time before researchers realize the intimate connections between multiple imputation and ensemble learning, so that more work along these lines may follow.

3.6 Categorical data

3.6.1 Generalized linear model

Imputation of missing categorical data is possible under the broad class of generalized linear models (McCullagh and Nelder, 1989). For incomplete binary variables we use *logistic regression*, where the outcome probability is modeled as

$$\Pr(y_i = 1 | X_i, \beta) = \frac{\exp(X_i\beta)}{1 + \exp(X_i\beta)} \tag{3.4}$$

A categorical variable with K unordered categories is imputed under the *multinomial logit model*

$$\Pr(y_i = k | X_i, \beta) = \frac{\exp(X_i\beta_k)}{\sum_{k=1}^{K} \exp(X_i\beta_k)} \tag{3.5}$$

for $k = 1, \ldots, K$, where β_k varies over the categories and where $\beta_1 = 0$ to identify the model. A categorical variable with K ordered categories is

Algorithm 3.5: Imputation of a binary variable by means of Bayesian logistic regression.♠

---

1. Estimate regression weights $\hat{\beta}$ from $(y_{\mathrm{obs}}, X_{\mathrm{obs}})$ by iteratively reweighted least squares.

2. Obtain V, the unscaled estimated covariance matrix of $\hat{\beta}$.

3. Draw q independent $N(0,1)$ variates in vector $\dot{z}_1$.

4. Calculate $V^{1/2}$ by Cholesky decomposition.

5. Calculate $\dot{\beta} = \hat{\beta} + \dot{z}_1 V^{1/2}$.

6. Calculate n_0 predicted probabilities $\dot{p} = 1/(1 + \exp(-X_{\mathrm{mis}}\dot{\beta}))$.

7. Draw n_0 random variates from the uniform distribution $U(0,1)$ in the vector u.

8. Calculate imputations $\dot{y}_j = 1$ if $u_j \leq \dot{p}_j$, and $\dot{y}_j = 0$ otherwise, where $j = 1, \ldots, n_0$.

---

imputed by the *ordered logit model*, or *proportional odds model*

$$\Pr(y_i \leq k | X_i, \beta, \tau_k) = \frac{\exp(\tau_k - X_i\beta)}{1 + \exp(\tau_k - X_i\beta)} \tag{3.6}$$

with the slope β is identical across categories, but the intercepts τ_k differ. For identification, we set $\tau_1 = 0$. The probability of observing category k is written as

$$\Pr(y_i = k | X_i) = \Pr(y_i \leq k | X_i) - \Pr(y_i \leq k - 1 | X_i) \tag{3.7}$$

where the model parameters β, τ_k and τ_{k-1} are suppressed for clarity. Scott Long (1997) is a very readable introduction to these methods. The practical application of these techniques in R is treated in Aitkin et al. (2009). The general idea is to estimate the probability model on the subset of the observed data, and draw synthetic data according to the fitted probabilities to impute the missing data. The parameters are typically estimated by iteratively reweighted least squares. As before, the variability of the model parameters β and $\tau_2, \ldots, \tau_K$ introduces additional uncertainty that needs to be incorporated into the imputations.

Algorithm 3.5 provides the steps for an approximate Bayesian imputation method using logistic regression. The method assumes that the parameter vector β follows a multivariate normal distribution. Although this is true in large samples, the distribution can in fact be far from normal for modest n_1, for large q or for predicted probabilities close to 0 or 1. The procedure is also

Table 3.4: Artificial data demonstrating complete separation. Adapted from
White et al. (2010).

	Symptom	
Disease	Yes	No
Yes	100	100
No	0	100
Unknown	100	100

approximate in the sense that it does not draw the estimated covariance V matrix. It is possible to define an explicit Bayesian sampling for drawing β and V from their exact posteriors. This method is theoretically preferable, but as it requires more elaborate modeling, it does not easily extend to other regression situations. In `mice` the algorithm is implemented as the method `logreg`.

It is easy to construct a bootstrap version that avoids some of the difficulties in Algorithm 3.5. Prior to estimating $\hat{\beta}$, we include a step that draws a bootstrap sample from Y_{obs} and X_{obs}. Steps 2–5 can then be replaced by equating $\dot{\beta} = \hat{\beta}$.

The algorithms for imputation of variables with more than two categories follow the same structure. In `mice` the multinomial logit model in method `polyreg` is estimated by the `nnet::multinom()` function in the `nnet` package. The ordered logit model in method `polr` is estimated by the `polr()` function of the `MASS` package. Even though the ordered model uses fewer parameters, it is often more difficult to estimate. In cases where `MASS::polr()` fails to converge, `nnet::multinom()` will take over its duties. See Venables and Ripley (2002) for more details on both functions.

3.6.2 Perfect prediction♠

There is a long-standing technical problem in models with categorical outcomes, known as *separation* or *perfect prediction* (Albert and Anderson, 1984; Lesaffre and Albert, 1989). The standard work by Hosmer and Lemeshow (2000, pp. 138–141) discussed the problem, but provided no solution. The problem occurs, for example, when predicting the presence of a disease from a set of symptoms. If one of the symptoms (or a combination of symptoms) always leads to the disease, then we can perfectly predict the disease for any patient who has the symptom(s).

Table 3.4 contains an artificial numerical example. Having the symptom always implies the disease, so knowing that the patient has the symptom will allow perfect prediction of the disease status. When such data are analyzed, most software will print out a warning message and produce unusually large standard errors.

Now suppose that in a new group of 200 patients (100 in each symptom group) we know only the symptom and impute disease status. Under MAR, we should impute all 100 cases with the symptom to the diseased group, and divide the 100 cases without the symptom randomly over the diseased and non-diseased groups. However, this is not what happens in Algorithm 3.5. The estimate of V will be very large as a result of separation. If we naively use this V then $\dot{\beta}$ in step 5 effectively covers both positive and negative values equally likely. This results in either correctly 100 imputations in Yes or incorrectly 100 imputations in No, thereby resulting in bias in the disease probability.

The problem has recently gained attention. There are at least six different approaches to perfect prediction:

1. Eliminate the variable that causes perfect prediction.

2. Take $\hat{\beta}$ instead of $\dot{\beta}$.

3. Use penalized regression with Jeffreys prior in step 2 of Algorithm 3.5 (Firth, 1993; Heinze and Schemper, 2002).

4. Use the bootstrap, and then apply method 1.

5. Use data augmentation, a method that concatenates pseudo-observations with a small weight to the data, effectively prohibiting infinite estimates (Clogg et al., 1991; White et al., 2010).

6. Apply the explicit Bayesian sampling with a suitable weak prior. Gelman et al. (2008) recommend using independent Cauchy distributions on all logistic regression coefficients.

Eliminating the most predictive variable is generally undesirable in the context of imputation, and may in fact bias the relation of interest. Option 2 does not yield proper imputations, and is therefore not recommended. Option 3 provides finite estimates, but has been criticized as not being well interpretable in a regression context (Gelman et al., 2008) and computationally inefficient (White et al., 2010). Option 4 corrects method 1, and is simple to implement. Options 5 and 6 have been recommended by White et al. (2010) and Gelman et al. (2008), respectively.

Methods 4, 5 and 6 all solve a major difficulty in the construction of automatic imputation techniques. It is not yet clear whether one of these methods is superior. The `logreg`, `polr` and `polyreg` methods in `mice` implement option 5.

3.6.3   Evaluation

The methods are based on the elegant generalized linear models. Simulations presented in Van Buuren et al. (2006) show that these methods performed quite well in the lab. When used in practice however, the methods may be unstable, slow and exhibit poor performance. Hardt et al. (2013) intentionally pushed the logistic methods to their limits, and observed that most methods break down relatively quick, i.e., if the proportion of missing values exceeds 0.4. Van der Palm et al. (2016a) found that `logreg` failed to pick up a three-way association in the data, leading to biased estimates. Likewise, Vidotto et al. (2015) observed that `logreg` did not recover the structure in the data as well as latent class models. Wu et al. (2015) found poor results for all three methods (i.e., binary, multinomial and proportional odds), and advise against their application. Akande et al. (2017) reported difficulties with fitting multinomial variables having many categories. The performance of the procedures suffered when variables with probabilities nearly equal to one (or zero) are included in the models. Methods based on the generalized linear model were found to be inferior to method `cart` (cf. Section 3.5) and to latent class models for categorical data (cf. Section 4.4). Audigier et al. (2017) found that logistic regression presented difficulties on the datasets with a high number of categories, resulting in undercoverage on several quantities.

Imputation of categorical data is more difficult than continuous data. As a rule of thumb, in logistic regression we need at least *10 events per predictor* in order to get reasonably stable estimates of the regression coefficients (Van Belle, 2002, p. 87). So if we impute 10 binary outcomes, we need 100 events, and if the events occur with a probability of 0.1, then we need $n > 1000$ cases. If we impute outcomes with more categories, the numbers rapidly increase for two reasons. First, we have more possible outcomes, and we need 10 events for each category. Second, when used as predictor, each nominal variable is expanded into dummy variables, so the number of predictors multiplies by the number of categories minus 1. The defaults `logreg`, `polyreg` and `polr` tend to preserve the main effects well provided that the parameters are identified and can be reasonably well estimated. In many datasets, especially those with many categories, the ratio of the number of fitted parameters relative to the number of events easily drops below 10, which may lead to estimation problems. In those cases, the advice is to specify more robust methods, like `pmm`, `cart` or `rf`.

3.7 Other data types

3.7.1 Count data

Examples of count data include the number of children in a family or the number of crimes committed. The minimum value is zero. Imputing incomplete count data should produce non-negative synthetic replacement values. Count data can be imputed in various ways:

1. Predictive mean matching (cf. Section 3.4).

2. Ordered categorical imputation (cf. Section 3.6).

3. (Zero-inflated) Poisson regression (Raghunathan et al., 2001).

4. (Zero-inflated) negative binomial regression (Royston, 2009).

Poisson regression is a class of models that is widely applied in biostatistics. The Poisson model can be thought of as the sum of the outcomes from a series of independent flips of the coin. The negative binomial is a more flexible model that is often applied an as alternative to account for over-dispersion. Zero-inflated versions of both models can be used if the number of zero values is larger than expected. The models are special cases of the generalized linear model, and do not bring new issues compared to, say, logistic regression imputation.

Kleinke and Reinecke (2013) developed methods for zero-inflated and over-dispersed data, using both Bayesian and bootstrap approaches. Methods are available in the `countimp` package for the Poisson model (`pois`, `pois.boot`), quasi Poission model (`qpois`, `qpois.boot`), the negative binomial model (`nb`), the zero-inflated Poisson (`2l.zip`, `2l.zip.boot`), and the zero-inflated negative binomial (`2l.zinb`, `2l.zinb.boot`). Note that, despite their naming, these 2l methods are for single-level imputation. An alternative is the `ImputeRobust` package (Salfran and Spiess, 2017), which implements the following `mice` methods for count data: `gamlssPO` (Poisson), `gamlssZIBI` (zero-inflated binomial) and `gamlssZIP` (zero-inflated Poisson)gamlssZIP. Kleinke (2017) evaluated the use of predictive mean matching as a multipurpose missing data tool. By and large, the simulations illustrate that the method is robust against violations of its assumptions, and can be recommended for imputation of mildly to moderately skewed variables when sample size is sufficiently large.

3.7.2 Semi-continuous data

Semi-continuous data have a high mass at one point (often zero) and a continuous distribution over the remaining values. An example is the number of cigarettes smoked per day, which has a high mass at zero because of the non-smokers, and an often highly skewed unimodal distribution for the smokers. The difference with count data is gradual. Semi-continuous data are typically treated as continuous data, whereas count data are generally considered discrete.

Imputation of semi-continuous variables needs to reproduce both the point mass and continuously varying part of the data. One possibility is to apply a general-purpose method that preserves distributional features, like predictive mean matching (cf. Section 3.4).

An alternative is to model the data in two parts. The first step is to determine whether the imputed value is zero or not. The second step is only done for those with a non-zero value, and consists of drawing a value from the continuous part. Olsen and Schafer (2001) developed an imputation technique by combining a logistic model for the discrete part, and a normal model for the continuous part, possibly after a normalizing transformation. A more general two-part model was developed by Javaras and Van Dyk (2003), who extended the standard general location model (Olkin and Tate, 1961) to impute partially observed semi-continuous data.

Yu et al. (2007) evaluated nine different procedures. They found that predictive mean matching performs well, provided that a sufficient number of data points in the neighborhood of the incomplete data are available. Vink et al. (2014) found that generic predictive mean matching is at least as good as three dedicated methods for semi-continuous data: the two-part models as implemented in `mi` (Su et al., 2011) and `irmi` (Templ et al., 2011b), and the blocked general location model by Javaras and Van Dyk (2003). Vroomen et al. (2016) investigated imputation of cost data, and found that predictive mean matching of the log-transformed outperformed plain predictive mean matching, a two-step method and complete-case analysis, and hence recommend log-transformed method for monetary data.

3.7.3 Censored, truncated and rounded data

An observation y_i is censored if its value is only partly known. In *right-censored* data we only know that $y_i > a_i$ for a censoring point a_i. In *left-censored* data we only know that $y_i \leq b_i$ for some known censoring point b_i, and in *interval censoring* we know $a_i \leq y_i \leq b_i$. Right-censored data arise when the true value is beyond the maximum scale value, for example, when body weight exceeds the scale maximum, say 150 kg. When y_i is interpreted as time taken to some event (e.g., death), right-censored data occur when the observation period ends before the event has taken place. Left and right censoring may cause floor and ceiling effects. Rounding data to fewer decimal places results in interval-censored data.

Truncation is related to censoring, but differs from it in the sense that value below (left truncation) or above (right truncation) the truncation point is not recorded at all. For example, if persons with a weight in excess of 150 kg are removed from the sample, we speak of truncation. The fact that observations are entirely missing turns the truncation problem into a missing data problem. Truncated data are less informative than censored data, and consequently truncation has a larger potential to distort the inferences of interest.

The usual approach for dealing with missing values in censored and truncated data is to delete the incomplete records, i.e., complete-case analysis. In the event that time is the censored variable, consider the following two problems:

- *Censored event times.* What would have been the uncensored event time if no censoring had taken place?

- *Missing event times.* What would have been the event time and the censoring status if these had been observed?

The problem of censored event times has been studied extensively. There are many statistical methods that can analyze left- or right-censored data directly, collectively known as *survival analysis*. Kleinbaum and Klein (2005), Hosmer et al. (2008) and Allison (2010) provide useful introductions into the field. Survival analysis is the method of choice if censoring is restricted to the single outcomes. The approach is, however, less suited for censored predictors or for multiple interdependent censored outcomes. Van Wouwe et al. (2009) discuss an empirical example of such a problem. The authors are interested in knowing time interval between resuming contraception and cessation of lactation in young mothers who gave birth in the last 6 months. As the sample was cross-sectional, both contraception and lactation were subject to censoring. Imputation could be used to impute the hypothetically uncensored event times in both durations, and this allowed a study of the association between the uncensored event times.

The problem of missing event times is relevant if the event time is unobserved. The censoring status is typically also unknown if the event time is missing. Missing event times may be due to happenstance, for example, resulting from a technical failure of the instrument that measures event times. Alternatively, the missing data could have been caused by truncation, where all event times beyond the truncation point are set to missing. It will be clear that the optimal way to deal with the missing events data depends on the reasons for the missingness. Analysis of the complete cases will systematically distort the analysis of the event times if the data are truncated.

Imputation of right-censored data has received most attention to date. In general, the method aims to find new (longer) event times that would have been observed had the data not been censored. Let n_1 denote the number of observed failure times, let $n_0 = n - n_1$ denote the number of censored event times and let $t_1, \ldots, t_n$ be the ordered set of failure and censored times. For some time point t, the *risk set* $R(t) = t_i > t$ for $i = 1, \ldots, n$ is the set of event and censored times that is longer than t. Taylor et al. (2002) proposed two imputation strategies for right-censored data:

1. *Risk set imputation.* For a given censored value t construct the risk set $R(t)$, and randomly draw one case from this set. Both the failure time and censoring status from the selected case are used to impute the data.

2. *Kaplan–Meier imputation.* For a given censored value t construct the risk set $R(t)$ and estimate the Kaplan–Meier curve from this set. A randomly drawn failure time from the Kaplan–Meier curve is used for imputation.

Both methods are asymptotically equivalent to the Kaplan–Meier estimator after multiple imputation with large m. The adequacy of imputation procedures will depend on the availability of possible donor observations, which diminishes in the tails of the survival distribution. The Kaplan–Meier method has the advantage that nearly all censored observations are replaced by im-

Algorithm 3.6: Imputation of right-censored data using predictive mean matching, Kaplan–Meier estimation and the bootstrap.♠

---

1. Estimate $\hat{\beta}$ by a proportional hazards model of y given X, where $y = (t, \phi)$ consists of time t and censoring indicator ϕ ($\phi_i = 0$ if t_i is censored).

2. Draw a bootstrap sample $(\dot{y}, \dot{X})$ of size n from (y, X).

3. Estimate $\dot{\beta}$ by a proportional hazards model of $\dot{y}$ given $\dot{X}$.

4. Calculate $\dot{\eta}(i, j) = |X_{[i]}\hat{\beta} - X_{[j]}\dot{\beta}|$ with $i = 1, \ldots, n$ and $j = 1, \ldots, n_0$, where $[j]$ indexes the cases with censored times.

5. Construct n_0 sets Z_j, each containing d candidate donors such that $t_i > t_j$ and $\sum_d \dot{\eta}(i, j)$ is minimum for each $j = 1, \ldots, n_0$. Break ties randomly.

6. For each Z_j, estimate the Kaplan–Meier curve $\hat{S}_j(t)$.

7. Draw n_0 uniform random variates u_j, and take $\dot{t}_j$ from the estimated cumulative distribution function $1 - \hat{S}_j(t)$ at u_j for $j = 1, \ldots, n_0$.

8. Set $\phi_j = 0$ if $\dot{t}_j = t_n$ and $\phi_{t_n} = 0$, else set $\phi_j = 1$.

---

puted failure times. In principle, both Bayesian sampling and bootstrap methods can be used to incorporate model uncertainty, but in practice only the bootstrap has been used.

Hsu et al. (2006) extended both methods to include covariates. The authors fitted a proportional hazards model and calculated a risk score as a linear combination of the covariates. The key adaptation is to restrict the risk set to those cases that have a risk score that is similar to the risk score of censored case, an idea similar to predictive mean matching. A donor group size with $d = 10$ was found to perform well, and Kaplan–Meier imputation was superior to risk set imputation across a wide range of situations.

Algorithm 3.6 is based on the KIMB method proposed by Hsu et al. (2006). The method assumes that censoring status is known, and aims to impute plausible event times for censored observations. Hsu et al. (2006) actually suggested fitting two proportional hazards models, one with survival time as outcome and one with censoring status as outcome, but in order to keep in line with the rest of this chapter, here we only fit the model for survival time. The way in which predictive mean matching is done differs slightly from Hsu et al. (2006).

The literature on imputation methods for censored and rounded data is rapidly evolving. Alternative methods for right-censored data have also been proposed (Wei and Tanner, 1991; Geskus, 2001; Lam et al., 2005; Liu et al., 2011). Lyles et al. (2001), Lynn (2001), Hopke et al. (2001) and Lee et al. (2018) concentrated on left-censored data. Imputation of interval-censored data (rounded data) has been discussed quite extensively (Heitjan and Rubin, 1990; Dorey et al., 1993; James and Tanner, 1995; Pan, 2000; Bebchuk and Betensky, 2000; Glynn and Rosner, 2004; Hsu, 2007; Royston, 2007; Chen and Sun, 2010; Hsu et al., 2015). Imputation of double-censored data, where both the initial and the final times are interval censored, is treated by Pan (2001) and Zhang et al. (2009). Delord and Génin (2016) extended Pan's approach to interval-censored competing risks data, thus allowing estimation of the survival function, cumulative incidence function, Cox and Fine & Gray regression coefficients. These methods are available in the MIICD package. Jackson et al. (2014) used multiple imputation to study departures from the independent censoring assumption in the Cox model.

By comparison, very few methods have been developed to deal with truncation. Methods for imputing a missing censoring indicator have been proposed by Subramanian (2009, 2011) and Wang and Dinse (2010).

3.8 Nonignorable missing data

3.8.1 Overview

All methods described thus far assume that the missing data mechanism is ignorable. In this case, there is no need for an explicit model of the missing data process (cf. Section 2.2.6). In reality, the mechanism may be nonignorable, even after accounting for any measurable factors that govern the response probability. In such cases, we can try to adapt the imputed data to make them more realistic. Since such adaptations are based on unverifiable assumptions, it is recommended to study carefully the impact of different possibilities on the final inferences by means of sensitivity analysis.

When is the assumption of ignorability suspect? It is hard to provide cut-and-dried criteria, but the following list illustrates some typical situations:

- If important variables that govern the missing data process are not available;

- If there is reason to believe that responders differ from non-responders, even after accounting for the observed information;

- If the data are truncated.

If ignorability does not hold, we need to model the distribution $P(Y, R)$ instead of $P(Y)$. For nonignorable missing data mechanisms, $P(Y, R)$ do not factorize into independent parts. Two main strategies to decompose $P(Y, R)$ are known as the *selection model* (Heckman, 1976) and the *pattern-mixture model* (Glynn et al., 1986). Little and Rubin (2002, ch. 15) and Little (2009) provide in-depth discussions of these models.

Imputations are created most easily under the pattern-mixture model. Herzog and Rubin (1983, pp. 222–224) proposed a simple and general family of nonignorable models that accounts for shift bias, scale bias and shape bias. Suppose that we expect that the nonrespondent data are shifted relative to the respondent data. Adding a simple shift parameter δ to the imputations creates a difference in the means of a δ. In a similar vein, if we suspect that the nonrespondents and respondents use different scales, we can multiply each imputation by a scale parameter. Likewise, if we suspect that the shapes of both distributions differ, we could redraw values from the candidate imputations with a probability proportional to the dissimilarity between the two distributions, a technique known as the SIR algorithm (Rubin, 1987b). We only discuss the shift parameter δ.

In practice, it may be difficult to specify the distribution of the nonrespondents, e.g., to provide a sensible specification of δ. One approach is to compare the results under different values of δ by sensitivity analysis. Though helpful, this puts the burden on the specification of realistic scenarios, i.e., a set of plausible δ-values. The next sections describe the selection model and pattern mixture in more detail, as a way to evaluate the plausibility of δ.

3.8.2   Selection model

The selection model (Heckman, 1976) decomposes the joint distribution $P(Y, R)$ as

$$P(Y, R) = P(Y)P(R|Y). \tag{3.8}$$

The selection model multiplies the marginal distribution $P(Y)$ in the population with the response weights $P(R|Y)$. Both $P(Y)$ and $P(R|Y)$ are unknown, and must be specified by the user. The model where $P(Y)$ is normal and where $P(R|Y)$ is a probit model is known as the Heckman model. This model is widely used in economics to correct for selection bias.

Numerical example. The column labeled Y in Table 3.5 contains the midpoints of 11 categories of systolic blood pressure. The column $P(Y)$ contains a hypothetically complete distribution of systolic blood pressure. It is specified here as symmetric with a mean of 150 mmHg (millimeters mercury). This distribution should be a realistic description of the *combined* observed and missing blood pressure values in the population of interest. The column $P(R = 1|Y)$ specifies the probability that blood pressure is actually observed at different levels of blood pressure. Thus, at a systolic blood pressure of 100 mmHg, we expect that 65% of the data will be observed. On the other hand, we expect that no missing data occur for those with a blood pressure of 200 mmHg. This specification produces 12.2% of missing data. The variability in the missingness probability is large, and reflects an extreme scenario where

Table 3.5: Numerical example of a nonignorable nonresponse mechanism, where more missing data occur in groups with lower blood pressures.

Y	$P(Y)$	$P(R=1|Y)$	$P(Y|R=1)$	$P(Y|R=0)$
100	0.02	0.65	0.015	0.058
110	0.03	0.70	0.024	0.074
120	0.05	0.75	0.043	0.103
130	0.10	0.80	0.091	0.164
140	0.15	0.85	0.145	0.185
150	0.30	0.90	0.307	0.247
160	0.15	0.92	0.157	0.099
170	0.10	0.94	0.107	0.049
180	0.05	0.96	0.055	0.016
190	0.03	0.98	0.033	0.005
200	0.02	1.00	0.023	0.000
$\bar{Y}$	150.00		151.58	138.60

the missing data are created mostly at the lower blood pressures. Section 9.2.1 discusses why more missing data in the lower levels are plausible. When taken together, the columns $P(Y)$ and $P(R=1|Y)$ specify a selection model.

3.8.3   Pattern-mixture model

The pattern-mixture model (Glynn et al., 1986; Little, 1993) decomposes the joint distribution $P(Y,R)$ as

$$P(Y,R) \quad = \quad P(Y|R)P(R) \tag{3.9}$$
$$= \quad P(Y|R=1)P(R=1) + P(Y|R=0)P(R=0) \tag{3.10}$$

Compared to Equation 3.8 this model only reverses the roles of Y and R, but the interpretation is quite different. The pattern-mixture model emphasizes that the combined distribution is a mix of the distributions of Y in the responders and nonresponders. The model needs a specification of the distribution $P(Y|R=1)$ of the responders (which can be conveniently modeled after the data), and of the distribution $P(Y|R=0)$ of the nonresponders (for which we have no data at all). The joint distribution is the mixture of these two distributions, with mixing probabilities $P(R=1)$ and $P(R=0) = 1 - P(R=1)$, the overall proportions of observed and missing data, respectively.

Numerical example. The columns labeled $P(Y|R=1)$ and $P(Y|R=0)$ in Table 3.5 contain the probability per blood pressure category for the respondents and nonrespondents. Since more missing data are expected to occur at lower blood pressures, the mass of the nonresponder distribution has shifted toward the lower end of the scale. As a result, the mean of the nonresponder distribution is equal to 138.6 mmHg, while the mean of the responder distribution equals 151.58 mmHg.

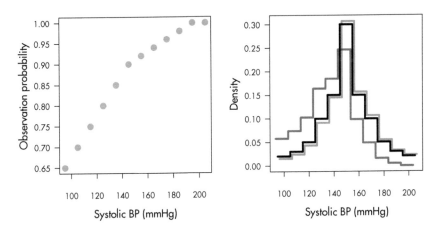

Figure 3.9: Graphic representation of the response mechanism for systolic blood pressure in Table 3.5. See text for explanation.

3.8.4 Converting selection and pattern-mixture models

The pattern-mixture model and the selection model are connected via Bayes rule. Suppose that we have a mixture model specified as the probability distributions $P(Y|R = 0)$ and $P(Y|R = 1)$ plus the overall response probability $P(R)$. The corresponding selection model can be calculated as

$$P(R = 1|Y = y) = P(Y = y|R = 1)P(R = 1)/P(Y = y) \qquad (3.11)$$

where the marginal distribution of Y is

$$P(Y = y) = P(Y = y|R = 1)P(R = 1) + P(Y = y|R = 0)P(R = 0) \qquad (3.12)$$

Reversely, the pattern-mixture model can be calculated from the selection model as follows:

$$P(Y = y|R = r) = P(R = r|Y = y)P(Y = y)/P(R = r) \qquad (3.13)$$

where the overall probability of observed ($r = 1$) or missing ($r = 0$) data is equal to

$$P(R = r) = \sum_y P(R = r|Y = y)P(Y = y) \qquad (3.14)$$

Numerical example. In Table 3.5 we calculate $P(Y = 100) = 0.015 \times 0.878 + 0.058 \times 0.122 = 0.02$. Likewise, we find $P(R = 1|Y) = 0.015 \times 0.878/0.02 = 0.65$ and $P(R = 0|Y) = 0.058 \times 0.122/0.02 = 0.35$. The reverse calculation is left as an exercise to the reader.

Figure 3.9 is an illustration of the posited missing data mechanism. The left-hand figure displays the missingness probabilities $P(R|Y)$ of the selection

δ	Interpretation
0 mmHg	MCAR, δ too small
−5 mmHg	Small effect
−10 mmHg	Large effect
−15 mmHg	Extreme effect
−20 mmHg	Too extreme effect

model. The right-hand plot provides the distributions $P(Y|R)$ in the observed (blue) and missing (red) data in the corresponding pattern-mixture model. The hypothetically complete distribution is given by the black curve. The distribution of blood pressure in the group with missing blood pressures is quite different, both in form and location. At the same time, observe that the effect of missingness on the combined distribution is only slight. The reason is that 87% of the information is actually observed.

The mean of the distribution of the observed data remains almost unchanged (151.6 mmHg instead of 150 mmHg), but the mean of the distribution of the missing data is substantially lower at 138.6 mmHg. Thus, under the assumed selection model we expect that the mean of the imputed data should be $151.6 - 138.6 = 13$ mmHg lower than in the observed data.

3.8.5   Sensitivity analysis

Sections 3.8.2–3.8.4 provide different, though related, views on the assumed response model. A fairly extreme response model where the missingness probability increases from 0% to 35% in the outcome produces a mean difference of 13 mmHg. The effect in the combined distribution is much smaller: 1.6 mmHg.

Section 3.8.1 discussed the idea of adding some extra mmHg to the imputed values, a method known as δ-adjustment. It is important to form an idea of what reasonable values for δ could be. Under the posited model, $\delta = 0$ mmHg is clearly too small (as it assumes MCAR), whereas $\delta = -20$ mmHg is too extreme (as it can only occur if nearly all missing values occur in the lowest blood pressures). Table 3.6 provides an interpretation of various values for δ. The most likely scenarios would yield $\delta = -5$ or $\delta = -10$ mmHg.

In practice, part of δ may be realized through the predictors needed under MAR. It is useful to decompose δ as $\delta = \delta_{MAR} + \delta_{MNAR}$, where δ_{MAR} is the mean difference caused by the predictors in the imputation models, and where δ_{MNAR} is the mean difference caused by an additional nonignorable part of the imputation model. If candidate imputations are produced under MAR, we only need to add a constant δ_{MNAR}. Section 9.2 continues this application.

Adding a constant may seem overly simple, but it is actually quite powerful. In cases where no one model will be obviously more realistic than any other, Rubin (1987a, p. 203) stressed the need for easily communicated models, like a "20% increase over the ignorable value." Little (2009, p. 49) warned that it is easy to be enamored of complicated models for $P(Y, R)$ so that we may be "lulled into a false sense of complacency about the fundamental lack of identification," and suggested simple methods:

> The idea of adding offsets is simple, transparent, and can be readily accomplished with existing software.

Adding a constant or multiplying by a value are in fact the most direct ways to specify nonignorable models.

3.8.6   Role of sensitivity analysis

Nonignorable models are only useful after the possibilities to make the data "more MAR" have been exhausted. A first step is always to create the best possible imputation model based on the available data. Section 6.3.2 provides specific advice on how to build imputation models.

The MAR assumption has been proven defensible for intentional missing data. In general, however, we can never rule out the possibility that the data are MNAR. In order to cater for this possibility, many advise performing a sensitivity analysis on the final result. This is voiced most clearly in recommendation 15 of the National Research Council's advice on clinical trials (National Research Council, 2010):

> Recommendation 15: Sensitivity analysis should be part of the primary reporting of findings from clinical trials. Examining sensitivity to the assumptions about the missing data mechanism should be a mandatory component of reporting.

While there is much to commend this rule, we should refrain from doing sensitivity analysis just for the sake of it. The proper execution of a sensitivity analysis requires us to specify plausible scenarios. An extreme scenario like "suppose that all persons who leave the study die" can have a major impact on the study result, yet it could be highly improbable and therefore of limited interest.

Sensitivity analysis on factors that are already part of the imputation model is superfluous. Preferably, before embarking on a sensitivity analysis, there should be reasonable evidence that the MAR assumption is (still) inadequate after the available data have been taken into account. Such evidence is also crucial in formulating plausible MNAR mechanisms. Any decisions about scenarios for sensitivity analysis should be taken in discussion with subject-matter specialists. There is no purely statistical solution to the problem of nonignorable missing data. Sensitivity analysis can increase our insight into the stability of the results, but in my opinion we should only use it if we have a firm idea of which scenarios for the missingness would be reasonable.

In practice, we may lack such insights. In such instances, I would prefer a carefully constructed imputation model (which is based on all available data) over a poorly constructed sensitivity analysis.

3.8.7   Recent developments

The literature on nonignorable models is large and diverse. The research in this area is active and recent. This section provides some pointers into the recent literature.

The historic overview by Kenward and Molenberghs (2015) provides an in-depth treatment of the selection, pattern-mixture and shared parameter models, including their connnections. The Handbook of Missing Data Methodology (Molenberghs et al., 2015) contains five chapters that discuss sensitivity analysis from all angles. The handbook should be the starting point for anyone considering models for data that are MNAR. Little et al. (2017) developed an alternative strategy based on selecting a subset of parameters of substantive interest. In particular cases, the conditions for ignoring the missing-data mechanism are more relaxed than under MAR.

Regulators prefer simple methods that impute the missing outcomes under MAR, and then add an adjustment δ to the imputes, while varying δ over a plausible range and independently for each treatment group (Permutt, 2016). The most interesting scenarios will be those where the difference between the δ's correspond to the size of the treatment effect in the completers. Contours of the p-values may be plotted on a graph as a function of the δ's to assist in a tipping-point analysis (Liublinska and Rubin, 2014).

Kaciroti and Raghunathan (2014) relate the identifying parameters from the pattern-mixture model to the corresponding missing data mechanism in the selection model. This dual interpretation provides a unified framework for performing sensitivity analysis. Galimard et al. (2016) proposed an imputation method under MNAR based on the Heckman model. The random indicator method (Jolani, 2012) is an experimental iterative method that redraws the missing data indicator under a selection model, and imputes the missing data under a pattern-mixture model, with the objective of estimating δ from the data under relaxed assumptions. Initial simulation results look promising. The algorithm is available as the `ri` method in `mice`.

3.9   Exercises

1. *MAR.* Reproduce Table 3.1 and Table 3.2 for MARRIGHT, MARMID and MARTAIL missing data mechanisms of Section 3.2.4.

    (a) Are there any choices that you need to make? If so, which?

(b) Consider the six possibilities to combine the missing data mechanism and missingness in x or y. Do you expect complete-case analysis to perform well in each case?

(c) Do the Bayesian sampling and bootstrap methods also work under the three MAR mechanisms?

2. *Parameter uncertainty.* Repeat the simulations of Section 3.2 on the `whiteside` data for different samples sizes.

(a) Use the method of Section 3.2.3 to generate an artificial population of 10000 synthetic gas consumption observations. Re-estimate the parameter from the artificial population. How close are they to the "true" values?

(b) Draw random samples from the artificial population. Systematically vary sample size. Is there some sample size at which `norm.nob` is as good as the Bayesian sampling and bootstrap methods?

(c) Is the result identical for missing y and missing x?

(d) Is the result the same after including insulation status in the model?

Chapter 4

Multivariate missing data

> *Conditional modeling allows enormous flexibility in dealing with practical problems.*
> Andrew Gelman, Trivellore Raghunathan

Chapter 3 dealt with univariate missing data. In practice, missing data may occur anywhere. This chapter discusses potential problems created by multivariate missing data, and outlines several approaches to deal with these issues.

4.1 Missing data pattern

4.1.1 Overview

Let the data be represented by the $n \times p$ matrix Y. In the presence of missing data Y is partially observed. Notation Y_j is the jth column in Y, and Y_{-j} indicates the complement of Y_j, that is, all columns in Y except Y_j. The *missing data pattern* of Y is the $n \times p$ binary response matrix R, as defined in Section 2.2.3.

For both theoretical and practical reasons, it is useful to distinguish various types of missing data patterns:

1. *Univariate and multivariate.* A missing data pattern is said to be univariate if there is only one variable with missing data.

2. *Monotone and non-monotone (or general).* A missing data pattern is said to be *monotone* if the variables Y_j can be ordered such that if Y_j is missing then all variables Y_k with $k > j$ are also missing. This occurs, for example, in longitudinal studies with drop-out. If the pattern is not monotone, it is called *non-monotone* or *general*.

3. *Connected and unconnected.* A missing data pattern is said to be *connected* if any observed data point can be reached from any other observed data point through a sequence of horizontal or vertical moves (like the rook in chess).

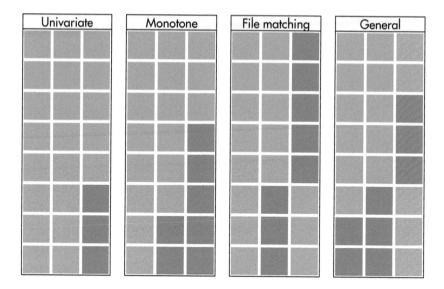

Figure 4.1: Some missing data patterns in multivariate data. Blue is observed, red is missing.

Figure 4.1 illustrates various data patterns in multivariate data. Monotone patterns can occur as a result of drop-out in longitudinal studies. If a pattern is monotone, the variables can be sorted conveniently according to the percentage of missing data. Univariate missing data form a special monotone pattern. Important computational savings are possible if the data are monotone.

All patterns displayed in Figure 4.1 are connected. The file matching pattern is connected since it is possible to travel to all blue cells by horizontal or vertical moves. This pattern will become unconnected if we remove the first column. In contrast, after removing the first column from the general pattern in Figure 4.1 it is still connected through the first two rows.

Connected patterns are needed to identify unknown parameters. For example, in order to be able to estimate a correlation coefficient between two variables, they need to be connected, either directly by a set of cases that have scores on both, or indirectly through their relation with a third set of connected data. Unconnected patterns may arise in particular data collection designs, like data combination of different variables and samples, or potential outcomes.

Missing data patterns of longitudinal data organized in the "long format" (cf. Section 11.1) are more complex than the patterns in Figure 4.1. See Van Buuren (2011, p. 179) for some examples.

4.1.2 Summary statistics

The missing data pattern influences the amount of information that can be transferred between variables. Imputation can be more precise if other variables are non-missing for those cases that are to be imputed. The reverse is also true. Predictors are potentially more powerful if they have are non-missing in rows that are vastly incomplete. This section discusses various measures of the missing data pattern.

The function `md.pattern()` in `mice` calculates the frequencies of the missing data patterns. For example, the frequency pattern of the dataset `pattern4` in Figure 4.1 is

```
md.pattern(pattern4, plot = FALSE)

  A B C
2 1 1 1 0
3 1 1 0 1
1 1 0 1 1
2 0 0 1 2
  2 3 3 8
```

The columns A, B and C are either 0 (missing) or 1 (observed). The first column provides the frequency of each pattern. The last column lists the number of missing entries per pattern. The bottom row provides the number of missing entries per variable, and the total number of missing cells. In practice, `md.pattern()` is primarily useful for datasets with a small number of columns.

Alternative measures start from pairs of variables. A pair of variables (Y_j, Y_k) can have four missingness patterns:

1. both Y_j and Y_k are observed (pattern `rr`);

2. Y_j is observed and Y_k is missing (pattern `rm`);

3. Y_j is missing and Y_k is observed (pattern `mr`);

4. both Y_j and Y_k are missing (pattern `mm`).

For example, for the monotone pattern in Figure 4.1 the frequencies are:

```
p <- md.pairs(pattern4)
p

$rr
  A B C
A 6 5 3
B 5 5 2
C 3 2 5
```

```
$rm
   A B C
A  0 1 3
B  0 0 3
C  2 3 0

$mr
   A B C
A  0 0 2
B  1 0 3
C  3 3 0

$mm
   A B C
A  2 2 0
B  2 3 0
C  0 0 3
```

Thus, for pair (A,B) there are five completely observed pairs (in rr), no pairs in which A is observed and B missing (in rm), one pair in which A is missing and B is observed (in mr) and two pairs with both missing A and B. Note that these numbers add up to the total sample size.

The *proportion of usable cases* (Van Buuren et al., 1999) for imputing variable Y_j from variable Y_k is defined as

$$I_{jk} = \frac{\sum_i^n (1 - r_{ij})r_{ik}}{\sum_i^n 1 - r_{ij}} \tag{4.1}$$

This quantity can be interpreted as the number of pairs (Y_j, Y_k) with Y_j missing and Y_k observed, divided by the total number of missing cases in Y_j. The proportion of usable cases I_{jk} equals 1 if variable Y_k is observed in all records where Y_j is missing. The statistic can be used to quickly select potential predictors Y_k for imputing Y_j based on the missing data pattern. High values of I_{jk} are preferred. For example, we can calculate I_{jk} in the dataset pattern4 in Figure 4.1 for all pairs (Y_j, Y_k) by

```
p$mr/(p$mr+p$mm)

        A B C
A 0.000 0 1
B 0.333 0 1
C 1.000 1 0
```

The first row contains $I_{AA} = 0$, $I_{AB} = 0$ and $I_{AC} = 1$. This informs us that B is not relevant for imputing A since there are no observed cases in B where A is missing. However, C is observed for both missing entries in A, and may thus be a relevant predictor. The I_{jk} statistic is an *inbound* statistic that measures how well the missing entries in variable Y_j are connected to the rest of the data.

The *outbound* statistic O_{jk} measures how observed data in variable Y_j connect to missing data in the rest of the data. The statistic is defined as

$$O_{jk} = \frac{\sum_i^n r_{ij}(1 - r_{ik})}{\sum_i^n r_{ij}} \qquad (4.2)$$

This quantity is the number of observed pairs (Y_j, Y_k) with Y_j observed and Y_k missing, divided by the total number of observed cases in Y_j. The quantity O_{jk} equals 1 if variable Y_j is observed in all records where Y_k is missing. The statistic can be used to evaluate whether Y_j is a potential predictor for imputing Y_k. We can calculate O_{jk} in the dataset `pattern4` in Figure 4.1 for all pairs (Y_j, Y_k) by

```
p$rm/(p$rm+p$rr)
```

```
    A     B    C
A 0.0 0.167 0.5
B 0.0 0.000 0.6
C 0.4 0.600 0.0
```

Thus A is potentially more useful to impute C (3 out of 6) than B (1 out of 6).

4.1.3 Influx and outflux

The inbound and outbound statistics in the previous section are defined for variable pairs (Y_j, Y_k). This section describes two overall measures of how each variable connects to others: influx and outflux.

The *influx coefficient* I_j is defined as

$$I_j = \frac{\sum_j^p \sum_k^p \sum_i^n (1 - r_{ij}) r_{ik}}{\sum_k^p \sum_i^n r_{ik}} \qquad (4.3)$$

The coefficient is equal to the number of variable pairs (Y_j, Y_k) with Y_j missing and Y_k observed, divided by the total number of observed data cells. The value of I_j depends on the proportion of missing data of the variable. Influx of a completely observed variable is equal to 0, whereas for completely missing variables we have $I_j = 1$. For two variables with the same proportion of missing data, the variable with higher influx is better connected to the observed data, and might thus be easier to impute.

The *outflux coefficient* O_j is defined in an analogous way as

$$O_j = \frac{\sum_j^p \sum_k^p \sum_i^n r_{ij}(1 - r_{ik})}{\sum_k^p \sum_i^n 1 - r_{ij}} \qquad (4.4)$$

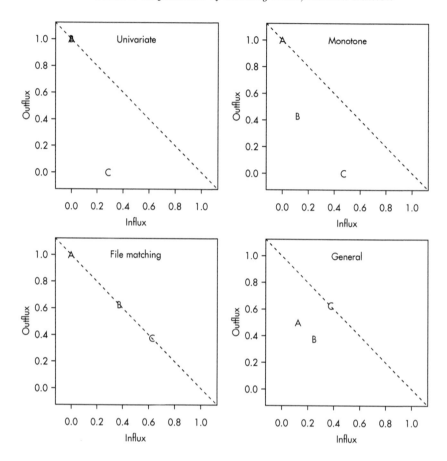

Figure 4.2: Fluxplot: Outflux versus influx in the four missing data patterns from Figure 4.1. The influx of a variable quantifies how well its missing data connect to the observed data on other variables. The outflux of a variable quantifies how well its observed data connect to the missing data on other variables. In general, higher influx and outflux values are preferred.

The quantity O_j is the number of variable pairs with Y_j observed and Y_k missing, divided by the total number of incomplete data cells. Outflux is an indicator of the potential usefulness of Y_j for imputing other variables. Outflux depends on the proportion of missing data of the variable. Outflux of a completely observed variable is equal to 1, whereas outflux of a completely missing variable is equal to 0. For two variables having the same proportion of missing data, the variable with higher outflux is better connected to the missing data, and thus potentially more useful for imputing other variables.

The function `flux()` in `mice` calculates I_j and O_j for all variables. For example, for `pattern4` we obtain

```
flux(pattern4)[,1:3]

  pobs influx outflux
A 0.750  0.125   0.500
B 0.625  0.250   0.375
C 0.625  0.375   0.625
```

The rows correspond to the variables. The columns contain the proportion of observed data, I_j and O_j. Figure 4.2 shows the influx-outflux pattern of the four patterns in Figure 4.1 produced by fluxplot(). In general, variables that are located higher up in the display are more complete and thus potentially more useful for imputation. It is often (but not always) true that $I_j + O_j \leq 1$, so in practice variables closer to the subdiagonal are typically better connected than those farther away. The fluxplot can be used to spot variables that clutter the imputation model. Variables that are located in the lower regions (especially near the lower-left corner) *and* that are uninteresting for later analysis are better removed from the data prior to imputation.

Influx and outflux are summaries of the missing data pattern intended to aid in the construction of imputation models. Keeping everything else constant, variables with high influx and outflux are preferred. Realize that outflux indicates the potential (and not actual) contribution to impute other variables. A variable with high O_j may turn out to be useless for imputation if it is unrelated to the incomplete variables. On the other hand, the usefulness of a highly predictive variable is severely limited by a low O_j. More refined measures of usefulness are conceivable, e.g., multiplying O_j by the average proportion of explained variance. Also, we could specialize to one or a few key variables to impute. Alternatively, analogous measures for I_j could be useful. The further development of diagnostic summaries for the missing data pattern is a promising area for further investigation.

4.2 Issues in multivariate imputation

Most imputation models for Y_j use the remaining columns Y_{-j} as predictors. The rationale is that conditioning on Y_{-j} preserves the relations among the Y_j in the imputed data. Van Buuren et al. (2006) identified various practical problems that can occur in multivariate missing data:

- The predictors Y_{-j} themselves can contain missing values;

- "Circular" dependence can occur, where Y_j^{mis} depends on Y_h^{mis} and Y_h^{mis} depends on Y_j^{mis} with $h \neq j$, because in general Y_j and Y_h are correlated, even given other variables;

- Variables are often of different types (e.g., binary, unordered, ordered, continuous), thereby making the application of theoretically convenient models, such as the multivariate normal, theoretically inappropriate;

- Especially with large p and small n, collinearity or empty cells can occur;

- The ordering of the rows and columns can be meaningful, e.g., as in longitudinal data;

- The relation between Y_j and predictors Y_{-j} can be complex, e.g., nonlinear, or subject to censoring processes;

- Imputation can create impossible combinations, such as pregnant fathers.

This list is by no means exhaustive, and other complexities may appear for particular data. The next sections discuss three general strategies for imputing multivariate data:

1. *Monotone data imputation.* For monotone missing data patterns, imputations are created by a sequence of univariate methods;

2. *Joint modeling.* For general patterns, imputations are drawn from a multivariate model fitted to the data;

3. *Fully conditional specification*, also known as *chained equations* and *sequential regressions*. For general patterns, a multivariate model is implicitly specified by a set of conditional univariate models. Imputations are created by drawing from iterated conditional models.

4.3   Monotone data imputation

4.3.1   Overview

Imputations of monotone missing data can be generated by specifying a sequence of univariate methods (one for each incomplete column), followed by drawing sequentially synthetic observations under each method.

Suppose that variables $Y_1, \ldots, Y_p$ are ordered into a monotone missing data pattern. The general recommended procedure is as follows (Rubin, 1987a, p. 172). The missing values of Y_1 are imputed from a (possibly empty) set of complete covariates X ignoring $Y_2, \ldots, Y_p$. Next, the missing values of Y_2 are imputed from (Y_1, X) ignoring $Y_3, \ldots, Y_p$, and so on. The procedure ends after Y_p is imputed from $(X, Y_1, \ldots, Y_{p-1})$. The univariate imputation methods as discussed in Chapter 3 can be used as building blocks. For example, Y_1 can be imputed by logistic regression, Y_2 by predictive mean matching and so on.

Numerical example. The first three columns of the data frame `nhanes2` in `mice` have a monotone missing data pattern. In terms of the above notation, X contains the complete variable `age`, Y_1 is the variable `hyp` and Y_2 is the variable `bmi`. Monotone data imputation can be applied to generate $m = 2$ complete datasets by:

```
data <- nhanes2[, 1:3]
md.pattern(data, plot = FALSE)

   age hyp bmi
16  1   1   1  0
1   1   1   0  1
8   1   0   0  2
    0   8   9 17

imp <- mice(data, visit = "monotone", maxit = 1, m = 2,
            print = FALSE)
```

The `md.pattern()` function outputs the three available data patterns in `data`. There are 16 complete rows, one row with missing `bmi`, and eight rows where both `bmi` and `hyp` are missing. The argument `visit = "monotone"` specifies that the visit sequence should be equal to the number of missing data per variable (so first `hyp` and then `bmi`). Since one iteration is enough, we use `maxit = 1` to limit the calculations. This code imputes `hyp` by logistic regression and `bmi` by predictive mean matching, the default methods for binary and continuous data, respectively.

Monotone data imputation requires that the missing data pattern is monotone. In addition, there is a second, more technical requirement: the parameters of the imputation models should be *distinct* (Rubin, 1987a, pp. 174–178). Let the jth imputation model be denoted by $P(Y_j^{\text{mis}}|X, Y_1, \ldots, Y_{p-1}, \phi_j)$, where ϕ_j represents the unknown parameters of the imputation model. For valid likelihood inferences, $\phi_1, \ldots, \phi_p$ should be distinct in the sense that the parameter space $\phi = (\phi_1, \ldots, \phi_p)$ in the multivariate model for the data is the cross-product of the individual parameter spaces (Schafer, 1997, p. 219). For Bayesian inference, it is required that the prior density of all parameters $\pi(\phi)$ factors into p independent densities $\pi(\phi) = \pi_1(\phi_1)\pi_2(\phi_2), \ldots, \pi_p(\phi_p)$ (Schafer, 1997, p. 224). In most applications these requirements are unlikely to limit the practical usefulness of the method because the parameters are typically unrelated and allowed to vary freely. We need to be aware, however, that monotone data imputation may fail if the parameters of imputation models for different Y_j somehow depend on each other.

4.3.2 Algorithm

Algorithm 4.1 provides the main steps of monotone data imputation. We order the variables according to their missingness, and impute from left to

Algorithm 4.1: Monotone data imputation of multivariate missing data.♠

1. Sort the data Y_j^{obs} with $j = 1, \ldots, p$ according to their missingness.

2. Draw $\dot\phi_1 \sim P(Y_1^{\text{obs}}|X)$.

3. Impute $\dot Y_1 \sim P(Y_1^{\text{mis}}|X, \dot\phi_1)$.

4. Draw $\dot\phi_2 \sim P(Y_2^{\text{obs}}|X, \dot Y_1)$.

5. Impute $\dot Y_2 \sim P(Y_2^{\text{mis}}|X, \dot Y_1, \dot\phi_2)$.

6. $\vdots$

7. Draw $\dot\phi_p \sim P(Y_p^{\text{obs}}|X, \dot Y_1, \ldots, \dot Y_{p-1})$.

8. Impute $\dot Y_p \sim P(Y_p^{\text{mis}}|X, \dot Y_1, \ldots, \dot Y_{p-1}, \dot\phi_p)$.

right. In practice, a pair of "draw-impute" steps is executed by one of the univariate methods of Chapter 3. Both Bayesian sampling and bootstrap imputation methods can be used, and can in fact be mixed. There is no need to iterate, and convergence is immediate. The algorithm is replicated m times from different starting points to obtain m multiply imputed datasets.

Monotone data imputation is fast and flexible, but requires a monotone pattern. In practice, a dataset may be near-monotone, and may become monotone if a small fraction of the missing data were imputed. For example, some subjects may drop out of the study resulting in a monotone pattern. There could be some unplanned missing data that destroy the monotone pattern. In such cases it can be computationally efficient to impute the data in two steps. First, fill in the missing data in a small portion of the data to restore the monotone pattern, and then apply the monotone data imputation (Li, 1988; Rubin and Schafer, 1990; Liu, 1993; Schafer, 1997; Rubin, 2003). There are often more ways to impute toward monotonicity, so a choice is necessary. Rubin and Schafer (1990) suggested ordering the variables according to the missing data rate.

Numerical example. The **nhanes2** data in **mice** contains 3 out of 27 missing values that destroy the monotone pattern: one for **hyp** (in row 6) and two for **bmi** (in rows 3 and 6). The following algorithm first imputes these 3 values by a simple random sample, and then fills in the remaining missing data by monotone data multiple imputation.

```
where <- make.where(nhanes2, "none")
where[6, "hyp"] <- TRUE
where[c(3, 6), "bmi"] <- TRUE
imp1 <- mice(nhanes2, where = where, method = "sample",
```

```
                seed = 21991, maxit = 1, print = FALSE)
data <- mice::complete(imp1)
imp2 <- mice(data, maxit = 1, visitSequence = "monotone",
             print = FALSE)
```

The primary advantage is speed. We need to make only two passes through the data. Since the method uses single imputation in the first step, it should be done only if the number of missing values that destroy the monotone pattern is small.

Observe that the imputed values for the missing `hyp` data in row 3 could also depend on `bmi` and `chl`, but in the procedure both predictors are ignored. In principle, we can improve the method by incorporating `bmi` and `chl` into the model, and then iterate. We will explore this technique in more detail in Section 4.5, but first we study the theoretically nice alternative.

4.4   Joint modeling

4.4.1   Overview

Joint modeling (JM) starts from the assumption that the data can be described by a multivariate distribution. Assuming ignorability, imputations are created as draws from the fitted distribution. The model can be based on any multivariate distribution. The multivariate normal distribution is most widely applied.

The general idea is as follows. For a general missing data pattern, missing data can occur anywhere in Y, so in practice the distribution from which imputations are to be drawn varies from row to row. For example, if the missingness pattern of row i is $r_{[i]} = (0, 0, 1, 1)$, then we need to draw imputations from the bivariate distribution $P_i(Y_1^{\text{mis}}, Y_2^{\text{mis}} | Y_3, Y_4, \phi_{1,2})$, whereas if $r_{[i']} = (0, 1, 1, 1)$ we need draws from the univariate distribution $P_{i'}(Y_1^{\text{mis}} | Y_1, Y_3, Y_4, \phi_1)$.

4.4.2   Continuous data

Under the assumption of multivariate normality $Y \sim N(\mu, \Sigma)$, the ϕ-parameters of these imputation models are functions of $\theta = (\mu, \Sigma)$ (Schafer, 1997, p. 157). The *sweep operator* transforms θ into ϕ by converting outcome variables into predictors, while the *reverse sweep operator* allows for the inverse operation (Beaton, 1964). The sweep operators allow rapid calculation of the ϕ parameters for imputation models that pertain to different missing data patterns. For reasons of efficiency, rows can be grouped along the missing data pattern. See Little and Rubin (2002, pp. 148–156) and Schafer (1997, pp. 157–163) for computational details.

The θ-parameters are usually unknown. For non-monotone missing data, however, it is generally difficult to estimate θ from Y_{obs} directly. The solution is to iterate imputation and parameter estimation using a general algorithm known as *data augmentation* (Tanner and Wong, 1987). At step t, the algorithm draws Y_{mis} and θ by alternating the following steps:

$$\dot{Y}^t_{\mathrm{mis}} \quad \sim \quad P(Y_{\mathrm{mis}}|Y_{\mathrm{obs}}, \dot{\theta}^{t-1}) \tag{4.5}$$
$$\dot{\theta}^t \quad \sim \quad P(\theta|Y_{\mathrm{obs}}, \dot{Y}^t_{\mathrm{mis}}) \tag{4.6}$$

where imputations from $P(Y_{\mathrm{mis}}|Y_{\mathrm{obs}}, \dot{\theta}^{t-1})$ are drawn by the method as described in the previous section, and where draws from the parameter distribution $P(\theta|Y_{\mathrm{obs}}, \dot{Y}^t_{\mathrm{mis}})$ are generated according to the method of Schafer (1997, p. 184).

Algorithm 4.2 lists the major steps needed to impute multivariate missing data under the normal model. Additional background can be found in Li (1988), Rubin and Schafer (1990) and Schafer (1997). Song and Belin (2004) generated multiple imputations under the common factor model. The performance of the method was found to be similar to that of the multivariate normal distribution, the main pitfall being the danger of setting the numbers of factors too low. Audigier et al. (2016) proposed an imputation method based on Bayesian principal components analysis, and suggested it as an alternative to regularize data with more columns than rows.

Schafer (1997, pp. 211–218) reported simulations that showed that imputations generated under the multivariate normal model are robust to non-normal data. Demirtas et al. (2008) confirmed this claim in a more extensive simulation study. The authors conclude that "imputation under the assumption of normality is a fairly reasonable tool, even when the assumption of normality is clearly violated; the fraction of missing information is high, especially when the sample size is relatively large." It is often beneficial to transform the data before imputation toward normality, especially if the scientifically interesting parameters are difficult to estimate, like quantiles or variances. For example, we could apply a logarithmic transformation before imputation to remove skewness, and apply an exponential transformation after imputation to revert to the original scale. See Von Hippel (2013) for a cautionary note.

Some work on automatic transformation methods for joint models is available. Van Buuren et al. (1993) developed an iterative transformation-imputation algorithm that finds optimal transformations of the variables toward multivariate normality. The algorithm is iterative because the multiply imputed values contribute to define the transformation, and vice versa. Transformations toward normality have also been incorporated in `transcan()` and `aregImpute()` of the `Hmisc` package in R (Harrell, 2001).

If a joint model is specified, it is nearly always the multivariate normal model. Alternatives like the t-distribution (Liu, 1995) are hardly being developed or applied. The recent research effort in the area has focused on models for multilevel data. These developments are covered in a Chapter 7.

1. Sort the rows of Y into S missing data patterns $Y_{[s]}, s = 1, \ldots, S$.

2. Initialize $\theta^0 = (\mu^0, \Sigma^0)$ by a reasonable starting value.

3. Repeat for $t = 1, \ldots, M$.

4. Repeat for $s = 1, \ldots, S$.

5. Calculate parameters $\dot{\phi}_s = \text{SWP}(\dot{\theta}^{t-1}, s)$ by sweeping the predictors of pattern s out of $\dot{\theta}^{t-1}$.

6. Calculate p_s as the number of missing data in pattern s. Calculate $o_s = p - p_s$.

7. Calculate the Cholesky decomposition C_s of the $p_s \times p_s$ submatrix of $\dot{\phi}_s$ corresponding to the missing data in pattern s.

8. Draw a random vector $z \sim N(0, 1)$ of length p_s.

9. Take $\dot{\beta}_s$ as the $o_s \times p_s$ submatrix of $\dot{\phi}_s$ of regression weights.

10. Calculate imputations $\dot{Y}^t_{[s]} = Y^{\text{obs}}_{[s]} \dot{\beta}_s + C'_s z$, where $Y^{\text{obs}}_{[s]}$ is the observed data in pattern s.

11. End repeat s.

12. Draw $\dot{\theta}^t = (\dot{\mu}, \dot{\Sigma})$ from the normal-inverse-Wishart distribution according to Schafer (1997, p. 184).

13. End repeat t.

4.4.3 Categorical data

The multivariate normal model is often applied to categorical data. Schafer (1997, p. 148) suggested rounding off continuous imputed values in categorical data to the nearest category "to preserve the distributional properties as fully as possible and to make them intelligible to the analyst." This advice was questioned by Horton et al. (2003), who showed that simple rounding may introduce bias in the estimates of interest, in particular for binary variables. Allison (2005) found that it is usually better not to round the data, and preferred methods specifically designed for categorical data, like logistic regression imputation or discriminant analysis imputation. Bernaards et al. (2007) confirmed the results of Horton et al. (2003) for simple rounding, and proposed two improvements to simple rounding: *coin flip* and *adaptive rounding*. Their simulations showed that "adaptive rounding seemed to provide the best performance, although its advantage over simple rounding was sometimes slight." Further work has been done by Yucel et al. (2008), who proposed rounding such that the marginal distribution in the imputations is similar to that of the observed data. Alternatively, Demirtas (2009) proposed two rounding methods based on logistic regression and an additional drawing step that makes rounding dependent on other variables in the imputation model. Another proposal is to model the indicators of the categorical variables (Lee et al., 2012). A single best rounding rule for categorical data has yet to be identified. Demirtas (2010) encourages researchers to avoid rounding altogether, and apply methods specifically designed for categorical data.

Several joint models for categorical variables have been proposed that do not rely on rounding. Schafer (1997) proposed several techniques to impute categorical data and mixed continuous-categorical data. Missing data in contingency tables can be imputed under the log-linear model. The model preserves higher-order interactions, and works best if the number of variables is small, say, up to six. Mixed continuous-categorical data can be imputed under the general location model originally developed by Olkin and Tate (1961). This model combines the log-linear and multivariate normal models by fitting a restricted normal model to each cell of the contingency table. Further extensions have been suggested by Liu and Rubin (1998) and Peng et al. (2004). Belin et al. (1999) pointed out some limitations of the general location model for a larger dataset with 16 binary and 18 continuous variables. Their study found substantial differences between the imputed and follow-up data, especially for the binary data.

Alternative imputation methods based on joint models have been developed. Van Buuren and Van Rijckevorsel (1992) maximized internal consistency by the k-means clustering algorithm, and outlined methods to generate multiple imputations. This is a single imputation method which artificially strengthens the relations in the data. The MIMCA imputation technique (Audigier et al., 2017) uses a similar underlying model, and derives variability in imputations by taking bootstrap samples under a chosen number of dimensions.

Van Ginkel et al. (2007) proposed *two-way imputation*, a technique for imputing incomplete categorical data by conditioning on the row and column sum scores of the multivariate data. This method has applications for imputing missing test item responses. Chen et al. (2011) proposed a class of models that specifies the conditional density by an odds ratio representation relative to the center of the distribution. This allows for separate models of the odds ratio function and the conditional density at the center.

Vermunt et al. (2008) pioneered the use of the latent class analysis for imputing categorical data. The latent class (or finite mixture) model describes the joint distribution as the product of locally independent categorical variables. When the number of classes is large, the model can be used as a generic density estimation tool that captures the relations between the variables by a highly parameterized model. The relevant associations in the data need not to be specified a-priori, and the main modeling effort consists of setting the number of latent classes. Unlike the saturated log-linear models advocated by Schafer (1997), latent models can handle a large number of variables. Vidotto et al. (2015) surveyed several different implementations of the latent class model for imputation, both frequentistic (Vermunt et al., 2008; Van der Palm et al., 2016b) and Bayesian (Si and Reiter, 2013), which differ in the ways to select the number of classes. Vidotto (2018) proposed an extension to longitudinal data.

Joint models for nested (multilevel) data have been intensively studied. Section 7.4 discusses these developments in more detail.

4.5 Fully conditional specification

4.5.1 Overview

Fully conditional specification (FCS) imputes multivariate missing data on a variable-by-variable basis (Van Buuren et al., 2006; Van Buuren, 2007a). The method requires a specification of an imputation model for each incomplete variable, and creates imputations per variable in an iterative fashion.

In contrast to joint modeling, FCS specifies the multivariate distribution $P(Y, X, R|\theta)$ through a set of conditional densities $P(Y_j|X, Y_{-j}, R, \phi_j)$. This conditional density is used to impute Y_j given X, Y_{-j} and R. Starting from simple random draws from the marginal distribution, imputation under FCS is done by iterating over the conditionally specified imputation models. The methods of Chapter 3 may act as building blocks. FCS is a natural generalization of univariate imputation.

Rubin (1987a, pp. 160–166) subdivided the work needed to create imputations into three tasks. The *modeling task* chooses a specific model for the data, the *estimation task* formulates the posterior parameters distribution given the model and the *imputation task* takes a random draws for the missing data by drawing successively from parameter and data distributions. FCS directly specifies the conditional distributions from which draws should be made, and hence bypasses the need to specify a multivariate model for the data.

The idea of conditionally specified models is quite old. Conditional probability distributions follow naturally from the theory of stochastic Markov chains (Bartlett, 1978, pp. 34–41, pp. 231–236). In the context of spatial data, Besag preferred the use of conditional probability models over joint probability models, since "the conditional probability approach has greater intuitive appeal to the practising statistician" (Besag, 1974, p. 223).

In the context of missing data imputation, similar ideas have surfaced under a variety of names: stochastic relaxation (Kennickell, 1991), variable-by-

Algorithm 4.3: MICE algorithm for imputation of multivariate missing data.

1. Specify an imputation model $P(Y_j^{\mathrm{mis}}|Y_j^{\mathrm{obs}}, Y_{-j}, R)$ for variable Y_j with $j = 1, \ldots, p$.

2. For each j, fill in starting imputations $\dot{Y}_j^0$ by random draws from Y_j^{obs}.

3. Repeat for $t = 1, \ldots, M$.

4. Repeat for $j = 1, \ldots, p$.

5. Define $\dot{Y}_{-j}^t = (\dot{Y}_1^t, \ldots, \dot{Y}_{j-1}^t, \dot{Y}_{j+1}^{t-1}, \ldots, \dot{Y}_p^{t-1})$ as the currently complete data except Y_j.

6. Draw $\dot{\phi}_j^t \sim P(\phi_j^t|Y_j^{\mathrm{obs}}, \dot{Y}_{-j}^t, R)$.

7. Draw imputations $\dot{Y}_j^t \sim P(Y_j^{\mathrm{mis}}|Y_j^{\mathrm{obs}}, \dot{Y}_{-j}^t, R, \dot{\phi}_j^t)$.

8. End repeat j.

9. End repeat t.

variable imputation (Brand, 1999), switching regressions (Van Buuren et al., 1999), sequential regressions (Raghunathan et al., 2001), ordered pseudo-Gibbs sampler (Heckerman et al., 2001), partially incompatible MCMC (Rubin, 2003), iterated univariate imputation (Gelman, 2004), chained equations (Van Buuren and Groothuis-Oudshoorn, 2000) and fully conditional specification (FCS) (Van Buuren et al., 2006).

4.5.2   The MICE algorithm

There are several ways to implement imputation under conditionally specified models. Algorithm 4.3 describes one particular instance: the MICE algorithm (Van Buuren and Groothuis-Oudshoorn, 2000, 2011). The algorithm starts with a random draw from the observed data, and imputes the incomplete data in a variable-by-variable fashion. One iteration consists of one cycle through all Y_j. The number of iterations M can often be low, say 5 or 10. The MICE algorithm generates multiple imputations by executing Algorithm 4.3 in parallel m times.

The MICE algorithm is a Markov chain Monte Carlo (MCMC) method, where the state space is the collection of all imputed values. More specifically, if the conditionals are compatible (cf. Section 4.5.3), the MICE algorithm is a Gibbs sampler, a Bayesian simulation technique that samples from the conditional distributions in order to obtain samples from the joint distribution

(Gelfand and Smith, 1990; Casella and George, 1992). In conventional applications of the Gibbs sampler, the full conditional distributions are derived from the joint probability distribution (Gilks, 1996). In the MICE algorithm, the conditional distributions are under direct control of the user, and so the joint distribution is only implicitly known, and may not actually exist. While the latter is clearly undesirable from a theoretical point of view (since we do not know the joint distribution to which the algorithm converges), in practice it does not seem to hinder useful applications of the method (cf. Section 4.5.3).

In order to converge to a stationary distribution, a Markov chain needs to satisfy three important properties (Roberts, 1996; Tierney, 1996):

- *irreducible*, the chain must be able to reach all interesting parts of the state space;

- *aperiodic*, the chain should not oscillate between different states;

- *recurrence*, all interesting parts can be reached infinitely often, at least from almost all starting points.

Do these properties hold for the MICE algorithm? Irreducibility is generally not a problem since the user has large control over the interesting parts of the state space. This flexibility is actually the main rationale for FCS instead of a joint model.

Periodicity is a potential problem, and can arise in the situation where imputation models are clearly inconsistent. A rather artificial example of an oscillatory behavior occurs when Y_1 is imputed by $Y_2\beta + \epsilon_1$ and Y_2 is imputed by $-Y_1\beta + \epsilon_2$ for some fixed, nonzero β. The sampler will oscillate between two qualitatively different states, so the correlation between Y_1 and Y_2 after imputing Y_1 will differ from that after imputing Y_2. In general, we would like the statistical inferences to be independent of the stopping point. A way to diagnose the *ping-pong* problem, or *order effect*, is to stop the chain at different points. The stopping point should not affect the statistical inferences. The addition of noise to create imputations is a safeguard against periodicity, and allows the sampler to "break out" more easily.

Non-recurrence may also be a potential difficulty, manifesting itself as explosive or non-stationary behavior. For example, if imputations are made by deterministic functions, the Markov chain may lock up. Such cases can sometimes be diagnosed from the trace lines of the sampler. See Section 6.5.2 for an example. As long as the parameters of imputation models are estimated from the data, non-recurrence is mild or absent.

The required properties of the MCMC method can be translated into conditions on the eigenvalues of the matrix of transition probabilities (MacKay, 2003, pp. 372–373). The development of practical tools that put these conditions to work for multiple imputation is still an ongoing research problem.

4.5.3   Compatibility♠

Gibbs sampling is based on the idea that knowledge of the conditional distributions is sufficient to determine a joint distribution, if it exists. Two conditional densities $p(Y_1|Y_2)$ and $p(Y_2|Y_1)$ are said to be *compatible* if a joint distribution $p(Y_1, Y_2)$ exists that has $p(Y_1|Y_2)$ and $p(Y_2|Y_1)$ as its conditional densities. More precisely, the two conditional densities are compatible if and only if their density ratio $p(Y_1|Y_2)/p(Y_2|Y_1)$ factorizes into the product $u(Y_1)v(Y_2)$ for some integrable functions u and v (Besag, 1974). So, the joint distribution either exists and is unique, or does not exist.

If the joint density itself is of genuine scientific interest, we should carefully evaluate the effect that imputations might have on the estimate of the distribution. For example, incompatible conditionals could produce a ridge (or spike) in an otherwise smooth density, and the location of the ridge may actually depend on the stopping point. If such is the case, then we should have a reason to favor a particular stopping point. Alternatively, we might try to reformulate the imputation model so that the order effect disappears.

Arnold and Press (1989) and Arnold et al. (1999) provide necessary and sufficient conditions for the existence of a joint distribution given two conditional densities. Gelman and Speed (1993) concentrate on the question whether an arbitrary mix of conditional and marginal distribution yields a unique joint distribution. Arnold et al. (2002) describe near-compatibility in discrete data.

Several papers are now available on the conditions under which imputations created by conditionally specified models are draws from the implicit joint distribution. According to Hughes et al. (2014), two conditions must hold for this to occur. First, the conditionals must be compatible, and second the margin must be noninformative. Suppose that $p(\phi_j)$ is the prior distribution of the set of parameters that relate Y_j to Y_{-j}, and that $p(\tilde{\phi}_j)$ is prior distribution of the set of parameters that describes that relations among the Y_{-j}. The noninformative margins condition states that if two sets of parameters are distinct (i.e., their joint parameter space is the product of their separate parameter spaces), and their joint distribution $p(\phi_j, \tilde{\phi}_j)$ are independent and factorizes as $p(\phi_j, \tilde{\phi}_j) = p(\phi_j)p(\tilde{\phi}_j)$. Independence is a property of the prior distributions, whereas distinctness is a property of the model. Hughes et al. (2014) show that distinctness holds for the saturated multinomial distribution with a Dirichlet prior, so imputations from this joint distribution can be achieved by a set of conditionally specified models. However, for the log-linear model with only two-way factor interaction (and no higher-order terms) distinctness only holds for a maximum of three variables. The noninformative marginal condition is sufficient, but not necessary. In most practical cases we are unable to show that the noninformative marginal condition holds, but we can stop the algorithms at different points and inspect the estimates for order effect. Simulations by Hughes et al. (2014) show that such order effects exist, but in general they are small. Liu et al. (2013) made the same division in the parameter space of compatible models, and showed that imputation created by conditional specification is asymptotically equivalent to full Bayesian imputation for an assumed joint model. Asymptotic equivalence assumes infinite m and infinite n, and holds when the joint model is misspecified. The order effect disappear with increasing sample size. Zhu and Raghunathan (2015) observed that the parameters of the conditionally specified models typically span a larger space than the space occupied by the implied joint model. A set of imputation models is *possibly compatible* if the conditional density for each variable j according to some joint distribution is a special case of the corresponding imputation model for j. If the parameters of the joint model can be separated, then iteration over the possible compatible conditional imputation models will provide draws from the conditional densities of the implied joint distribution.

Several methods for identifying compatibility from actual data have been developed (Tian et al., 2009; Ip and Wang, 2009; Wang and Kuo, 2010; Chen, 2011; Yao et al., 2014; Kuo et al., 2017). However, the application of these methods is challenging because of the many possible choices of conditional models. What happens when the joint distribution does not exist? The MICE algorithm is ignorant of the non-existence of the joint distribution, and happily produces imputations whether the joint distribution exists or not. Can the imputed data be trusted when we cannot find a joint distribution $p(Y_1, Y_2)$ that has $p(Y_1|Y_2)$ and $p(Y_2|Y_1)$ as its conditionals?

Incompatibility easily arises if deterministic functions of the data are imputed along with their originals. For example, the imputation model may contain interaction terms, data summaries or nonlinear functions of the data. Such terms introduce feedback loops and impossible combinations into the system, which can invalidate the imputations (Van Buuren and Groothuis-Oudshoorn, 2011). It is important to diagnose this behavior and eliminate feedback loops from the system. Chapter 6 describes the tools to do this.

Van Buuren et al. (2006) described a small simulation study using strongly incompatible models. The adverse effects on the estimates after multiple imputation were only minimal in the cases studied. Though FCS is only guaranteed to work if the conditionals are compatible, these simulations suggested that the results may be robust against violations of compatibility. Li et al. (2012) presented three examples of problems with MICE. However, their examples differ from the usual sequential regression setup in various ways, and do not undermine the validity of the approach (Zhu and Raghunathan, 2015). Liu et al. (2013) pointed out that application of incompatible conditional models cannot provide imputations from any joint model. However, they also found that Rubin's rules provide consistent point estimates for incompatible models under fairly general conditions, as long as each conditional model was correctly specified. Zhu and Raghunathan (2015) showed that incompatibility does not need to lead to divergence. While there is no joint model to converge to, the algorithm can still converge. The key in achieving convergence is that the imputation models should closely model the data. For example, include the skewness of the residuals, or ideally, generate the imputations from the underlying (but usually unknown) mechanism that generated the data.

The interesting point is that the last two papers have shifted the perspective from the user's joint model to the data producer's data generating model. With incompatible models, the most important condition is the validity of each conditional model. As long as the conditional models are able to replay the missing data according to the mechanism that generated the data, we might not be overly concerned with issues of compatibility.

In the majority of cases, scientific interest will focus on quantities that are more remote to the joint density, such as regression weights, factor loadings, prevalence estimates and so on. In such cases, the joint distribution is more like a nuisance factor that has no intrinsic value.

Apart from potential feedback problems, it appears that incompatibility seems like a relatively minor problem in practice, especially if the missing data rate is modest and the imputation models fit the data well. In order to evaluate these aspects, we need to inspect convergence and assess the fit of the imputations.

4.5.4   Congeniality or compatibility?

Meng (1994) introduced the concept of *congeniality* to refer to the relation between the imputation model and the analysis model. Some recent papers have used the term *compatibility* to refer to essentially the same concept, and this alternative use of the term compatibility may generate confusion. This section explains the two different meanings attached to the compatibility.

Compatibility refers to the property that the conditionally specified models together specify some joint distribution from which imputations are to be drawn. Compatibility is a theoretical requirement of the Gibbs sampler. The evidence obtained thus far indicated that mutual incompatibility of conditionals will only have a minor impact on the final inferences, as long as the conditional models are well specified to fit the data. See Section 4.5.3 for more detail.

Another use of compatibility refers to the relation between the substantive model and the imputation model. It is widely accepted that the imputation model should be more general than the substantive model. Meng (1994) stated that the analysis procedure should be congenial to the imputation model, where congeniality is a property of the analysis procedure. Bartlett et al. (2015) connected congeniality to compatibility by extending the joint distribution of the imputation model to include the substantive model. Bartlett et al. (2015) reasoned that an imputation model is congenial to the substantive model if the two models are compatible. In that case, a joint model exists whose conditionals include both the imputation and substantive model. Models that are incompatible may lead to biased estimates of parameters in the substantive model. Hence, incompatibility is a bad thing that should be prevented. Technically speaking, the use of the term compatibility is correct, but the interpretation and implications of this form of incompatibility are very different, and in fact close in spirit to Meng's congeniality.

This book reserves the term *compatibility* to refer to the property of the imputation model whether its parts make up a joint distribution (irrespective of an analysis model), and use the term *congeniality* to refer to the relation between the imputation model and the substantive model.

4.5.5 Model-based and data-based imputation

This section highlights an interesting new development for setting up imputation models.

Observe that Bartlett et al. (2015) reversed the direction of the relation between the imputation and substantive models. Meng (1994) takes a given imputation model, and then asks whether the analysis model is congenial to it, whereas Bartlett et al. (2015) start from the complete-data model, and ask whether the imputation model is congenial/compatible. These are complementary perspectives, leading to different strategies in setting up imputation models. If there is a strong scientific model, then it is natural to use model-based imputation, which puts the substantive model in the driver's seat, and ensures that the distribution from which imputations are generated is compatible to the substantive model. The challenge here is to create imputations that remain faithful to the data, and do not amplify aspects assumed in the model that are not supported by the data. If there is weak scientific theory, or if a wide range of models is fitted, then apply data-based imputation, where imputations are generated that closely represent the features of the data, without any particular analysis model in mind. Then the challenge is to create imputations that will accommodate for a wide range of substantive models.

The model-based approach is theoretically well grounded, and procedures are available for substantive models based on normal regression, discrete outcomes and proportional hazards. Related work can be found in Wu (2010), Goldstein et al. (2014), Erler et al. (2016), Erler et al. (2018) and Zhang and Wang (2017). Some simulations showed promising results (Grund et al., 2018b). Implementations in R are available as the smcfcs package (Bartlett and Keogh, 2018), and the mdmb package (Robitzsch and Lüdtke, 2018), as well as Blimp (Enders et al., 2018). One potential drawback is that the imputations might be specific to the model at hand, and need to be redone if the model changes. Another potential issue could be the calculations needed, which requires the use of rejection samplers. There is not yet much experience with such practicalities, but a method that approaches the imputation problem "from the other side" is an interesting and potentially useful addition to the imputer's toolbox.

4.5.6 Number of iterations

When m sampling streams are calculated in parallel, monitoring convergence is done by plotting one or more statistics of interest in each stream against iteration number t. Common statistics to be plotted are the mean and standard deviation of the synthetic data, as well as the correlation between different variables. The pattern should be free of trend, and the variance within a chain should approximate the variance between chains.

In practice, a low number of iterations appears to be enough. Brand (1999) and (Van Buuren et al., 1999) set the number of iterations M quite low, usually somewhere between 5 to 20 iterations. This number is much lower than in other applications of MCMC methods, which often require thousands of iterations.

Why can the number of iterations in MICE be so low? First of all, realize that the imputed data $\dot{Y}_{\text{mis}}$ form the only memory in the the MICE algorithm. Chapter 3 explained that imputed data can have a considerable amount of random noise, depending on the strength of the relations between the variables. Applications of MICE with lowly correlated data therefore inject a lot of noise into the system. Hence, the autocorrelation over t will be low, and convergence will be rapid, and in fact immediate if all variables are independent. Thus, the incorporation of noise into the imputed data has pleasant side-effect of speeding up convergence. Reversely, situations to watch out for occur if:

- the correlations between the Y_j's are high;

- the missing data rates are high; or

- constraints on parameters across different variables exist.

The first two conditions directly affect the amount of autocorrelation in the system. The latter condition becomes relevant for customized imputation models. We will see some examples in Section 6.5.2.

In the context of missing data imputation, our simulations have shown that unbiased estimates and appropriate coverage usually requires no more than just five iterations. It is, however, important not to rely automatically on this result as some applications can require considerably more iterations.

4.5.7 Example of slow convergence

Consider a small simulation experiment with three variables: one complete covariate X and two incomplete variables Y_1 and Y_2. The data consist of draws from the multivariate normal distribution with correlations $\rho(X, Y_1) = \rho(X, Y_2) = 0.9$ and $\rho(Y_1, Y_2) = 0.7$. The variables are ordered as $[X, Y_1, Y_2]$. The complete pattern is $R_1 = (1, 1, 1)$. Missing data are randomly created in two patterns: $R_2 = (1, 0, 1)$ and $R_3 = (1, 1, 0)$. Variables Y_1 and Y_2 are jointly observed on $n_{(1,1,1)}$ complete cases. The following code defines the function to generate the incomplete data.

```
generate <- function(n = c(1000, 4500, 4500, 0),
                      cor = matrix(c(1.0, 0.9, 0.9,
                                     0.9, 1.0, 0.7,
                                     0.9, 0.7, 1.0), nrow = 3)) {
  require(MASS)
  nt <- sum(n)
  cs <- cumsum(n)
  data <- mvrnorm(nt, mu = rep(0,3), Sigma = cor)
  dimnames(data) <- list(1:nt, c("X", "Y1", "Y2"))
  if (n[2] > 0) data[(cs[1]+1):cs[2],"Y1"] <- NA
  if (n[3] > 0) data[(cs[2]+1):cs[3],"Y2"] <- NA
  if (n[4] > 0) data[(cs[3]+1):cs[4],c("Y1","Y2")] <- NA
  return(data)
}
```

As an imputation model, we specified compatible linear regressions $Y_1 = \beta_{1,0} + \beta_{1,2}Y_2 + \beta_{1,3}X + \epsilon_1$ and $Y_2 = \beta_{2,0} + \beta_{2,1}Y_1 + \beta_{2,3}X + \epsilon_2$ to impute Y_1 and Y_2. The following code defines the function used for imputation.

```
impute <- function(data, m = 5, method = "norm",
                   print = FALSE, maxit = 10, ...) {
  statistic <- matrix(NA, nrow = maxit, ncol = m)
  for (iter in 1:maxit) {
    if (iter==1) imp <- mice(data, m = m, method = method,
                             print = print, maxit = 1, ...)
    else imp <- mice.mids(imp, maxit = 1, print = print, ...)
    statistic[iter, ] <- unlist(with(imp, cor(Y1, Y2))$analyses)
  }
  return(list(imp = imp, statistic = statistic))
}
```

The difficulty in this particular problem is that the correlation $\rho(Y_1, Y_2)$ under the conditional independence of Y_1 and Y_2 given X is equal to $0.9 \times 0.9 = 0.81$, whereas the true value equals 0.7. It is thus of interest to study how the correlation $\rho(Y_1, Y_2)$ develops over the iterations, but this is not a standard function in mice(). As an alternative, the impute() function repeatedly calls mice.mids() with maxit = 1, and calculates $\rho(Y_1, Y_2)$ after each iteration from the complete data.

The following code defines six scenarios where the number of complete cases is varied as $n_{(1,1,1)} \in \{1000, 500, 250, 100, 50, 0\}$, while holding the total sample size constant at $n = 10000$. The proportion of complete rows thus varies between 10% and 0%.

```
simulate <- function(
  ns = matrix(c(1000, 500, 250, 100, 50, 0,
                rep(c(4500, 4750, 4875, 4950, 4975, 5000), 2),
                rep(0, 6)), nrow = 6),
  m = 5, maxit = 10, seed = 1, ...) {
  if (!missing(seed)) set.seed(seed)
  s <- cbind(rep(1:nrow(ns), each = maxit * m),
             apply(ns, 2, rep, each = maxit * m),
             rep(1:maxit, each = m), 1:m, NA)
  colnames(s) <- c("k", "n111", "n101", "n110", "n100",
                   "iteration", "m", "rY1Y2")
  for (k in 1:nrow(ns)) {
    data <- generate(ns[k, ], ...)
    r <- impute(data, m = m, maxit = maxit, ...)
    s[s[,"k"] == k, "rY1Y2"] <- t(r$statistic)
  }
  return(data.frame(s))
}
```

The `simulate()` function code collects the correlations $\rho(Y_1, Y_2)$ per iteration in the data frame s. Now call the function with

```
slow.demo <- simulate(maxit = 150, seed = 62771)
```

Figure 4.3 shows the development of $\rho(Y_1, Y_2)$ calculated on the completed data after every iteration of the MICE algorithm. At iteration 1, $\rho(Y_1, Y_2)$ is approximately 0.81, the value expected under independence of Y_1 and Y_2, conditional on X. The influence of the complete records with both Y_1 and Y_2 observed percolates into the imputations, so that the chains slowly move into the direction of the population value of 0.7. The speed of convergence heavily depends on the number of missing cases. For 90% or 95% missing data, the streams are essentially flat after about 15–20 iterations. As the percentage of missing data increases, more and more iterations are needed before the true correlation of 0.7 trickles through. In the extreme cases with 100% missing data, the correlation $\rho(Y_1, Y_2)$ cannot be estimated due to lack of information in the data. In this case, the different streams do not converge at all, and wander widely within the Cauchy–Schwarz bounds (0.6 to 1.0 here). But even here we could argue that the sampler has essentially converged. We could stop at iteration 200 and take the imputations from there. From a Bayesian perspective, this still would yield an essentially correct inference about $\rho(Y_1, Y_2)$, being that it could be anywhere within the Cauchy–Schwarz bounds. So even in this pathological case with 100% missing data, the results look sensible as long as we account for the wide variability.

The lesson we can learn from this simulation is that we should be careful about convergence in missing data problems with high correlations and high

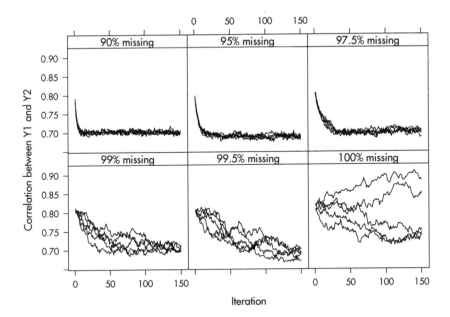

Figure 4.3: Correlation between Y_1 and Y_2 in the imputed data per iteration in five independent runs of the MICE algorithm for six levels of missing data. The true value is 0.7. The figure illustrates that convergence can be slow for high percentages of missing data.

missing data rates. At the same time, observe that we really have to push the MICE algorithm to its limits to see the effect. Over 99% of real data will have lower correlations and lower missing data rates. Of course, it never hurts to do a couple of extra iterations, but my experience is that good results can often be obtained with a small number of iterations.

4.5.8 Performance

Each conditional density has to be specified separately, so FCS requires some modeling effort on the part of the user. Most software provides reasonable defaults for standard situations, so the actual effort required may be small. A number of simulation studies provide evidence that FCS generally yields estimates that are unbiased and that possess appropriate coverage (Brand, 1999; Raghunathan et al., 2001; Brand et al., 2003; Tang et al., 2005; Van Buuren et al., 2006; Horton and Kleinman, 2007; Yu et al., 2007).

4.6 FCS and JM

4.6.1 Relations between FCS and JM

FCS is related to JM in some special cases. If $P(X, Y)$ has a multivariate normal model distribution, then all conditional densities are linear regressions with a constant normal error variance. So, if $P(X, Y)$ is multivariate normal then $P(Y_j|X, Y_{-j})$ follows a linear regression model. The reverse is also true: If the imputation models $P(Y_j|X, Y_{-j})$ are all linear with constant normal error variance, then the joint distribution will be multivariate normal. See Arnold et al. (1999, p. 186) for a description of the precise conditions. Thus, imputation by FCS using all linear regressions is identical to imputation under the multivariate normal model.

Another special case occurs for binary variables with only two-way interactions in the log-linear model. In the special case $p = 3$ suppose that $Y_1, \ldots, Y_3$ are modeled by the log-linear model that has the three-way interaction term set to zero. It is known that the corresponding conditional distribution $P(Y_1|Y_2, Y_3)$ is the logistic regression model $\log(P(Y_1)/1 - P(Y_1)) = \beta_0 + \beta_2 Y_2 + \beta_3 Y_3$ (Goodman, 1970). Analogous definitions exist for $P(Y_2|Y_1, Y_3)$ and $P(Y_3|Y_1, Y_2)$. This means that if we use logistic regressions for Y_1, Y_2 and Y_3, we are effectively imputing under the multivariate "no three-way interaction" log-linear model. Hughes et al. (2014) showed that this relation does not extend to more than three variables.

4.6.2 Comparisons

FCS cannot use computational shortcuts like the sweep operator, so the calculations per iterations are more intensive than under JM. Also, JM has better theoretical underpinnings.

On the other hand, FCS allows tremendous flexibility in creating multivariate models. One can easily specify models that are outside any known standard multivariate density $P(X, Y, R|\theta)$. FCS can use specialized imputation methods that are difficult to formulate as a part of a multivariate density $P(X, Y, R|\theta)$. Imputation methods that preserve unique features in the data, e.g., bounds, skip patterns, interactions, bracketed responses and so on can be incorporated. It is possible to maintain constraints between different variables in order to avoid logical inconsistencies in the imputed data that would be difficult to do as part of a multivariate density $P(X, Y, R|\theta)$.

Lee and Carlin (2010) found that JM performs as well as FCS, even in the presence of binary and ordinal variables. These authors also observed substantial improvements for skewed variables by transforming the variable to symmetry (for JM) or by using predictive mean matching (for FCS). Kropko et al. (2014) found that JM and FCS performed about equally well for continuous and binary variable, but FCS outperforms JM on every metric when the variable of interest is categorical. With predictive mean matching, FCS outperforms JM "for every metric and variable type, including the continuous variable." Seaman and Hughes (2018) compared FCS to a restricted general location model. As expected, the latter model is more efficient when correctly specified, but the gains are small unless the relations between the variables are very strong. As FCS was found to be more robust under misspecification, the authors advise FCS over JM.

4.6.3 Illustration

The Fourth Dutch Growth Study by Fredriks et al. (2000a) collected data on 14500 Dutch children between 0 and 21 years. The development of secondary pubertal characteristics was measured by the so-called Tanner stages, which divides the continuous process of maturation into discrete stages for the ages between 8 and 21 years. Pubertal stages of boys are defined for genital development (gen: five ordered stages G1–G5), pubic hair development (phb: six ordered stages P1–P6) and testicular volume (tv: 1–25 ml).

We analyze the subsample of 424 boys in the age range 8–21 years using the boys data in mice. There were 180 boys (42%) for which scores for genital development were missing. The missingness was strongly related to age, rising from about 20% at ages 9–11 years to 60% missing data at ages 17–20 years.

The data consist of three complete covariates: age (age), height (hgt) and weight (wgt), and three incomplete outcomes measuring maturation. The following code block creates $m = 10$ imputations by the normal model, by predictive mean matching and by the proportional odds model.

```
select <- with(boys, age >= 8 & age <= 21.0)
djm <- boys[select, -4]
djm$gen <- as.integer(djm$gen)
djm$phb <- as.integer(djm$phb)
djm$reg <- as.integer(djm$reg)
dfcs <- boys[select, -4]

## impute under jm and fcs
jm <- mice(djm, method = "norm", seed = 93005, m = 10,
           print = FALSE)
pmm <- mice(djm, method = "pmm", seed = 71332, m = 10,
            print = FALSE)
fcs <- mice(dfcs, seed = 81420, m = 10, print = FALSE)
```

Figure 4.4 plots the results of the first five imputations from the normal model. It was created by the following statement:

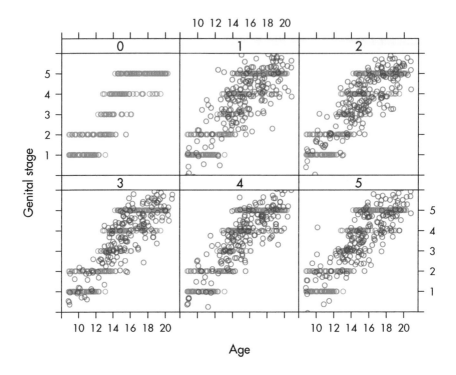

Figure 4.4: Joint modeling: Imputed data for genital development (Tanner stages G1–G5) under the multivariate normal model. The panels are labeled by the imputation numbers 0–5, where 0 is the observed data and 1–5 are five multiply imputed datasets.

```
xyplot(jm, gen ~ age | as.factor(.imp), subset = .imp < 6,
       xlab = "Age", ylab = "Genital stage",
       col = mdc(1:2), ylim = c(0, 6))
```

The figure portrays how genital development depends on age for both the observed and imputed data. The spread of the synthetic values in Figure 4.4 is larger than the observed data range. The observed data are categorical while the synthetic data vary continuously. Note that there are some negative values in the imputations. If we are to do categorical data analysis on the imputed data, we need some form of rounding to make the synthetic values comparable with the observed values.

Imputations for the proportonal odds model in Figure 4.5 differ markedly from those in Figure 4.4. This model yields imputations that are categorical, and hence no rounding is needed.

The complete-data model describes the probability of achieving each Tan-

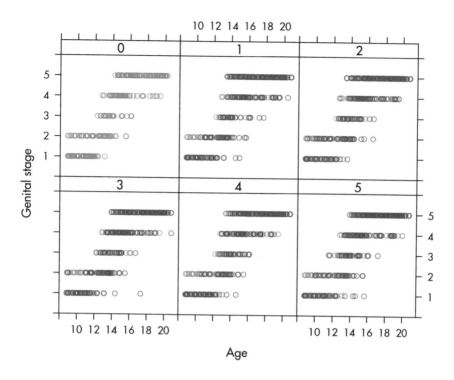

Figure 4.5: Fully conditional specification: Imputed data of genital development (Tanner stages G1–G5) under the proportional odds model.

ner stage as a nonlinear function of age according to the model proposed in Van Buuren and Ooms (2009). The calculations are done with `gamlss` (Stasinopoulos and Rigby, 2007). Under the assumption of ignorability, analysis of the complete cases will not be biased, so the complete-case analysis provides a handle to the appropriate solution. The blue lines in Figure 4.6 indicate the model fitted on the complete cases, whereas the thin black lines correspond to the analyses of the 10 imputed datasets.

The different panels of Figure 4.6 corresponds to different imputation methods. The panel labeled *JM: multivariate normal* contains the model fitted to the unprocessed imputed data produced under the multivariate normal model. There is a large discrepancy between the complete-case analysis and the models fitted to the imputed data, especially for the older boys. The fit improves in the panel labeled *JM: rounded*, where imputed data are rounded to the nearest category. There is considerable misfit, and the behavior of the imputed data around the age of 10 years is a bit curious. The panel labeled *FCS: predictive mean matching* applied Algorithm 3.3 as a component of the MICE algorithm. Though this technique improves upon the previous two methods,

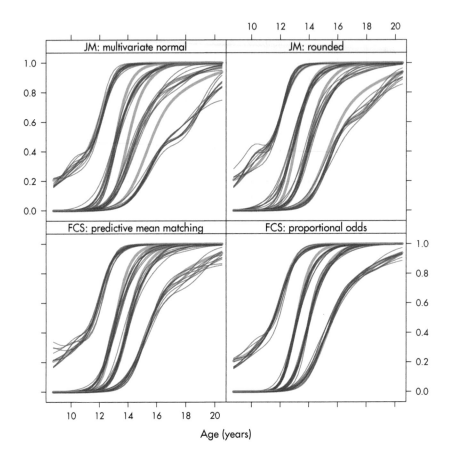

Figure 4.6: Probability of achieving stages G2–G5 of genital development by age (in years) under four imputation methods ($m = 10$).

some discrepancies for the older boys remain. The panel labeled *FCS: proportional odds* displays the results after applying the method for ordered categorical data as discussed in Section 3.6. The imputed data essentially agree with the complete-case analysis, perhaps apart from some minor deviations around the probability level of 0.9.

Figure 4.6 shows clear differences between FCS and JM when data are categorical. Although rounding may provide reasonable results in particular datasets, it seems that it does more harm than good here. There are many ways to round, rounding may require unrealistic assumptions and it will attenuate correlations. Horton et al. (2003), Ake (2005) and Allison (2005) recommend against rounding when data are categorical. See Section 4.4.3. Horton et al. (2003) expected that bias problems of rounding would taper off if variables have more than two categories, but the analysis in this section suggests that JM may also be biased for categorical data with more than two categories. Even though it may sound a bit trivial, my recommendation is: Impute categorical data by methods for categorical data.

4.7 MICE extensions

The MICE algorithm listed in box 4.3 can be extended in several ways.

4.7.1 Skipping imputations and overimputation

By default, the MICE algorithm imputes the missing data, and leaves the observed data untouched. In some cases it may also be useful to skip imputation of certain cells. For example, we wish to skip imputation of quality of life for the deceased, or not impute customer satisfaction for people who did not buy the product. The primary difficulty with this option is that it creates missing data in the predictors, so the imputer should either remove the predictor from all imputation models, or have the missing values propagated through the algorithm. Another use case involves imputing cells with observed data, a technique called *overimputation*. For example, it may be useful to evaluate whether the observed point data fit the imputation model. If all is well, we expect the observed data point in the center of the multiple imputations. The primary difficulty with this option is to ensure that only the observed data (and not the imputed data) are used as an outcome in the imputation model. Version 3.0 of `mice` includes the `where` argument, a matrix with with logicals that has the same dimensions as the data, that indicates where in the data the imputations should be created. This matrix can be used to specify for each cell whether it should be imputed or not. The default is that the missing data are imputed.

4.7.2 Blocks of variables, hybrid imputation

An important difference between JM and FCS is that JM imputes all variables at once, whereas FCS imputes each variable separately. JM and FCS are the extremes scenarios of the much wider range of *hybrid imputation models*. In actual data analysis sets of variables are often connected in some way. Examples are:

- A set of scale items and its total score;

- A variable with one or more transformations;

- Two variables with one or more interaction terms;

- A block of normally distributed Z-scores;

- Compositions that add up to a total;

- Set of variables that are collected together.

Instead of specifying the steps for each variable separately, it is more user-friendly to impute these as a block. Version `3.0` of `mice` includes a new `block` argument that partitions the complete set of variables into blocks. All variables within the same block are jointly imputed, which provides a strategy to specify hybrids of JM and FCS. The joint models need to be open to accept external covariates. One possibility is to use predictive mean matching to impute multivariate nonresponse, where the donor values for the variables within the block come from the same donor (Little, 1988). The main algorithm in `mice 3.0` iterates over the blocks rather than the variables. By default, each variable is its own block, which gives the familiar behavior.

4.7.3 Blocks of units, monotone blocks

Another way to partition the data is to define blocks of units. One weakness of the algorithm in box 4.3 is that it may become unstable when many of the predictors are imputed. Zhu (2016) developed a solution called "Block sequential regression multivariate imputation", where units are partitioned into blocks according to the missing data pattern. The imputation model for a given variable is modified for each block, such that only the observed data with the block can serve as predictor. The method generalizes the monotone block approach of Li et al. (2014).

4.7.4 Tile imputation

The block-wise partitioning methods are complementary strategies to multivariate imputation. The methods in Section 4.7.2 partition the columns and apply one model to many outcomes, whereas the methods in Section 4.7.3 partition the rows and apply many models to one outcome. These operations can be freely combined into a whole new class of algorithms based on *tiles*, i.e., combinations of row and column partitions. This is a vast and yet unexplored field. I expect that it will be possible to develop imputation algorithms that are user-friendly, stable and automatic. A major new application of such tile algorithm will be in the imputation of combined data. The problem of automatic detection of "optimal tiles" provides both enormous challenges and substantial pay-offs.

4.8   Conclusion

Multivariate missing data lead to analytic problems caused by mutual dependencies between incomplete variables. The missing data pattern provides important information for the imputation model. The influx and outflux measures are useful to sift out variables that cannot contribute to the imputations. For general missing data patterns, both JM and FCS approaches can be used to impute multivariate missing data. JM is the model of choice if the data conform to the modeling assumptions because it has better theoretical properties. The FCS approach is much more flexible, easier to understand and allows for imputations close to the data. Automatic tile imputation algorithms with simultaneous partitions of rows and columns of the data form a vast and unexplored field.

4.9   Exercises

1. *MAR.* Repeat Exercise 3.1 for a multivariate missing data mechanism.

2. *Convergence.* Figure 4.3 shows that convergence can take longer for very high amounts of missing data. This exercise studies an even more extreme situation.

   (a) The default argument `ns` of the `simulate()` function in Section 4.5.7 defines six scenarios with different missing data patterns. Define a 6×4 matrix `ns2`, where patterns R_2 and R_3 are replaced by pattern $R_4 = (1, 0, 0)$. How many more missing values are there in each scenario?

   (b) For the new scenarios, do you expect convergence to be slower or faster? Explain.

   (c) Change the scenario in which all data in Y_1 and Y_2 are missing so that there are 20 complete cases. Then run

   ```
   slow2 <- simulate(ns = ns2, maxit = 50, seed = 62771)
   ```

   and create a figure similar to Figure 4.3.

   (d) Compare your figure with Figure 4.3. Are there any major differences? If so, which?

   (e) Did the figure confirm your idea about convergence speed you had formulated in (b)?

   (f) How would you explain the behavior of the trace lines?

3. *Binary data.* Perform the simulations of Section 4.5.7 with binary Y_1 and Y_2. Use the odds ratio instead of the correlation to measure the association between Y_1 and Y_2. Does the same conclusion hold?

Chapter 5

Analysis of imputed data

You must use a computer to do data science: you cannot do it in your head, or with pencil and paper.
Hadley Wickham

Creating plausible imputations is the most challenging activity in multiple imputation. Once we have the multiply imputed data, we can estimate the parameters of scientific interest from each of the m imputed datasets, but now without the need to deal with the missing data, as all data are now complete. These repeated analyses produce m results.

The m results will feed into step 3 (pooling the results). The pooling step to derive the final statistical inferences is relatively straightforward, but its application in practice is not entirely free of problems. First of all, the complete-data analyses are nontrivial. Historically, the imputation literature (including the first edition of this book) has concentrated on step 1 (creating the imputations) and on step 3 (pooling the results), and has worked from the notion that step 2 (estimating the parameters) is well-specified and easy to execute once the data are complete. In practice step 2 can be quite involved. The step often includes model searching, optimization, validation, prediction, assessment of the quality of model fit, in fact, step 2 may embrace almost any aspect of machine learning and data science. All the analyses need to be repeated for each of the m datasets, which may put a considerable burden on the data analyst.

Fortunately, thanks to tremendous advances in recent computational technology, the use of modern data science techniques in step 2 is now becoming feasible. This chapter focuses on step 2. The next chapter addresses issues related to step 3.

5.1 Workflow

Figure 5.1 outlines the three main steps in any multiple imputation analysis. In step 1, we create several m complete versions of the data by replacing the missing values by plausible data values. The task of step 2 is to estimate the parameters of scientific or commercial interest from each imputed

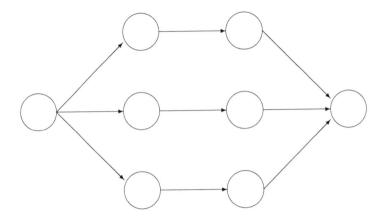

Incomplete data   Imputed data   Analysis results   Pooled results

Figure 5.1: Scheme of main steps in multiple imputation.

Table 5.1: Overview of the classes in the `mice` package.

Class	Name	Produced by	Description
mids	imp	`mice()`	multiply imputed dataset
mild	idl	`complete()`	multiply imputed list of data
mira	fit	`with()`	Multiple imputation repeated analyses
mipo	est	`pool()`	Multiple imputation pooled results

dataset. Step 3 involves pooling the m parameter estimates into one estimate, and obtaining an estimate of its variance. The results allow us to arrive at valid decisions from the data, accounting for the missing data and having the correct type I error rate.

5.1.1   Recommended workflows

There is more than one way to divide the work that implements the steps of Figure 5.1. The classic workflow in `mice` runs functions `mice()`, `with()` and `pool()` in succession, each time saving the intermediate result:

```
# mids workflow using saved objects
library(mice)
imp <- mice(nhanes, seed = 123, print = FALSE)
fit <- with(imp, lm(chl ~ age + bmi + hyp))
est1 <- pool(fit)
```

The objects `imp`, `fit` and `est` have classes `mids`, `mira` and `mipo`, respectively. See Table 5.1 for an overview. The classic workflow works because `mice` contains a `with()` function that understands how to deal with a `mids`-object. The classic `mids` workflow has been widely adopted, but there are more possibilities.

The `magrittr` package introduced the pipe operator to R. This operator removes the need to save and reread objects, resulting in more compact and better readable code:

```
# mids workflow using pipes
library(magrittr)
est2 <- nhanes %>%
  mice(seed = 123, print = FALSE) %>%
  with(lm(chl ~ age + bmi + hyp)) %>%
  pool()
```

The `with()` function handles two tasks: to fill in the missing data and to analyze the data. Splitting these over two separate functions provided the user easier access to the imputed data, and hence is more flexible. The following code uses the `complete()` function to save the imputed data as a list of dataset (i.e., as an object with class `mild`), and then executes the analysis on each dataset by the `lapply()` function.

```
# mild workflow using base::lapply
est3 <- nhanes %>%
  mice(seed = 123, print = FALSE) %>%
  mice::complete("all") %>%
  lapply(lm, formula = chl ~ age + bmi + hyp) %>%
  pool()
```

If desired, we may extend the `mild` workflow by recycling through multiple arguments by means of the `Map` function.

```
# mild workflow using pipes and base::Map
est4 <- nhanes %>%
  mice(seed = 123, print = FALSE) %>%
  mice::complete("all") %>%
  Map(f = lm, MoreArgs = list(f = chl ~ age + bmi + hyp)) %>%
  pool()
```

RStudio has been highly successful with the introduction of the free and open `tidyverse` ecosystem for data acquisition, organization, analysis, visualization and reproducible research. The book by Wickham and Grolemund (2017) provides an excellent introduction to data science using `tidyverse`. The `mild` workflow can be written in `tidyverse` as

```
# mild workflow using purrr::map
library(purrr)
est5 <- nhanes %>%
  mice(seed = 123, print = FALSE) %>%
  mice::complete("all") %>%
  map(lm, formula = chl ~ age + bmi + hyp) %>%
  pool()
```

Manipulating the imputed data is easy if we store the imputed data in `long` format.

```
# long workflow using base::by
est6 <- nhanes %>%
  mice(seed = 123, print = FALSE) %>%
  mice::complete("long")  %>%
  by(as.factor(.$.imp), lm, formula = chl ~ age + bmi + hyp) %>%
  pool()
```

The `long` format can be processed by the `dplyr::do()` function into a list-column and pooled, as follows:

```
# long workflow using a dplyr list-column
library(dplyr)
est7 <- nhanes %>%
  mice(seed = 123, print = FALSE) %>%
  mice::complete("long") %>%
  group_by(.imp) %>%
  do(model = lm(formula = chl ~ age + bmi + hyp, data = .)) %>%
  as.list() %>%
  .[[-1]] %>%
  pool()
```

These workflows yield identical estimates, but allow for different extensions.

5.1.2 Not recommended workflow: Averaging the data

Researchers are often tempted to average the multiply imputed data, and analyze the averaged data as if it were complete. This method yields incorrect standard errors, confidence intervals and p-values, and thus should not be used if any form of statistical testing or uncertainty analysis is to be done on the imputed data. The reason is that the procedure ignores the between-imputation variability, and hence shares all the drawbacks of single imputation. See Section 1.3.

Averaging the data and analyzing the aggregate is easy to do with `dplyr`:

```
# incorrect workflow: averaging data, no pooling
ave <- nhanes %>%
  mice(seed = 123, print = FALSE) %>%
  mice::complete("long") %>%
  group_by(.id) %>%
  summarise_all(.funs = mean) %>%
  select(-.id, -.imp)
est8 <- lm(formula = chl ~ age + bmi + hyp, data = ave)
```

This workflow is faster and easier than the methods in Section 5.1.1, since there is no need to replicate the analyses m times. In the words of Dempster and Rubin (1983), this workflow is

> ... seductive because it can lull the user into the pleasurable state of believing that the data are complete after all.

The ensuing statistical analysis does not know which data are observed and which are missing, and treats all data values as real, which will underestimate the uncertainty of the parameters. The reported standard errors and p-values after data-averaging are generally too low. The correlations between the variables of the averaged data will be too high. For example, the correlation matrix is the average data

```
cor(ave)
```

```
          age      bmi     hyp    chl
age     1.000  -0.4079  0.5262  0.478
bmi    -0.408   1.0000  0.0308  0.313
hyp     0.526   0.0308  1.0000  0.381
chl     0.478   0.3127  0.3812  1.000
```

are more extreme than the average of the m correlation matrices[1]

```
cor <- nhanes %>%
  mice(seed = 123, print = FALSE) %>%
  mice::complete("all") %>%
  lapply(cor)
Reduce("+", cor) / length(cor)
```

```
          age      bmi     hyp    chl
age     1.000  -0.3676  0.4741  0.442
bmi    -0.368   1.0000  0.0377  0.278
hyp     0.474   0.0377  1.0000  0.299
chl     0.442   0.2782  0.2985  1.000
```

---

[1]Section 5.2.2 describes a better way to combine correlations.

which is an example of ecological fallacy. As researchers tend to like low p-values and high correlations, there is a cynical reward for the analysis of the average data. However, analysis of the average data cannot give a fair representation of the uncertainties associated with the underlying data, and hence is not recommended.

5.1.3   Not recommended workflow: Stack imputed data

A variation on this theme is to stack the imputed data, thus creating $m \times n$ complete records. Each record is weighted by a factor $1/m$, so that the total sample size is equal to n. The statistical analysis amounts to performing a weighted linear regression. If the scientific interest is restricted to point estimates and if the complete-data model is linear, this analysis of the stacked imputed data will yield unbiased estimates. Be aware that routine methods for calculating test statistics, confidence intervals or p-values will provide invalid answers if applied to the stacked imputed data.

Creating and analyzing a stacked imputed dataset is easy:

```
est9 <- nhanes2 %>%
  mice(seed = 123, print = FALSE) %>%
  mice::complete("long") %>%
  lm(formula = chl ~ age + bmi + hyp)
```

While the estimated regression coefficients are unbiased, we cannot trust the standard errors, t-values and so on. An advantage of stacking over averaging is that it is easier to analyze categorical data. Although stacking can be useful in specific contexts, like variable selection, in general it is not recommended.

5.1.4   Repeated analyses

The appropriate way to analyze multiply imputed data is to fit the complete-data model on each imputed dataset separately. In `mice` we can use the `with()` command for this purpose. This function takes two main arguments. The first argument of the call is a `mids` object produced by `mice()`. The second argument is an expression that is to be applied to each completed dataset. The `with()` function implements the following loop ($\ell = 1, \ldots, m$):

1. it creates the ℓth imputed dataset

2. it runs the expression on the imputed dataset

3. it stores the result in the list `fit$analyses`

For example, we fit a regression model to each dataset and print out the estimates from the first and second completed datasets by

```
fit <- with(imp, lm(chl~bmi+age))
coef(fit$analyses[[1]])
```

(Intercept)	bmi	age
33.55	4.08	27.37

```
coef(fit$analyses[[2]])
```

(Intercept)	bmi	age
-51.47	6.77	35.44

Note that the estimates differ from each other because of the uncertainty created by the missing data. Applying the standard pooling rules is done by

```
est <- pool(fit)
summary(est)
```

	estimate	std.error	statistic	df	p.value
(Intercept)	6.26	64.29	0.0974	9.62	0.92394
bmi	4.99	2.10	2.3762	9.24	0.03370
age	29.17	9.15	3.1890	12.89	0.00718

which shows the correct estimates after multiple imputation.

Any R expression produced by `expression()` can be evaluated on the multiply imputed data. For example, suppose we want to calculate the difference in frequencies between categories 1 and 2 of `hyp`. This is conveniently done by the following statements:

```
expr <- expression(freq <- table(hyp), freq[1] - freq[2])
fit <- with(imp, eval(expr))
unlist(fit$analyses)
```

```
1  1  1  1  1
9  9 11 15 15
```

All the major software packages nowadays have ways to execute the m repeated analyses to the imputed data.

5.2 Parameter pooling

5.2.1 Scalar inference of normal quantities

Section 2.4 describes Rubin's rules for pooling the results from the m complete-data analyses. These rules are based on the assumption that the

Table 5.2: Suggested transformations toward normality for various types of statistics. The transformed quantities can be pooled by Rubin's rules.

Statistic	Transformation	Source
Correlation	Fisher z	Schafer (1997)
Odds ratio	Logarithm	Agresti (1990)
Relative risk	Logarithm	Agresti (1990)
Hazard ratio	Logarithm	Marshall et al. (2009)
Explained variance R^2	Fisher z on root	Harel (2009)
Survival probabilities	Complementary log-log	Marshall et al. (2009)
Survival distribution	Logarithm	Marshall et al. (2009)

parameter estimates $\hat{Q}$ are normally distributed around the population value Q with a variance of U. Many types of estimates are approximately normally distributed, e.g., means, standard deviations, regression coefficients, proportions and linear predictors. Rubin's pooling rules can be applied directly to such quantities (Schafer, 1997; Marshall et al., 2009).

5.2.2   Scalar inference of non-normal quantities

How should we combine quantities with non-normal distributions: correlation coefficients, odds ratios, relative risks, hazard ratios, measures of explained variance and so on? The quality of the pooled estimate and the confidence intervals can be improved when pooling is done in a scale for which the distribution is close to normal. Thus, transformation toward normality and back-transformation into the original scale improves statistical inference.

As an example, consider transforming a correlation coefficient ρ_ℓ for $\ell = 1, \ldots, m$ toward normality using the Fisher z transformation

$$z_\ell = \frac{1}{2} \ln \frac{1 + \rho_\ell}{1 - \rho_\ell} \tag{5.1}$$

For large samples, the distribution of z_ℓ is normal with variance $\sigma^2 = 1/(n-3)$. It is straightforward to calculate the pooled correlation $\bar{z}$ and its variance by Rubin's rules. The result can be back-transformed by the inverse Fisher transformation

$$\bar{\rho} = \frac{e^{2\bar{z}} - 1}{e^{2\bar{z}} + 1} \tag{5.2}$$

The confidence interval of $\bar{\rho}$ is calculated in the z-scale as usual, and then back-transformed by Equation 5.2.

Table 5.2 suggests transformations toward approximate normality for various types of statistics. There are quantities for which the distribution is complex or unknown. Examples include the Cramér C statistic (Brand, 1999) and the discrimination index (Marshall et al., 2009). Ideally, the entire sampling distribution should be pooled in such cases, but the corresponding pooling methods have yet to be developed. The current advice is to search for ad hoc transformations to make the sampling distribution close to normality, and then apply Rubin's rules.

5.3 Multi-parameter inference

There are many situations where we need to test whether a set of parameters is significantly different from zero. For example, if a categorical variable enters the analysis through a set of dummy variables, all parameters related to this set should be tested simultaneously. More generally, ANOVA type of designs can be formulated and tested as a multi-parameter inference regression problem. Van Ginkel and Kroonenberg (2014) provide many practical examples with missing data.

Schafer (1997) distinguished three types of statistics in multi-parameter tests: D_1 (multivariate Wald test), D_2 (combining test statistics) and D_3 (likelihood ratio test). The next sections outline the idea of each approach, and demonstrate how these tests can be performed as part of the repeated data analyses .

5.3.1 D_1 Multivariate Wald test

The multivariate Wald test is an extension of the procedure for scalar quantities as described in Section 2.4.2. The procedure tests whether $Q = Q_0$, where Q_0 is a k-vector of values under the null hypothesis (typically all zeros). The multivariate Wald test requires an estimate of the variance-covariance matrix U of $\bar{Q}$. We could use T from Equation 2.20, but this estimate can be unreliable. The problem is that for small m, the estimate of the between-imputation variance B is unstable, and if $m \leq k$, it is not even full rank. Thus T can be unreliable if B is a substantial component of T.

Li et al. (1991b) proposed an estimate of T in which B and $\bar{U}$ are assumed to be proportional to each other. A more stable estimate of the total variance is then $\tilde{T} = (1 + r_1)\bar{U}$, where $r_1 = \bar{r}$ (from Equation 2.29) is the average fraction of missing information.

Under the assumption that $(Q_0 - \bar{Q})$ is sufficiently close to a normal distribution, the D_1-statistic

$$D_1 = (\bar{Q} - Q_0)'\tilde{T}^{-1}(\bar{Q} - Q_0)/k \tag{5.3}$$

follows an F-distribution F_{k,ν_w} with k and ν_1 degrees of freedom, where

$$\nu_1 = \begin{cases} 4 + (t - 4)[1 + (1 - 2t^{-1})r_1^{-1}]^2 & \text{if } t = k(m - 1) > 4 \\ t(1 + k^{-1})(1 + r_1^{-1})^2/2 & \text{otherwise} \end{cases} \tag{5.4}$$

The p-value for D_1 is

$$P_1 = \Pr[F_{k,\nu_1} > D_1] \tag{5.5}$$

The assumption that the fraction of missing information is the same across all variables and statistics is unlikely to hold in practice. However, Li et al. (1991b) provide encouraging simulation results for situations where this assumption is violated. Except for some extreme cases, the level of the procedure was close to the nominal level, while the loss of power from such violations was modest.

The work of Li et al. (1991b) is based on large samples. Reiter (2007) developed a small-sample version for the degrees of freedom using ideas similar to Barnard and Rubin (1999). Reiter's ν_f spans several lines of text, and is not given here. A simulation study conducted by Reiter showed marked improvements over the earlier formulation, especially in smaller samples. Simulation work by Grund et al. (2016b) and Liu and Enders (2017) confirmed that for small samples (say $n < 50$) ν_f is more conservative than ν_1, and produced type I errors rates closer to their nomimal value. Raghunathan (2015) recently provided an elegant alternative based on Equation 2.32 with ν_{obs} substituted as $\nu_{\mathrm{obs}} = (\nu_{\mathrm{com}} + 1)\nu_{\mathrm{com}}/(\nu_{\mathrm{com}} + 3)(1 + \bar{r})$. It is not yet known how this correction compares to ν_1 and ν_f.

The `mice` package implements the multivariate Wald test as the `D1()` function. Let us impute the **nhanes2** data, and fit the linear regression of **chl** on **age** and **bmi**.

```
library(mice)
imp <- mice(nhanes2, m = 10, print = FALSE, seed = 71242)
m2 <- with(imp, lm(chl ~ age + bmi))
pool(m2)
```

```
Class: mipo      m = 10
              estimate      ubar        b        t dfcom     df    riv
(Intercept)      -2.98   2896.71  1266.44  4289.79    21  11.28  0.481
age40-59         45.73    296.37   152.46   464.07    21  10.43  0.566
age60-99         65.62    342.71   229.04   594.66    21   9.08  0.735
bmi               6.40      3.26     1.41     4.81    21  11.32  0.477
             lambda    fmi
(Intercept)   0.325  0.419
age40-59      0.361  0.456
age60-99      0.424  0.519
bmi           0.323  0.418
```

We want to simplify the model by testing for **age**. Since **age** is a categorical variable with three categories, removing it involves deleting two columns at the same time, hence the univariate Wald test does not apply. The solution is to fit the model without **age**, and run the multivariate Wald statistic to test whether the model estimates are different.

```
m1 <- with(imp, lm(chl ~ bmi))
summary(D1(m2, m1))


Models:
  model            formula
      1 chl ~ age + bmi
      2     chl ~ bmi

Comparisons:
    test statistic df1  df2 df.com p.value   riv
  1 ~~ 2      4.23    2 14.4     21   0.036 0.63

Number of imputations:   10    Method D1
```

Since the Wald test is significant, removing **age** from the model reduces its predictive power.

5.3.2 D_2 Combining test statistics♠

The multivariate Wald test may become cumbersome when k is large, or when many tests are to be done. Some analytic models may not produce an estimate of Q, or of its variance-covariance matrix. For example, nonparametric tests like the sign test or Wilcoxon-Mann-Whitney produce a p-value, and no estimate of Q. In cases like these, we can still calculate a combined significance test using the m test statistics (e.g., p-values, Z-values, χ^2-values, t-values) as input.

Rubin (1987a, p. 87) and Li et al. (1991a) describe a procedure for pooling the values of the test statistics. Suppose that d_ℓ is the test statistic obtained from the analysis of the ℓth imputed dataset Y_ℓ, $\ell = 1, \ldots, m$. Let $\bar{d} = m^{-1} \sum_\ell d_\ell$ be the average test statistic. The statistic for the combined test is

$$D_2 = \frac{\bar{d}k^{-1} - (m+1)(m-1)^{-1}r_2}{1 + r_2} \tag{5.6}$$

where the relative increase of the variance is calculated as

$$r_2 = \left(1 + \frac{1}{m}\right) \frac{1}{m-1} \sum_{\ell=1}^{m} \left(\sqrt{d_\ell} - \overline{\sqrt{d}}\right)^2 \tag{5.7}$$

with $\overline{\sqrt{d}} = m^{-1} \sum_\ell \sqrt{d_\ell}$, so that r_2 equals the sample variance of $\sqrt{d_1}, \sqrt{d_2}, \ldots, \sqrt{d_m}$ multiplied by $(1 + 1/m)$. The p-value for testing the null hypothesis is

$$P_2 = \Pr[F_{k,\nu_2} > D_2] \tag{5.8}$$

where

$$\nu_2 = k^{-3/m}(m-1)(1 + r_2^{-1})^2 \tag{5.9}$$

The procedure assumes that the test statistic is approximately normally distributed. This is clearly not the case for p-values, which follow a uniform distribution under the null. One may transform the p-values to approximate normality, combine and back-transform afterwards. Based on this idea, Licht (2010, pp. 40–43) proposed a method for obtaining significance levels from repeated p-values similar to Equation 5.6 with custom r_2 and ν_2.

In context of significance testing for logistic regression, Eekhout et al. (2017) suggest taking the median of the m p-values as the combined p-value, an exceedingly simple method. It nevertheless appears to outperform more sophisticated techniques if the variable to be tested is categorical with more than two categories. It would be useful to explore whether this *median P rule* has wider validity.

Let us continue with the previous example. Suppose that our software cannot export the variance-covariance matrix in each repeated analysis, but it does provide a table with the Wald statistics for testing age. The D2 function calculates the D_2-statistic and its degrees of freedom as

```
D2(m2, m1)
```

```
    test statistic df1  df2 df.com p.value   riv
1 ~~ 2      2.26    2 17.6     NA   0.134  1.81
```

In contrast to the previous analysis, observe that the D_2-statistic is not significant at an α-level of 0.05. The reason is that the D_2 test is less informed by the data, and hence less powerful than the D_1 test.

5.3.3 D_3 Likelihood ratio test♠

The likelihood ratio test (Meng and Rubin, 1992) is designed to handle situations where one cannot obtain the covariance matrices of the complete-data estimates. This could be the case if the dimensionality of Q is high, which can occur with partially classified contingency tables. For large n the procedure is equivalent to the method of Section 5.3.1. The likelihood ratio test is the preferred method for testing random effects (Singer and Willett, 2003), and connects to global model fit statistics in structural equation models (Enders and Mansolf, 2018).

Let the vector Q contain the parameters of interest. We wish to test the hypothesis $Q = Q_0$ for some given Q_0. The usual scenario is that we compare two models, one where Q can vary freely and one more restrictive model that constrains $Q = Q_0$.

The procedure for calculating the likelihood ratio test is as follows. First, estimate $\bar{Q}$ (for the full model) and $\bar{Q}_0$ (for the restricted model) from the m datasets by Rubin's rules. Calculate the value of the log-likelihood functions $l(\hat{Q}_\ell)$ (for the full model) and $l(\hat{Q}_{0,\ell})$ (for the restricted model), and determine the average of the likelihood ratio tests across the m datasets, i.e.,

$$\hat{d} = m^{-1} \sum_{\ell} -2(l(\hat{Q}_{0,\ell}) - l(\hat{Q}_{\ell})) \tag{5.10}$$

Then re-estimate the full and restricted models, with their model parameters fixed to $\bar{Q}$ and $\bar{Q}_0$, respectively, and average the corresponding likelihood ratio tests as

$$\bar{d} = m^{-1} \sum_{\ell} -2(l(\bar{Q}_{0,\ell}) - l(\bar{Q}_{\ell})) \tag{5.11}$$

The test statistic proposed by Meng and Rubin (1992) is

$$D_3 = \frac{\bar{d}}{k(1 + r_3)} \tag{5.12}$$

where

$$r_3 = \frac{m+1}{k(m-1)}(\hat{d} - \bar{d}) \tag{5.13}$$

estimates the average relative increase in variance due to nonresponse. The quantity r_3 is asymptotically equivalent to $\bar{r}$ from Equation 2.29. The p-value for D_3 is equal to

$$P_3 = \Pr[F_{k,\nu_3} > D_3] \tag{5.14}$$

where $\nu_3 = \nu_1$, or equal Reiter's correction for small samples.

The likelihood ratio test does not require normality. For complete data, the likelihood ratio test is invariant to scale changes, which is the reason that many prefer the likelihood ratio scale over the Wald test. However, Schafer (1997, p. 118) observed that the invariance property is lost in multiple imputation because the averaging operations in Equations 5.10 and 5.11 may yield somewhat different results under nonlinear transformations of $l(Q)$. He advised that the best results will be obtained if the distribution of Q is approximately normal. One may transform the parameters to achieve normality, provided that appropriate care is taken to infer that the result is still within the allowable parameter space.

Liu and Enders (2017) found in their simulations that D_3 can become negative, a nonsensical value, in some scenarios. They suggest that a value of $r_3 > 10$ or a 1000% increase in sampling variance due to missing data may act as warning signals for this anomaly.

Routine use of the likelihood ratio statistic has long been hampered by difficulties in calculating the likelihood ratio tests for the models with fixed parameters $\bar{Q}$ and $\bar{Q}_0$. With the advent of the `broom` package (Robinson, 2017), the calculations have become feasible for a wide class of models. The `D3()` function in `mice` can be used to calculate the likelihood ratio test. We apply it to the data from previous examples by

```
D3(m2, m1)
```

```
  test statistic df1 df2 df.com p.value riv
1 ~~ 2      9.57   2  18     Inf 0.00147 766
```

The D_3-statistic strongly indicates that age is a significant predictor. Note however the extremely large value for r_3 (column riv), so the result must be taken with a grain of salt. The likely cause for this anomaly could well be the lack of a small-sample correction for this test.

5.3.4 D_1, D_2 or D_3?

If the estimates are approximately normal and if the software can produce the required variance-covariance matrices, we recommend using D_1 with an adjustment for small samples if $n < 100$. D_1 is a direct extension of Rubin's rules to multi-parameter problems, theoretically convincing, mature and widely applied. D_1 is insensitive to the assumption of equal fractions of missing information, is well calibrated, works well with small m (unless the fractions of information are large and variable) and suffers only modest loss of power. The relevant literature (Rubin, 1987a; Li et al., 1991b; Reiter, 2007; Grund et al., 2016b; Liu and Enders, 2017) is quite consistent.

If only the test statistics are available for pooling, then the D_2-statistic is a good option, provided that the number of imputations $m > 20$. The test is easy to calculate and applies to different test statistics. For $m < 20$, the power may be low. D_2 tends to become optimistic for high fractions of missing information (> 0.3), and this effect unfortunately increases with sample size (Grund et al., 2016b). Thus, careless application of D_2 to large datasets with many missing values may yield high rates of false positives.

The likelihood ratio statistic D_3 is theoretically sound. Calculation of D_3 requires refitting the repeated analysis models with the estimates constrained to their pooled values. This was once an issue, but probably less so in the future. D_3 is asymptotically equivalent to D_1, and may be preferred for theoretical reasons: it does not require normality in the complete-data model, it is often more powerful and it may be more stable than if k is large (as $\bar{U}$ need not be inverted). Grund et al. (2016b), Liu and Enders (2017) and Eekhout et al. (2017) found that D_3 produces Type 1 error rates that were comparable to D_1. D_3 tends to be somewhat conservative in smaller samples, especially with high fractions of missing information and with high k. Also, D_3 has lower statistical power in some of the extreme scenarios. For small samples, D_1 has a slight edge over D_3, so given the current available evidence D_1 is the better option for $n < 200$. In larger samples ($n \geq 200$) D_1 and D_3 appear equally good, so the choice between them is mostly a matter of convenience.

5.4 Stepwise model selection

The standard multiple imputation scheme consists of three phases:

1. Imputation of the missing data m times;

2. Analysis of the m imputed datasets;

3. Pooling of the parameters across m analyses.

This scheme is difficult to apply if stepwise model selection is part of the statistical analysis in phase 2. Application of stepwise variable selection methods may result in sets of variables that differ across the m datasets. It is not obvious how phase 3 should be done.

5.4.1 Variable selection techniques

Brand (1999, chap. 7) was the first to recognize and treat the variable selection problem. He proposed a solution in two steps. The first step involves performing stepwise model selection separately on each imputed dataset, followed by the construction of a new supermodel that contains all variables that were present in at least half of the initial models. The idea is that this criterion excludes variables that were selected accidentally. Moreover, it is a rough correction for multiple testing. Second, a special procedure for backward elimination is applied to all variables present in the supermodel. Each variable is removed in turn, and the pooled likelihood ratio p-value (Equation 5.14) is calculated. If the largest p-value is larger than 0.05, the corresponding variable is removed, and the procedure is repeated on the smaller model. The procedure stops if all $p \leq 0.05$. The procedure was found to be a considerable improvement over complete-case analysis.

Yang et al. (2005) proposed variable selection techniques using Bayesian model averaging. The authors studied two methods. The first method, called "impute then select," applies Bayesian variable selection methods on the imputed data. The second method, called "simultaneously impute and select" combines selection and missing data imputation into one Gibbs sampler. Though the latter slightly outperformed the first method, the first method is more broadly applicable. Application of the second method seems to require equivalent imputation and analysis models, thus defeating one of the main advantages of multiple imputation.

Wood et al. (2008) and Vergouwe et al. (2010) studied several scenarios for variable selection. We distinguish three general approaches:

1. *Majority.* A method that selects variables in the final that appear in at least half of the models.

2. *Stack*. Stack the imputed datasets into a single dataset, assign a fixed weight to each record and apply the usual variable selection methods.

3. *Wald*. Stepwise model selection is based on the Wald statistic calculated from the multiply imputed data.

The majority method is identical to step 1 of Brand (1999), whereas the Wald test method is similar to Brand's step 2, with the likelihood ratio test replaced by the Wald test. The Wald test method is recommended since it is a well-established approach that follows Rubin's rules, whereas the majority and stack methods fail to take into account the uncertainty caused by the missing data. Indeed, Wood et al. (2008) found that the Wald test method is the only procedure that preserved the type I error.

Zhao and Long (2017) review recent work on variable selection on imputed data. These authors favor approaches based on the least absolute shrinkage and selection operator (LASSO) (Tibshirani, 1996). The MI-LASSO method by Chen and Wang (2013) tests the coefficients across all the stacked datasets, thus ensuring model consistency across different imputations. Marino et al. (2017) proposed an extension to select covariates in multilevel models.

In practice, it may be useful to combine methods. The Wald test method is computationally intensive, but is now easily available in `mice` as the `D1()` function. A strong point of the majority method is that it gives insight into the variability between the imputed datasets. An advantage of the stack method is that only one dataset needs to be analyzed. The discussion of Wood et al. (2008) contains additional simulations of a two-step method, in which a preselection made by the majority and stack methods is followed by the Wald test. This yielded a faster method with better theoretical properties. In practice, a judicious combination of approaches might turn out best.

5.4.2   Computation

The following steps illustrate the main steps involved by implementing a simple majority method to select variables in `mice`.

```
data <- boys[boys$age >= 8, -4]
imp <- mice(data, seed = 28382, m = 10, print = FALSE)
scope <- list(upper = ~ age + hgt + wgt + hc + gen + phb + reg,
              lower = ~1)
expr <- expression(f1 <- lm(tv ~ 1),
                   f2 <- step(f1, scope = scope))
fit <- with(imp, expr)
```

This code imputes the boys data $m = 10$ times, fits a stepwise linear model to predict `tv` (testicular volume) separately to each of the imputed dataset. The following code blocks counts how many times each variable was selected.

```
formulas <- lapply(fit$analyses, formula)
terms <- lapply(formulas, terms)
votes <- unlist(lapply(terms, labels))
table(votes)

votes
age gen  hc hgt phb reg wgt
 10   9   1   6   9  10   1
```

The `lapply()` function is used three times. The first statement extracts the model formulas fitted to the m imputed datasets. The second `lapply()` call decomposes the model formulas into pieces, and the third call extracts the names of the variables included in all m models. The `table()` function counts the number of times that each variable in the 10 replications. Variables `age`, `gen` and `reg` are always included, whereas `hc` was selected in only one of the models. Since `hgt` appears in more than 50% of the models, we can use the Wald test to determine whether it should be in the final model.

```
fit.without <- with(imp, lm(tv ~ age + gen + reg + phb))
fit.with <- with(imp, lm(tv ~ age + gen + reg + phb + hgt))
D1(fit.with, fit.without)

   test statistic df1  df2 df.com p.value   riv
1 ~~ 2       2.15  1 19.3    409   0.159 0.978
```

The p-value is equal to 0.173, so `hgt` is not needed in the model. If we go one step further, and remove `phb`, we obtain

```
fit.without <- with(imp, lm(tv ~ age + gen + reg))
fit.with <- with(imp, lm(tv ~ age + gen + reg + phb))
D1(fit.with, fit.without)

   test statistic df1  df2 df.com p.value  riv
1 ~~ 2       2.49  5 97.9    410  0.0362 1.29
```

The significant difference ($p = 0.029$) between the models implies that `phb` should be retained. We obtain similar results for the other three variables, so the final model contains `age`, `gen`, `reg` and `phb`.

5.4.3   Model optimism

The main danger of data-driven model building strategies is that the model found may depend highly on the sample at hand. For example, Viallefont et al. (2001) showed that of the variables declared to be "significant" with p-values between 0.01 and 0.05 by stepwise variable selection, only 49% actually were true risk factors. Various solutions have been proposed to counter such *model optimism*. A popular procedure is bootstrapping the model as developed in Sauerbrei and Schumacher (1992) and Harrell (2001). Although Austin (2008) found it ineffective to identify true predictors, this method has often been found to work well for developing predictive models. The method randomly draws multiple samples with replacement from the observed sample, thus mimicking the sampling variation in the population from which the sample was drawn. Stepwise regression analyses are replayed in each bootstrap sample. The proportion of times that each prognostic variable is retained in the stepwise regression model is known as the *inclusion frequency* (Sauerbrei and Schumacher, 1992). This proportion provides information about the strength of the evidence that an indicator is an independent predictor. In addition, each bootstrap model can be fitted to the original sample. The difference between the apparent performance and the bootstrap performance provides the basis for performance measures that correct for model optimism. Steyerberg (2009, p. 95) provides an easy-to-follow procedure to calculate such *optimism-corrected performance* measures.

Clearly, the presence of missing data adds uncertainty to the model building process, so optimism can be expected to be more severe with missing data. It is not yet clear what the best way is to estimate optimism from incomplete data. Heymans et al. (2007) explored the combination of multiple imputation and the bootstrap. There appear to be at least four general procedures:

1. *Imputation.* Multiple imputation generates 100 imputed datasets. Automatic backward selection is applied to each dataset. Any differences found between the 100 fitted models are due to the missing data.

2. *Bootstrap.* 200 bootstrap samples are drawn from one singly imputed completed data. Automatic backward selection is applied to each dataset. Any differences found between the 200 fitted models are due to sampling variation.

3. *Nested bootstrap.* The bootstrap method is applied on each of the multiply imputed datasets. Automatic backward selection is applied to each of the 100 × 200 datasets. Differences between the fitted model portray both sampling and missing data uncertainty.

4. *Nested imputation.* The imputation method is applied on each of the bootstrapped datasets.

Heymans et al. (2007) observed that the imputation method produced a wider range of inclusion frequencies than the bootstrap method. This is attractive since a better separation of strong and weak predictors may ease model building. The area under the curve is an overall index of predictive strength. Though the type of method had a substantial effect on the apparent c-index estimate, the optimism-corrected c-index estimate was quite similar. The optimism-corrected calibration slope estimates tended to be lower in the methods involving imputation, thus necessitating more shrinkage.

A drawback of the method is the use of classic stepwise variable selection techniques, which do not generalize well to high-dimensional data. Musoro et al. (2014) improved the methods of Heymans et al. (2007) through their use of the LASSO.

Long and Johnson (2015) developed a procedure, called bootstrap impu-tation and stability selection (BI-SS) , that generates bootstrap samples from the original data, imputes each bootstrap sample by single imputation, obtains the randomized LASSO estimate from each sample, and then selects the active set according to majority. The multiple imputation random LASSO (MIRL) method by Liu et al. (2016) first performs multiple imputation, obtains boot-strap samples from each imputed dataset, estimates regression weights under LASSO, and then selects the active set by majority. It is not yet known how BS-SS and MIRL compare to each other.

5.5 Parallel computation

Multiple imputation is a parallel technique. If there are m processors available, it is possible to generate the m imputed datasets, estimate the m complete-data statistics and store the m results by m independent parallel streams. The overhead needed is minimal since each stream requires the same amount of processor time. If more than m processors are available, a better alternative is to subdivide each stream into several substreams. Huge savings in execution time can be obtained in this way (Beddo, 2002).

Unfortunately, R is single-threaded, so the exploitation of the parallel na-ture of multiple imputation is not automatic, and requires some additional work. There are currently three alternatives to perform the calculation of `mice` in a parallel fashion.

1. Gordon (2014) presents a fully worked out example code that builds upon the `doParallel` library, and that combines `complete()` and `ibind()`. With some programming this example can be adapted to other datasets.

2. The `parlMICE()` function is a wrapper around `mice()` that can divide the imputations over multiple cores or CPUs. Schouten and Vink (2017) show that substantial gains are already possible with three free cores, especially for a combination of a large number of imputations m and a large sample size n.

3. The `par.mice()` function in the `micemd` package (Audigier and Resche-Rigon, 2018) takes the same arguments as the `mice()` function, plus two extra arguments related to the parallel calculations. It also builds on the `parallel` package.

The last two options are quite similar. Application of these methods is especially beneficial for simulation studies, where the same model needs to be replicated a large number of times. Support for multi-core processing is likely to grow, so keep an eye on the Internet.

5.6   Conclusion

The statistical analysis of the multiply imputed data involved repeated analysis followed by parameter pooling. Rubin's rules apply to a wide variety of quantities, especially if these quantities are transformed toward normality. Dedicated statistical tests and model selection technique are now available. Although many techniques for complete data now have their analogues for incomplete data, the present state-of-the-art does not cover all. As multiple imputation becomes more familiar and more routine, we will see new post-imputation methodology that will be progressively more refined.

5.7   Exercises

Allison and Cicchetti (1976) investigated the interrelationship between sleep, ecological and constitutional variables. They assessed these variables for 39 mammalian species. The authors concluded that slow-wave sleep is negatively associated with a factor related to body size. This suggests that large amounts of this sleep phase are disadvantageous in large species. Also, paradoxical sleep was associated with a factor related to predatory danger, suggesting that large amounts of this sleep phase are disadvantageous in prey species.

Allison and Cicchetti (1976) performed their analyses under complete-case analysis. In this exercise we will recompute the regression equations for slow wave ("nondreaming") sleep (hrs/day) and paradoxical ("dreaming") sleep (hrs/day), as reported by the authors. Furthermore, we will evaluate the imputations.

1. *Complete-case analysis.* Compute the regression equations (1) and (2) from the paper of Allison and Cicchetti (1976) under complete-case analysis.

2. *Imputation.* The `mammalsleep` data are part of the `mice` package. Impute the data with `mice()` under all the default settings. Recalculate the regression equations (1) and (2) on the multiply imputed data.

3. *Traces.* Inspect the trace plot of the MICE algorithm. Does the algorithm appear to converge?

4. *More iterations.* Extend the analysis with 20 extra iterations using `mice.mids()`. Does this affect your conclusion about convergence?

5. *Distributions.* Inspect the data with diagnostic plots for univariate data. Are the univariate distributions of the observed and imputed data similar? Can you explain why they do (or do not) differ?

6. *Relations.* Inspect the data with diagnostic plots for the most interesting bivariate relations. Are the relations similar in the observed and imputed data? Can you explain why they do (or do not) differ?

7. *Defaults.* Consider each of the seven default choices from Section 6.1 in turn. Do you think the default is appropriate for your data? Explain why.

8. *Improvement.* Do you have particular suggestions for improvement? Which? Implement one (or more) of your suggestions. Do the results now look more plausible or realistic? Explain. What happened to the regression equations?

9. *Multivariate analyses.* Repeat the factor analysis and the stepwise regression. Beware: There might be pooling problems.

Part II

Advanced techniques

Chapter 6

Imputation in practice

Ad hoc methods were designed to get past the missing data so that at least some analyses could be done.
John W. Graham

Chapters 3 and 4 describe methods to generate multiple imputations. The application of these techniques in practice should be done with appropriate care. This chapter focuses on practical issues that surround the methodology. This chapter assumes that multiple imputations are created by means of the MICE algorithm, as described in Section 4.5.2.

6.1 Overview of modeling choices

The specification of the imputation model is the most challenging step in multiple imputation. The imputation model should

- account for the process that created the missing data,

- preserve the relations in the data, and

- preserve the uncertainty about these relations.

The idea is that adherence to these principles will yield proper imputations (cf. Section 2.3.3), and thus result in valid statistical inferences. What are the choices that we need to make, and in what order? Van Buuren and Groothuis-Oudshoorn (2011) list the following seven choices:

1. First, we should decide whether the MAR assumption is plausible. See Sections 1.2 and 2.2.4 for an introduction to MAR and MNAR. FCS can handle both MAR and MNAR. Multiple imputation under MNAR requires additional modeling assumptions that influence the generated imputations. There are many ways to do this. Section 3.8 described one way to do so within the FCS framework. Section 6.2 deals with this issue in more detail.

2. The second choice refers to the form of the imputation model. The form encompasses both the structural part and the assumed error distribution. In FCS the form needs to be specified for each incomplete column in the data. The choice will be steered by the scale of the variable to be imputed, and preferably incorporates knowledge about the relation between the variables. Chapter 3 described many different methods for creating univariate imputations.

3. A third choice concerns the set of variables to include as predictors in the imputation model. The general advice is to include as many relevant variables as possible, including their interactions (Collins et al., 2001). This may, however, lead to unwieldy model specifications. Section 6.3 describes the facilities within the mice() function for setting the predictor matrix.

4. The fourth choice is whether we should impute variables that are functions of other (incomplete) variables. Many datasets contain derived variables, sum scores, interaction variables, ratios and so on. It can be useful to incorporate the transformed variables into the multiple imputation algorithm. Section 6.4 describes methods that we can use to incorporate such additional knowledge about the data.

5. The fifth choice concerns the order in which variables should be imputed. The visit sequence may affect the convergence of the algorithm and the synchronization between derived variables. Section 6.5.1 discusses relevant options.

6. The sixth choice concerns the setup of the starting imputations and the number of iterations M. The convergence of the MICE algorithm can be monitored in many ways. Section 6.5.2 outlines some techniques that assist in this task.

7. The seventh choice is m, the number of multiply imputed datasets. Setting m too low may result in large simulation error and statistical inefficiency, especially if the fraction of missing information is high. Section 2.8 provided guidelines for setting m.

Please realize that these choices are always needed. Imputation software needs to make default choices. These choices are intended to be useful across a wide range of applications. However, the default choices are not necessarily the best for the data at hand. There is simply no magical setting that always works, so often some tailoring is needed. Section 6.6 highlights some diagnostic tools that aid in determining the choices.

6.2   Ignorable or nonignorable?

Recall from Section 2.2.6 that the assumption of ignorability is essentially the belief that the available data are sufficient to correct the missing data. There are two main strategies that we might pursue if the response mechanism is nonignorable:

- Expand the data in the imputation model in the hope of making the missing data mechanism closer to MAR, or

- Formulate and fit a nonignorable imputation model and perform sensitivity analysis on the critical parameters.

Collins et al. (2001) remarked that it is a "safe bet" there will be lurking variables Z that are correlated both with the variables of interest Y and with the missingness of Y. The important question is, however, whether these correlations are strong enough to produce substantial bias if no measures are taken. Collins et al. (2001) performed simulations that provided some answers in the case of linear regression. If the missing data rate did not exceed 25% and if the correlation between the Z and Y was 0.4, omitting Z from the imputation model had a negligible effect. For more extreme situations, with 50% missing data and/or a correlation of 0.9, the effect depended strongly on the form of the missing data mechanism. When the probability to be missing was linear in Z (like MARRIGHT in Section 3.2.4), then omitting Z from the imputation model only affected the intercept, whereas the regression weights and variance estimates were unaffected. When more missing data were created in the extremes (like MARTAIL), the reverse occurred: omitting Z affected the regression coefficients and variance estimates, but the intercept was unbiased with the correct confidence interval. In summary, all estimates under multiple imputation were remarkably robust against MNAR in many instances. Beyond a correlation of 0.4 or a missing data rate over 25% the form of the missing data mechanism determines which parameters are affected.

Based on these results, we suggest the following guidelines. The MAR assumption is often a suitable starting point. If the MAR assumption is suspect for the data at hand, a next step is to find additional data that are strongly predictive of the missingness, and include these into the imputation model. If all possibilities for such data are exhausted and if the assumption is still suspect, perform a concise simulation study as in Collins et al. (2001) customized for the problem at hand with the goal of finding out how extreme the MNAR mechanism needs to be to influence the parameters of scientific interest. Finally, use a nonignorable imputation model (cf. Section 3.8) to correct the direction of imputations created under MAR. Vary the most critical parameters, and study their influence on the final inferences. Section 9.2 contains an example of how this can be done in practice.

Table 6.1: Built-in univariate imputation techniques in the mice package.

Method	Description	Scale Type
pmm	Predictive mean matching	Any*
midastouch	Weighted predictive mean matching	Any
sample	Random sample from observed values	Any
cart	Classification and regression trees	Any
rf	Random forest imputation	Any
mean	Unconditional mean imputation	Numeric
norm	Bayesian linear regression	Numeric
norm.boot	Normal imputation with bootstrap	Numeric
norm.nob	Normal imputation ignoring model error	Numeric
norm.predict	Normal imputation, predicted values	Numeric
quadratic	Imputation of quadratic terms	Numeric
ri	Random indicator for nonignorable data	Numeric
logreg	Logistic regression	Binary*
logreg.boot	Logistic regression with bootstrap	Binary
polr	Proportional odds model	Ordinal*
polyreg	Polytomous logistic regression	Nominal*
lda	Discriminant analysis	Nominal

* = default

6.3   Model form and predictors

6.3.1   Model form

The MICE algorithm requires a specification of a univariate imputation method separately for each incomplete variable. Chapter 3 discussed many possible methods. The measurement level largely determines the form of the univariate imputation model. The mice() function distinguishes numerical, binary, ordered and unordered categorical data, and sets the defaults accordingly.

Table 6.1 lists the built-in univariate imputation method in the mice package. The defaults have been chosen to work well in a wide variety of situations, but in particular cases different methods may be better. For example, if it is known that the variable is close to normally distributed, using norm instead of the default pmm may be more efficient. For large datasets where sampling variance is not an issue, it could be useful to select norm.nob, which does not draw regression parameters, and is thus simpler and faster. The norm.boot method is a fast non-Bayesian alternative for norm. The norm methods are an alternative to pmm in cases where pmm does not work well, e.g., when insufficient nearby donors can be found.

The mean method is included for completeness and should not be generally used. For sparse categorical data, it may be better to use method pmm instead of `logreg`, `polr` or `polyreg`. Method `logreg.boot` is a version of `logreg` that uses the bootstrap to emulate sampling variance. Method `lda` is generally inferior to `polyreg` (Brand, 1999), and should be used only as a backup when all else fails. Finally, `sample` is a quick method for creating starting imputations without the need for covariates.

6.3.2 Predictors

A useful feature of the `mice()` function is the ability to specify the set of predictors to be used for each incomplete variable. The basic specification is made through the `predictorMatrix` argument, which is a square matrix of size `ncol(data)` containing 0/1 data. Each row in `predictorMatrix` identifies which predictors are to be used for the variable in the row name. If `diagnostics=T` (the default), then `mice()` returns a `mids` object containing a `predictorMatrix` entry. For example, type

```
imp <- mice(nhanes, print = FALSE)
imp$predictorMatrix
```

```
    age bmi hyp chl
age   0   1   1   1
bmi   1   0   1   1
hyp   1   1   0   1
chl   1   1   1   0
```

The rows correspond to incomplete target variables, in the sequence as they appear in the data. A value of 1 indicates that the column variable is a predictor to impute the target (row) variable, and a 0 means that it is not used. Thus, in the above example, `bmi` is predicted from `age`, `hyp` and `chl`. Note that the diagonal is 0 since a variable cannot predict itself. Since `age` contains no missing data, `mice()` silently sets all values in the row to 0. The default setting of the `predictorMatrix` specifies that every variable predicts all others.

Conditioning on all other data is often reasonable for small to medium datasets, containing up to, say, 20–30 variables, without derived variables, interactions effects and other complexities. As a general rule, using every bit of available information yields multiple imputations that have minimal bias and maximal efficiency (Meng, 1994; Collins et al., 2001). It is often beneficial to choose as large a number of predictors as possible. Including as many predictors as possible tends to make the MAR assumption more plausible, thus reducing the need to make special adjustments for MNAR mechanisms (Schafer, 1997).

For datasets containing hundreds or thousands of variables, using all predictors may not be feasible (because of multicollinearity and computational problems) to include all these variables. It is also not necessary. In my experience, the increase in explained variance in linear regression is typically negligible after the best, say, 15 variables have been included. For imputation purposes, it is expedient to select a suitable subset of data that contains no more than 15 to 25 variables. Van Buuren et al. (1999) provide the following strategy for selecting predictor variables from a large database:

1. Include all variables that appear in the complete-data model, i.e., the model that will be applied to the data after imputation, including the outcome (Little, 1992; Moons et al., 2006). Failure to do so may bias the complete-data model, especially if the complete-data model contains strong predictive relations. Note that this step is somewhat counter-intuitive, as it may seem that imputation would artificially strengthen the relations of the complete-data model, which would be clearly undesirable. If done properly however, this is not the case. On the contrary, not including the complete-data model variables will tend to bias the results toward zero. Note that interactions of scientific interest also need to be included in the imputation model.

2. In addition, include the variables that are related to the nonresponse. Factors that are known to have influenced the occurrence of missing data (stratification, reasons for nonresponse) are to be included on substantive grounds. Other variables of interest are those for which the distributions differ between the response and nonresponse groups. These can be found by inspecting their correlations with the response indicator of the variable to be imputed. If the magnitude of this correlation exceeds a certain level, then the variable should be included.

3. In addition, include variables that explain a considerable amount of variance. Such predictors help reduce the uncertainty of the imputations. They are basically identified by their correlation with the target variable.

4. Remove from the variables selected in steps 2 and 3 those variables that have too many missing values within the subgroup of incomplete cases. A simple indicator is the percentage of observed cases within this subgroup, the percentage of usable cases (cf. Section 4.1.2).

Most predictors used for imputation are incomplete themselves. In principle, one could apply the above modeling steps for each incomplete predictor in turn, but this may lead to a cascade of auxiliary imputation problems. In doing so, one runs the risk that every variable needs to be included after all.

In practice, there is often a small set of key variables, for which imputations are needed, which suggests that steps 1 through 4 are to be performed for key variables only. This was the approach taken in Van Buuren and Groothuis-Oudshoorn (1999), but it may miss important predictors of predictors. A safer and more efficient, though more laborious, strategy is to perform the modeling steps also for the predictors of predictors of key variables. This is done in Groothuis-Oudshoorn et al. (1999). I expect that it is rarely necessary to go beyond predictors of predictors. At the terminal node, we can apply a simple method like `sample` that does not need any predictors for itself.

The `mice` package contains several tools that aid in automatic predictor selection. The `quickpred()` function is a quick way to define the predictor matrix using the strategy outlined above. The `flux()` function was introduced in Section 4.1.3. The `mice()` function detects multicollinearity, and solves the problem by removing one or more predictors for the model. Each removal is noted in the `loggedEvents` element of the `mids` object. For example,

```
imp <- mice(cbind(nhanes, chl2 = 2 * nhanes$chl),
            print = FALSE, maxit = 1, m = 3, seed = 1)

Warning: Number of logged events: 1

imp$loggedEvents

  it im dep       meth  out
1  0  0     collinear chl2
```

yields a warning that informs us that at initialization variable `chl2` was removed from the imputation model because it is collinear with `chl`. As a result, `chl` will be imputed, but `chl2` is not. We may override removal by

```
imp <- mice(cbind(nhanes, chl2 = 2 * nhanes$chl),
            print = FALSE, maxit = 1, m = 3, seed = 1,
            remove.collinear = FALSE)

Warning: Number of logged events: 3

imp$loggedEvents

  it im  dep meth out
1  1  1 chl2  pmm chl
2  1  2 chl2  pmm chl
3  1  3 chl2  pmm chl
```

Now, the algorithm detects multicollinearity during iterations, and removes `chl` from the imputation model for `chl2`. Although this imputes both `chl` and `chl2`, their relation is not maintained. See Figure 6.1. As a general rule, feedback between different versions of the same variable should be prevented. The next section describes a number of techniques that are useful in various situations. Another measure to control the algorithm is the `ridge` parameter, denoted by κ in Algorithms 3.1, 3.2 and 3.3. The `ridge` parameter is specified as an argument to `mice()`. Setting `ridge=0.001` or `ridge=0.01` makes the algorithm more robust at the expense of bias.

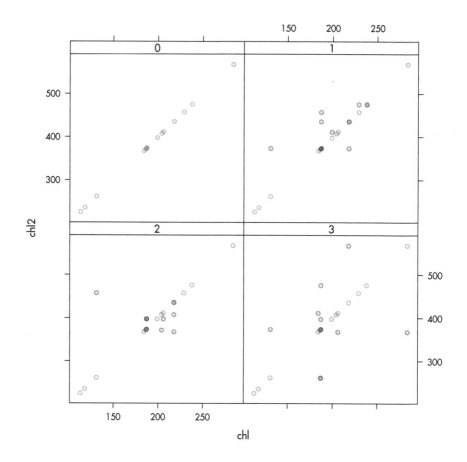

Figure 6.1: Scatterplot of chl2 against chl for $m = 3$. The observed data are linearly related, but the imputed data do not respect the relationship.

6.4　Derived variables

6.4.1　Ratio of two variables

In practice, there is often extra knowledge about the data that is not modeled explicitly. For example, consider the weight/height ratio whr, defined as wgt/hgt (kg/m). If any one of the triplet (hgt, wgt, whr) is missing, then the missing value can be calculated with certainty by a simple deterministic rule. Unless we specify otherwise, the imputation model is unaware of the relation between the three variables, and will produce imputations that are inconsistent with the rule. Inconsistent imputations are clearly undesirable since they yield combinations of data values that are impossible in the real world. Including knowledge about derived data in the imputation model prevents imputations from being inconsistent. Knowledge about the derived data can take many forms, and includes data transformations, interactions, sum scores, recoded versions, range restrictions, if-then relations and polynomials.

The easiest way to deal with the problem is to leave any derived data outside the imputation process. For example, we may impute any missing height and weight data, and append `whr` to the imputed data afterward. It is simple to do that in `mice` by

```
data <- boys[, c("age", "hgt", "wgt", "hc", "reg")]
imp <- mice(data, print = FALSE, seed = 71712)
long <- mice::complete(imp, "long", include = TRUE)
long$whr <- with(long, 100 * wgt / hgt)
imp.itt <- as.mids(long)
```

The approach is known as *Impute, then transform* (Von Hippel, 2009). While `whr` will be consistent, the obvious problem with this approach is that `whr` is not used by the imputation method, and hence biases the estimates of parameters related to `whr` towards zero. Note the use of the `as.mids` function, which transforms the imputed data `long` back into a `mids` object.

Another possibility is to create `whr` before imputation, and impute `whr` as just another variable, known as *JAV* (White et al., 2011b), or under the name *Transform, then impute* (Von Hippel, 2009). This is easy to do, as follows:

```
data$whr <- 100 * data$wgt / data$hgt
imp.jav1 <- mice(data, seed = 32093, print = FALSE)
```

```
Warning: Number of logged events: 45
```

The warning results from the linear dependencies among the predictors, which were introduced by adding `whr`. The `mice()` function checks for linear dependencies during the iterations, and temporarily removes predictors from the univariate imputation models where needed. Each removal action is documented in the the `loggedEvents` component of the `imp.jav1` object. The last three removal events are

```
tail(imp.jav1$loggedEvents, 3)
```

```
   it im dep meth out
43  5  4 whr  pmm wgt
44  5  5 wgt  pmm whr
45  5  5 whr  pmm wgt
```

which informs us that `wgt` was removed while imputing `whr`, and vice versa.

We may prevent automatic removal by setting the relevant entries in the predictorMatrix to zero.

```
pred <- make.predictorMatrix(data)
pred[c("wgt", "whr"), c("wgt", "whr")] <- 0
pred[c("hgt", "whr"), c("hgt", "whr")] <- 0
pred
```

	age	hgt	wgt	hc	reg	whr
age	0	1	1	1	1	1
hgt	1	0	1	1	1	0
wgt	1	1	0	1	1	0
hc	1	1	1	0	1	1
reg	1	1	1	1	0	1
whr	1	0	0	1	1	0

```
imp.jav2 <- mice(data, pred = pred, seed = 32093, print = FALSE)
```

This is a little faster (5-10%) and cleans out the warning.

A third approach is *Passive imputation*, where the transformation is done on-the-fly within the imputation algorithm. Since the transformed variable is available for imputation, the hope is that passive imputation removes the bias of the *Impute, then transform* methods, while restoring consistency among the imputations that was broken in *JAV*. Figure 6.2 visualizes the consistency of the three methods.

In mice passive imputation is invoked by specifying the tilde symbol as the first character of the imputation method. This provides a simple method for specifying dependencies among the variables, such as transformed variables, recodes, interactions, sum scores and so on. In the above example, we invoke passive imputation by

```
data <- boys[, c("age", "hgt", "wgt", "hc", "reg")]
data$whr <- 100 * data$wgt / data$hgt
meth <- make.method(data)
meth["whr"] <- "~I(100 * wgt / hgt)"
pred <- make.predictorMatrix(data)
pred[c("wgt", "hgt"), "whr"] <- 0
imp.pas <- mice(data, meth = meth, pred = pred,
                print = FALSE, seed = 32093)
```

The I() operator in the meth definitions instructs R to interpret the argument as literal. So I(100 * wgt / hgt) calculates whr by dividing wgt by hgt (in meters). The imputed values for whr are thus derived from hgt and wgt according to the stated transformation, and hence are consistent. Since whr is that last column in the data, it is updated after wgt and hgt are imputed. The changes to the default predictor matrix are needed to break any feedback

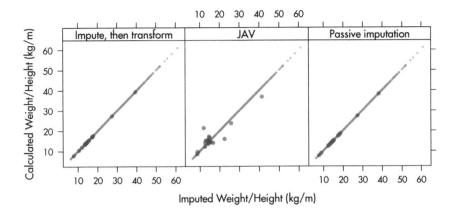

Figure 6.2: Three imputation models to impute weight/height ratio (whr). Methods *Impute, then transform* and *Passive imputation* respect the relation between whr and height (hgt) in the imputed data, whereas *JAV* does not.

loops between the derived variables and their originals. It is important to do this since otherwise `whr` may contain absurd imputations and the algorithm may have difficulties in convergence.

How well do these methods impute a ratio? The simulation studies by Von Hippel (2009) and Seaman et al. (2012) favored the use of *JAV*, but neither addressed imputation of the ratio of two variables. Let us look at a small simulation comparing the three methods. As the population data, take the 681 complete records of variables `age`, `hgt`, `wgt`, `hc` and `reg`, and create a model for predicting height circumference from `hc` from more easily measured variables, including `whr`.

```
pop <- na.omit(boys[, c("age", "hgt", "wgt", "hc", "reg")])
pop$whr <- with(pop, 100 * wgt / hgt)
broom::tidy(lm(hc ~ age + hgt + wgt + whr, data = pop))
```

	term	estimate	std.error	statistic	p.value
1	(Intercept)	23.714	0.65602	36.15	4.25e-160
2	age	-0.308	0.05383	-5.72	1.56e-08
3	hgt	0.176	0.00736	23.97	2.41e-92
4	wgt	-0.496	0.02843	-17.44	1.62e-56
5	whr	1.062	0.05827	18.22	1.22e-60

This is a simple linear model, but the proportion of explained variance is very high, about 0.9. The ratio variable `whr` explains about 5% of the variance on top of the other variables. Let us randomly delete 25% of `hgt` and 25% of

Table 6.2: Evaluation of parameter for whr with 25% MCAR missing in hgt and 25% MCAR missing in wgt using four imputation strategies ($n_{sim} = 200$).

Method	Bias	% Bias	Coverage	CI Width	RMSE
Impute, transform	-0.28	26.4	0.09	0.322	0.289
JAV	-0.90	84.5	0.00	0.182	0.897
Passive imputation	-0.28	26.8	0.06	0.328	0.293
smcfcs	-0.03	2.6	0.90	0.334	0.094
Listwise deletion	0.01	0.7	0.90	0.307	0.094

wgt, apply each of the three methods 200 times using $m = 5$, and evaluate the parameter for whr.

Table 6.2 shows that all three methods have substantial negative biases. Method *JAV* almost nullifies the parameter. The other two methods are better, but still far from optimal. Actually, none of these methods can be recommended for imputing a ratio.

Bartlett et al. (2015) proposed a novel rejection sampling method that creates imputations that are congenial in the sense of Meng (1994) with the substantive (complete-data) model. The method was applied to squared terms and interactions, and here we investigate whether it extends to ratios. The method has been implemented in the smcfcs package. The imputation method requires a specification of the complete-data model, as arguments smtype and smformula. An example of how to generate imputations, fit models, and pool the results is:

```
library(smcfcs)
data <- pop
data[sample(nrow(data), size = 100), "wgt"] <- NA
data[sample(nrow(data), size = 100), "hgt"] <- NA
data$whr <- 100 * data$wgt / data$hgt
meth <- c("", "norm", "norm", "", "", "norm")
imps <- smcfcs(originaldata = data, meth = meth, smtype = "lm",
               smformula = "hc ~ age + hgt + wgt + whr")
fit <- lapply(imps$impDatasets, lm,
              formula = hc ~ age + hgt + wgt + whr)
summary(pool(fit))
```

The results of the simulations are also in Table 6.2 under the heading of the smcfcs. The smcfcs method is far better than the three previous alternatives, and almost as good as one could wish for. Rejection sampling for imputation is still new and relatively unexplored, so this seems a promising area for further work.

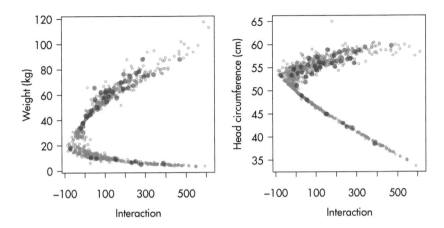

Figure 6.3: The relation between the interaction term wgt.hc (on the horizontal axes) and its components wgt and hc (on the vertical axes).

6.4.2 Interaction terms

The standard MICE algorithm only models main effects. Sometimes the interaction between variables is of scientific interest. For example, in a longitudinal study we could be interested in assessing whether the rate of change differs between two treatment groups, in other words, the treatment-by-group interaction. The standard algorithm does not take interactions into account, so the interactions of interest should be added to the imputation model.

The usual type of interactions between two continuous variables is to subtract the mean and take the product. The following code adds an interaction betwen wgt and hc to the boys data and imputes the data by passive imputation:

```
expr <- expression((wgt - 40) * (hc - 50))
boys$wgt.hc <- with(boys, eval(expr))
meth <- make.method(boys)
meth["wgt.hc"] <- paste("~I(", expr, ")", sep = "")
meth["bmi"] <- ""
pred <- make.predictorMatrix(boys)
pred[c("wgt", "hc"), "wgt.hc"] <- 0
imp.int <- mice(boys, m = 1, meth = meth, pred = pred,
                print = FALSE, seed = 62587, maxit = 10)
```

Figure 6.3 illustrates that the scatterplots of the real and synthetic values are similar. Furthermore, the imputations adhere to the stated recipe (wgt - 40) * (hc - 50). Interactions involving categorical variables can be done in similar ways (Van Buuren and Groothuis-Oudshoorn, 2011), for example by imputing the data in separate groups. One may do this in mice by splitting

Table 6.3: Evaluation of parameter for wgt.hc with 25% MCAR missing in hc and 25% MCAR missing in wgt using four imputation strategies ($n_{\text{sim}} = 200$).

Method	Bias	% Bias	Coverage	CI Width	RMSE
Impute, transform	0.20	22.7	0.17	0.290	0.207
JAV	0.63	71.5	0.00	0.229	0.632
Passive imputation	0.20	22.6	0.17	0.283	0.207
scmfcs	-0.01	0.8	0.92	0.306	0.083
Listwise deletion	-0.01	0.8	0.91	0.237	0.076

the dataset into two or more parts, run `mice()` on each part and then combine the imputed datasets with `rbind()`.

Other methods for imputing interactions are *JAV*, *Impute, then transform* and `smcfcs`. Table 6.3 contains the results of simulations similar to those in Section 6.4.1, but adapted to include the interaction effect shown in Figure 6.3 by using the complete-data model `lm(hgt  wgt + hc + wgt.hc)`. The results tell the same story as before, with `smcfcs` the best method, followed by *Passive imputation* and *Impute, then transform*.

Von Hippel (2009) stated that *JAV* would give consistent results under MAR, but Seaman et al. (2012) showed that consistency actually required MCAR. It is interesting that Seaman et al. (2012) found that *JAV* generally performed better than passive imputation, which is not confirmed in our simulations. It is not quite clear where the difference comes from, but the discussion *JAV* versus passive pales somewhat in the light of `smcfcs`.

Generic methods to preserve interactions include tree-based regression and classification (Section 3.5) as well as various joint modeling methods (Section 4.4). The relative strengths and limitations of these approaches still need to be sorted out.

6.4.3   Quadratic relations♠

One way to analyze nonlinear relations by a linear model is to include quadratic or cubic versions of the explanatory variables into the model. Creating imputed values under such models poses some challenges. Current imputation methodology either preserves the quadratic relation in the data and biases the estimates of interest, or provides unbiased estimates but does not preserve the quadratic relation (Von Hippel, 2009). It seems that we either have a congenial but misspecified model, or an uncongenial model that is specified correctly. This section describes an approach that aims to resolve this problem.

The model of scientific interest is

$$X = \alpha + Y\beta_1 + Y^2\beta_2 + \epsilon \tag{6.1}$$

with $\epsilon \sim N(0, \sigma^2)$. We assume that X is complete, and that $Y = (Y_{\text{obs}}, Y_{\text{mis}})$ is partially missing. The problem is to find imputations for Y such that estimates of α, β_1, β_2 and σ^2 based on the imputed data are unbiased, while ensuring that the quadratic relation between Y and Y^2 will hold in the imputed data.

Define the *polynomial combination* Z as $Z = Y\beta_1 + Y^2\beta_2$ for some β_1 and β_2. The idea is to impute Z instead of Y and Y^2, followed by a decomposition of the imputed data Z into components Y and Y^2. Imputing Z reduces the multivariate imputation problem to a univariate problem, which is easier to manage. Under the assumption that $P(X, Z)$ is multivariate normal, we can impute the missing part of Z by Algorithm 3.1. In cases where a normal residual distribution is suspect, we replace the linear model by predictive mean matching. The next step is to decompose Z into Y and Y^2. Under the model in Equation 6.1 the value Y has two roots:

$$Y_- = -\tfrac{1}{2\beta_2}\left(\sqrt{4\beta_2 Z + \beta_1^2} + \beta_1\right) \tag{6.2}$$

$$Y_+ = \tfrac{1}{2\beta_2}\left(\sqrt{4\beta_2 Z + \beta_1^2} - \beta_1\right) \tag{6.3}$$

where we assume that the discriminant $4\beta_2 Z + \beta_1^2$ is larger than zero. For a given Z, we can take either $Y = Y_-$ or $Y = Y_+$, and square it to obtain Y^2. Either root is consistent with $Z = Y\beta_1 + Y^2\beta_2$, but the choice among these two options requires care. Suppose we choose Y_- for all Z. Then all Y will correspond to points located on the left arm of the parabolic function. The minimum of the parabola is located at $Y_{\min} = -\beta_1/2\beta_2$, so all imputations will occur in the left-hand side of the parabola. This is probably not intended.

The choice between the roots is made by random sampling. Let V be a binary random variable defined as 1 if $Y > Y_{\min}$, and as 0 if $Y \leq Y_{\min}$. Let us model the probability $P(V = 1)$ by logistic regression as

$$\text{logit}(P(V = 1)) = X\psi_X + Z\psi_Z + XZ\psi_{YZ} \tag{6.4}$$

where the ψs are parameters in the logistic regression model. Under the assumption of ignorability, we calculate the predicted probability $P(V = 1)$ from $X_{\min}$ and $Z_{\min}$. As a final step, a random draw from the binomial distribution is made, and the corresponding (negative or positive) root is selected as the imputation. This is repeated for each missing value.

Algorithm 6.1 provides a detailed overview of all steps involved. The imputations $\dot{Z}$ satisfy $\dot{Z} = \dot{Y}\hat{\beta}_1 + \dot{Y}^2\hat{\beta}_2$, as required. The technique is available in `mice` as the method `quadratic`. The evaluation by Vink and Van Buuren (2013) showed that the method provided unbiased estimates under four types of extreme MAR mechanisms. The idea can be generalized to polynomial bases of higher orders.

6.4.4 Compositional data♠

Sometimes we know that a set of variables should add up to a given total. If one of the additive terms is missing, we can directly calculate its value

Algorithm 6.1: Multiple imputation of quadratic terms.♠

1. Calculate Y_{obs}^2 for the observed Y.

2. Multiply impute Y_{mis} and Y_{mis}^2 as if they were unrelated by linear regression or predictive mean matching, resulting in imputations $\dot{Y}$ and $\dot{Y}^2$.

3. Estimate $\hat{\beta}_1$ and $\hat{\beta}_2$ by pooled linear regression of X given $Y = (Y_{\text{obs}}, \dot{Y})$ and $Y^2 = (Y_{\text{obs}}^2, \dot{Y}^2)$.

4. Calculate the polynomial combination $Z = Y\hat{\beta}_1 + Y^2\hat{\beta}_2$.

5. Multiply impute Z_{mis} by linear regression or predictive mean matching, resulting in imputed $\dot{Z}$.

6. Calculate roots $\dot{Y}_-$ and $\dot{Y}_+$ given $\hat{\beta}_1$, $\hat{\beta}_2$ and $\dot{Z}$ using Equations 6.2 and 6.3.

7. Calculate the value on the horizontal axis at the parabolic minimum/maximum $Y_{\text{min}} = -\hat{\beta}_1/2\hat{\beta}_2$.

8. Calculate $V_{\text{obs}} = 0$ if $Y_{\text{obs}} \leq Y_{\text{min}}$, else $V_{\text{obs}} = 1$.

9. Impute V_{mis} by logistic regression of V given X, Z and XZ, resulting in imputed $\dot{V}$.

10. If $\dot{V} < 0$ then assign $\dot{Y} = \dot{Y}_-$, else set $\dot{Y} = \dot{Y}_+$.

11. Calculate $\dot{Y}^2$.

with certainty by deducting the known terms from the total. This is known as deductive imputation (De Waal et al., 2011). If two additive terms are missing, imputing one of these terms uses the available one degree of freedom, and hence implicitly determines the other term. Data of this type are known as compositional data, and they occur often in household and business surveys. Imputation of compositional data has only recently received attention (Tempelman, 2007; Hron et al., 2010; De Waal et al., 2011; Vink et al., 2015). Hron et al. (2010) proposed matching on the Aitchison distance, a measure specifically designed for compositional data. The method is available in R as the `robCompositions` package (Templ et al., 2011a).

This section suggests a somewhat different method for imputing compositional data. Let $Y_{123} = Y_1 + Y_2 + Y_3$ be the known total score of the three variables Y_1, Y_2 and Y_3. We assume that Y_3 is complete and that Y_1 and Y_2 are jointly missing or observed. The problem is to create multiple imputations in Y_1 and Y_2 such that the sum of Y_1, Y_2 and Y_3 equals a given total Y_{123}, and such that parameters estimated from the imputed data are unbiased and have appropriate coverage.

Since Y_3 is known, we write $Y_{12} = Y_{123} - Y_3$ for the sum score $Y_1 + Y_2$. The key to the solution is to find appropriate values for the ratio $P_1 = Y_1/Y_{12}$, or equivalently for $(1 - P_1) = Y_2/Y_{12}$. Let $P(P_1|Y_1^{\text{obs}}, Y_2^{\text{obs}}, Y_3, X)$ denote the posterior distribution of P_1, which is possibly dependent on the observed information. For each incomplete record, we make a random draw $\dot{P}_1$ from this distribution, and calculate imputations for Y_1 as $\dot{Y}_1 = \dot{P}_1 Y_{12}$. Likewise, imputations for Y_2 are calculated by $\dot{Y}_2 = (1 - \dot{P}_1)Y_{12}$. It is easy to show that $\dot{Y}_1 + \dot{Y}_2 = Y_{12}$, and hence $\dot{Y}_1 + \dot{Y}_2 + Y_3 = Y_{123}$, as required.

The best way in which the posterior should be specified has still to be determined. In this section we apply standard predictive mean matching. We study the properties of the method by a small simulation study. The first step is to create an artificial dataset with known properties as follows:

```
set.seed(43112)
n <- 400
Y1 <- sample(1:10, size = n, replace = TRUE)
Y2 <- sample(1:20, size = n, replace = TRUE)
Y3 <- 10 + 2 * Y1 + 0.6 * Y2 + sample(-10:10, size = n,
                                      replace = TRUE)
Y <- data.frame(Y1, Y2, Y3)
Y[1:100, 1:2] <- NA
md.pattern(Y, plot = FALSE)
```

```
      Y3  Y1  Y2
300    1   1   1   0
100    1   0   0   2
       0 100 100 200
```

Thus, Y is a 400×3 dataset with 300 complete records and with 100 records in which both Y_1 and Y_2 are missing. Next, define three auxiliary variables that are needed for imputation:

```
Y123 <- Y1 + Y2 + Y3
Y12 <- Y123 - Y[,3]
P1 <- Y[,1] / Y12
data <- data.frame(Y, Y123, Y12, P1)
```

where the naming of the variables corresponds to the total score Y_{123}, the sum score Y_{12} and the ratio P_1.

The imputation model specifies how Y_1 and Y_2 depend on P_1 and Y_{12} by means of passive imputation. The predictor matrix specifies that only Y_3 and Y_{12} may be predictors of P_1 in order to avoid linear dependencies.

```
meth <- make.method(data)
meth["Y1"]   <- "~ I(P1 * Y12)"
meth["Y2"]   <- "~ I((1 - P1) * Y12)"
meth["Y12"] <- "~ I(Y123 - Y3)"
pred <- make.predictorMatrix(data)
pred["P1", ] <- 0
pred[c("P1"), c("Y12", "Y3")] <- 1
imp1 <- mice(data, meth = meth, pred = pred, m = 10,
             print = FALSE)
```

The code $I(P1 * Y12)$ calculates Y_1 as the product of P_1 and Y_{12}, and so on. The pooled estimates are calculated as

```
round(summary(pool(with(imp1, lm(Y3 ~ Y1 + Y2))))[, 1:2], 2)
```

	estimate	std.error
(Intercept)	9.71	0.98
Y1	1.99	0.11
Y2	0.61	0.05

The estimates are reasonably close to their true values of 10, 2 and 0.6, respectively. A small simulation study with these data using 100 simulations and $m = 10$ revealed average estimates of 9.94 (coverage 0.96), 1.95 (coverage 0.95) and 0.63 (coverage 0.91). Though not perfect, the estimates are close to the truth, while the data adhere to the summation rule.

Figure 6.4 shows where the solution might be further improved. The distribution of P_1 in the observed data is strongly patterned. This pattern is only partially reflected in the imputed $\dot{P}_1$ after predictive mean matching on both Y_{12} and Y_3. It is possible to imitate the pattern perfectly by removing Y_3 as a predictor for P_1. However, this introduces bias in the parameter estimates. Evidently, some sort of compromise between these two options might further remove the remaining bias. This is an area for further research.

For a general missing data pattern, the procedure can be repeated for all pairs $(Y_j, Y_{j'})$ that have missing data. First create a consistent starting imputation that adheres to the rule of composition, then apply the above method to pairs $(Y_j, Y_{j'})$ that belong to the composition. This algorithm is a variation on the MICE algorithm with iterations occurring over pairs of variables rather than separate variables.

Vink (2015) extended these ideas to nested compositional data, where a given element of the composition is broken down into subelements. The method, called *predictive ratio matching*, calculates the ratio of two components, and then borrows imputations from donors that have a similar ratio. Component pairs are visited in an ingenious way and combined into an iterative algorithm.

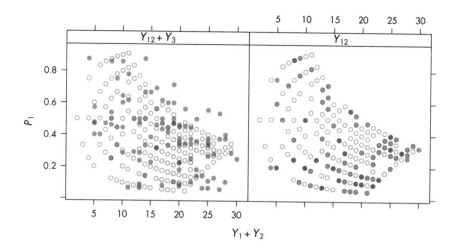

Figure 6.4: Distribution of P_1 (relative contribution of Y_1 to $Y_1 + Y_2$) in the observed and imputed data at different levels of $Y_1 + Y_2$. The strong geometrical shape in the observed data is partially reproduced in the model that includes Y_3.

6.4.5 Sum scores

The sum score is undefined if one of the variables to be added is missing. We can use sum scores of imputed variables within the MICE algorithm to economize on the number of predictors. For example, suppose we create a summary maturation score of the pubertal measurements gen, phb and tv, and use that score to impute the other variables instead of the three original pubertal measurements.

Another area of application is the imputation of test items that form a scale. When the number of items is small relative to the sample size, good results can be obtained by imputing the items in a full imputation model, where all items are used to impute others (Van Buuren, 2010). This method becomes unfeasible for a larger number of items. In that case, one may structure the imputation problem assuming one knows which items belong to which scale. Suppose that the data consist of an outcome variable out, a background variable bck, a scale a with ten items a1-a10, a scale b with twelve items b1-b12, and that all variables contain missing values. After filling in starting imputations, the imputation model would take the following steps:

1. Impute out given bck, a, b, where a and b are the summed scale scores from b1-b10 and b1-b12;

2. Impute bck given out, a and b;

3. Impute a1 given out, bck, b and a2-a10;

4. Impute a2 given out, bck, b and a1, a3-a10;

5. Impute a3-a10 along the same way;

6. Impute b1 given out, bck, a and b2-b12, where a is the updated summed scale score;

7. Impute b2-b12 along the same way.

This technique will condense the imputation model, so that it will become both faster and more stable. It is easy to specify such models in mice. See Van Buuren and Groothuis-Oudshoorn (2011) for examples.

Plumpton et al. (2016) found that the technique reduced the standard error on average by 39% compared to complete-case analysis, resulting in more precise conclusions. Eekhout et al. (2018) found that the technique outperforms existing techniques (complete-case analysis, imputing only total scores) with respect to bias and efficiency. As an alternative, one may use the mean of the observed items as the scale score (the parcel summary), which is easier than calculating the total score. Both studies obtained results that were comparable to using the total score. Note that the parcel summary mean requires the assumption that the missing items are MCAR, which may be problematic for speeded tests and for scales where items increase in difficulty. The magnitude of the effect of violations of the MCAR assumption on bias is still to be determined.

6.4.6   Conditional imputation

In some cases it makes sense to restrict the imputations, possibly conditional on other data. The method in Section 1.3.5 produced negative values for the positive-valued variable Ozone. One way of dealing with this mismatch between the imputed and observed values is to censor the values at some specified minimum or maximum value. The mice() function has an argument called post that takes a vector of strings of R commands. These commands are parsed and evaluated just after the univariate imputation function returns, and thus provide a way to post-process the imputed values. Note that post only affects the synthetic values, and leaves the observed data untouched. The squeeze() function in mice replaces values beyond the specified bounds by the minimum and maximal scale values. A hacky way to ensure positive imputations for Ozone under stochastic regression imputation is

```
data <- airquality[, 1:2]
post <- make.post(data)
post["Ozone"] <-
    "imp[[j]][, i] <- squeeze(imp[[j]][, i], c(1, 200))"
imp <- mice(data, method = "norm.nob", m = 1,
```

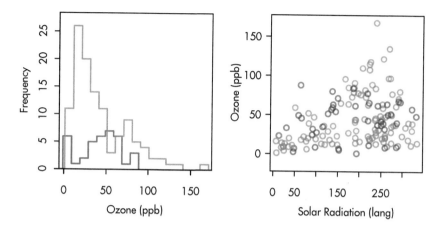

Figure 6.5: Stochastic regression imputation of Ozone, where the imputed values are restricted to the range 1–200. Compare to Figure 1.4.

```
maxit = 1, seed = 1, post = post)
```

Compare Figure 6.5 to Figure 1.4. The negative ozone value of -18.8 has now been replaced by a value of 1.

The previous syntax of the `post` argument is a bit cumbersome. The same result can be achieved by neater code:

```
post["Ozone"] <- "ifdo(c(Ozone < 1, Ozone > 200), c(1, 200))"
```

The `ifdo()` function is a convenient way to create conditional imputes. For example, in the `boys` data puberty is measured only for boys older than 8 years. Before this age it is unlikely that puberty has started. It is a good idea to bring this extra information into the imputation model to stabilize the solution. More precisely, we may restrict any imputations of `gen`, `phb` and `tv` to the lowest possible category for those boys younger than 8 years. This can be achieved by

```
post <- make.post(boys)
post["gen"] <- "ifdo(age < 8, levels(gen)[1])"
post["phb"] <- "ifdo(age < 8, levels(phb)[1])"
post["tv"] <- "ifdo(age < 8, 1)"
```

Figure 6.6 compares the scatterplot of genital development against age for the free and restricted solutions. Around infancy and early childhood, the imputations generated under the free solution are clearly unrealistic due to the severe extrapolation of the data between the ages 0–8 years. The restricted

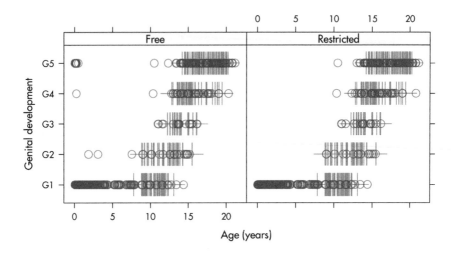

Figure 6.6: Genital development of Dutch boys by age. The free solution does not constrain the imputations, whereas the restricted solution requires all imputations below the age of 8 years to be at the lowest category.

solution remedies this situation by requiring that pubertal development does not start before the age of 8 years.

The post-processing facility provides almost limitless possibilities to customize the imputed values. For example, we could reset the imputed value in some subset of the missing data to NA, thus imputing only some of the variables. Of course, appropriate care is needed when using this partially imputed variable later on as a predictor. Another possibility is to add or multiply the imputed data by a given constant in the context of a sensitivity analysis for nonignorable missing data mechanisms (see Section 3.8). More generally, we might re-impute some entries in the dataset depending on their current value, thus opening up possibilities to specify methods for nonignorable missing data.

6.5 Algorithmic options

6.5.1 Visit sequence

The default MICE algorithm imputes incomplete columns in the data from left to right. Theoretically, the visit sequence of the MICE algorithm is irrelevant as long as each column is visited often enough, though some schemes

are more efficient than others. In practice, there are small order effects of
the MICE algorithm, where the parameter estimates depend on the sequence
of the variables. To date, there is little evidence that this matters in practice, even for clearly incompatible imputation models (Van Buuren et al.,
2006). For monotone missing data, convergence is immediate if variables are
ordered according to their missing data rate. Rather than reordering the data,
it is more convenient to change the visit sequence of the algorithm by the
`visitSequence` argument. In its basic form, the `visitSequence` argument
is a vector of names, or a vector of integers in the range `1:ncol(data)` of
arbitrary length, specifying the sequence of blocks (usually variables) for one
iteration of the algorithm. Any given block may be visited more than once
within the same iteration, which can be useful to ensure proper synchronization among blocks of variables. Consider the `mids` object `imp.int` created in
Section 6.4.2. The visit sequence is

```
imp.int$visitSequence
```

```
[1] "age"    "hgt"    "wgt"    "bmi"    "hc"     "gen"
[7] "phb"    "tv"     "reg"    "wgt.hc"
```

If the `visitSequence` is not specified, the `mice()` function imputes the
data from left to right. Thus here `wgt.hc` is calculated after `reg` is imputed,
so at this point `wgt.hc` is synchronized with both `wgt` and `hc`. Note, however,
that `wgt.hc` is not synchronized with `wgt` and `hc` when imputing `pub`, `gen`,
`tv` or `reg`, so `wgt.hc` is not representing the current interaction effect. This
could result in wrong imputations. We can correct this by including an extra
visit to `wgt.hc` after `wgt` or `hc` has been imputed as follows:

```
vis <- c("hgt", "wgt", "hc", "wgt.hc", "gen", "phb",
          "tv", "reg")
expr <- expression((wgt - 40) * (hc - 50))
boys$wgt.hc <- with(boys, eval(expr))
imp.int2 <- mice(boys, m = 1, maxit = 1, visitSequence = vis,
                 meth = imp.int$meth, pred = imp.int$pred,
                 seed = 23390)
```

```
  iter imp variable
   1   1  hgt wgt  hc  wgt.hc  gen  phb  tv  reg
```

When the missing data pattern is close to monotone, convergence may be
speeded by visiting the columns in increasing order of the number of missing
data. We can specify this order by the `"monotone"` keyword as

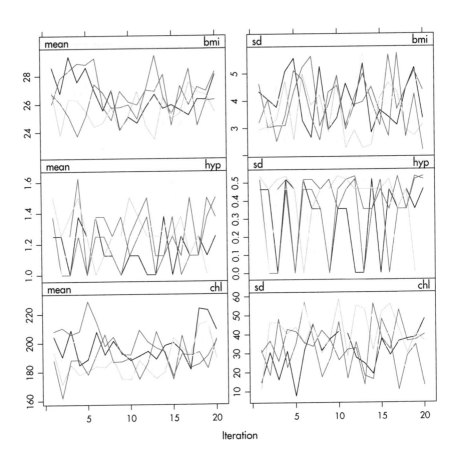

Figure 6.7: Mean and standard deviation of the synthetic values plotted against iteration number for the imputed nhanes data.

```
imp.int2 <- mice(boys, m = 1, maxit = 1, vis =  "monotone",
                 meth = imp.int$meth, pred = imp.int$pred,
                 seed = 23390)
```

```
iter imp variable
  1   1  reg wgt hgt hc wgt.hc gen phb tv
```

6.5.2 Convergence

There is no clear-cut method for determining when the MICE algorithm has converged. It is useful to plot one or more parameters against the iteration number. The mean and variance of the imputations for each parallel stream can be plotted by

```
imp <- mice(nhanes, seed = 62006, maxit = 20, print = FALSE)
plot(imp)
```

which produces Figure 6.7. On convergence, the different streams should be freely intermingled with one another, without showing any definite trends. Convergence is diagnosed when the variance between different sequences is no larger than the variance within each individual sequence. Inspection of the streams may reveal particular problems of the imputation model. A pathological case of non-convergence occurs with the following code:

```
meth <- make.method(boys)
meth["bmi"] <- "~I(wgt / (hgt / 100)^2)"
imp.bmi1 <- mice(boys, meth = meth, maxit = 20,
                 print = FALSE, seed = 60109)
```

Convergence is problematic here because imputations of bmi feed back into hgt and wgt. Figure 6.8 shows that the streams hardly mix and slowly resolve into a steady state. The problem is solved by breaking the feedback loop as follows:

```
pred <- make.predictorMatrix(boys)
pred[c("hgt", "wgt"), "bmi"] <- 0
imp.bmi2 <- mice(boys, meth = meth, pred = pred, maxit = 20,
                 print = FALSE, seed = 60109)
```

Figure 6.9 is the resulting plot for the same three variables. There is little trend and the streams mingle well.

The default plot() function for mids objects plots the mean and variance of the imputations. While these parameters are informative for the behavior of the MICE algorithm, they may not always be the parameter of greatest interest. It is easy to replace the mean and variance by other parameters, and monitor these. Schafer (1997, pp. 129–131) suggested monitoring the "worst linear function" of the model parameters, i.e., a combination of parameters that will experience the most problematic convergence. If convergence can be established for this parameter, then it is likely that convergence will also be achieved for parameters that converge faster. Alternatively, we may monitor some statistic of scientific interest, e.g., a correlation or a proportion. See Sections 4.5.7 (Pearson correlation) and 9.4.3 (Kendall's τ) for examples.

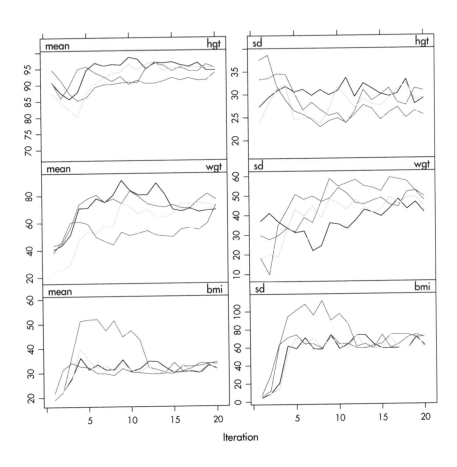

Figure 6.8: Non-convergence of the MICE algorithm caused by feedback of bmi into hgt and wgt.

It is possible to use formal convergence statistics. Several expository reviews are available that assess convergence diagnostics for MCMC methods (Cowles and Carlin, 1996; Brooks and Gelman, 1998; El Adlouni et al., 2006). Cowles and Carlin (1996) conclude that "automated convergence monitoring (as by a machine) is unsafe and should be avoided." No method works best in all circumstances. The consensus is to assess convergence with a combination of tools. The added value of using a combination of convergence diagnostics for missing data imputation has not yet been systematically studied.

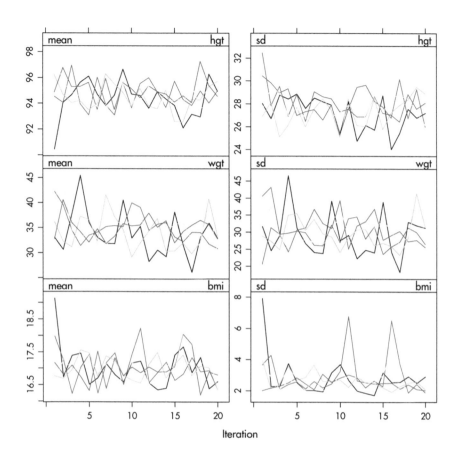

Figure 6.9: Healthy convergence of the MICE algorithm for hgt, wgt and bmi.

6.6 Diagnostics

Assessing model fit is also important for building trust by assessing the plausibility of the generated imputations. Diagnostics for statistical models are procedures to find departures from the assumptions underlying the model. Model evaluation is a huge field in which many special tools are available, e.g., Q-Q plots, residual and influence statistics, formal statistical tests, information criteria and posterior predictive checks. In principle, all these techniques can be applied to evaluate the imputation model. Conventional model evaluation concentrates on the fit between the data and the model. In imputation it is often more informative to focus on *distributional discrepancy*, the difference between the observed and imputed data. The next section illustrates this with an example.

6.6.1   Model fit versus distributional discrepancy

The MICE algorithm fits the imputation model to the records with observed Y_j^{obs}, and applies the fitted model to generate imputations for the records with unobserved Y_j^{mis}. The fit of the imputation model to the data can thus be studied from Y_j^{obs}.

The worm plot is a diagnostic tool to assess the fit of a nonlinear regression (Van Buuren and Fredriks, 2001; Van Buuren, 2007b). In technical terms, the worm plot is a detrended Q-Q plot conditional on a covariate. The model fits the data if the worms are close to the horizontal axis.

Figure 6.10 is the worm plot calculated from imputed data after predictive mean matching. The fit between the observed data and the imputation model is bad. The blue points are far from the horizontal axis, especially for the youngest children. The shapes indicate that the model variance is much larger than the data variance. In contrast to this, the red and blue worms are generally close, indicating that the distributions of the imputed and observed body weights are similar. Thus, despite the fact that the model does not fit the data, the distributions of the observed and imputed data are similar. This distributional similarity is more relevant for the final inferences than model fit per se.

6.6.2   Diagnostic graphs

One of the best tools to assess the plausibility of imputations is to study the discrepancy between the observed and imputed data. The idea is that good imputations have a distribution similar to the observed data. In other words, the imputations could have been real values had they been observed. Except under MCAR, the distributions do not need to be identical, since strong MAR mechanisms may induce systematic differences between the two distributions. However, any dramatic differences between the imputed and observed data should certainly alert us to the possibility that something is wrong.

This book contains many colored figures that emphasize the relevant contrasts. Graphs allow for a quick identification of discrepancies of various types:

- the points have different means (Figure 2.2);

- the points have different spreads (Figures 1.2, 1.3 and 1.5);

- the points have different scales (Figure 4.4);

- the points have different relations (Figure 6.2);

- the points do not overlap and they defy common sense (Figure 6.6).

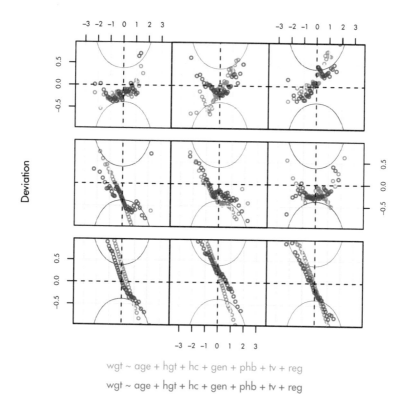

Figure 6.10: Worm plot of the predictive mean matching imputations for body weight. Different panels correspond to different age ranges. While the imputation model does not fit the data in many age groups, the distributions of the observed and imputed data often match up very well.

Differences between the densities of the observed and imputed data may suggest a problem that needs to be further checked. The `mice` package contains several graphic functions that can be used to gain insight into the correspondence of the observed and imputed data: `bwplot()`, `stripplot()`, `densityplot()` and `xyplot()`.

The `stripplot()` function produces the individual points for numerical variables per imputation as in Figure 6.11 by

```
imp <- mice(nhanes, seed = 29981)
stripplot(imp, pch = c(21, 20), cex = c(1, 1.5))
```

The stripplot is useful to study the distributions in datasets with a low

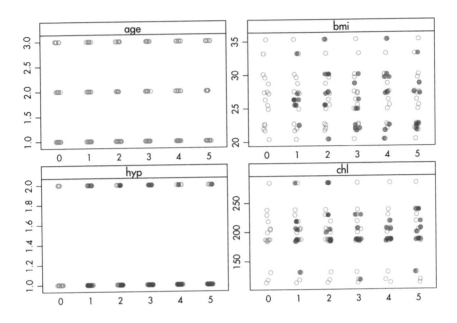

Figure 6.11: A stripplot of the multiply imputed nhanes data with $m = 5$.

number of data points. For large datasets it is more appropriate to use the function `bwplot()` that produces side-by-side box-and-whisker plots for the observed and synthetic data.

The `densityplot()` function produces Figure 6.12 by

```
densityplot(imp, layout = c(3, 1))
```

which shows kernel density estimates of the imputed and observed data. In this case, the distributions match up well.

Interpretation is more difficult if there are discrepancies. Such discrepancies may be caused by a bad imputation model, by a missing data mechanism that is not MCAR or by a combination of both. Bondarenko and Raghunathan (2016) proposed a more refined diagnostic tool that aims to compare the distributions of observed and imputed data conditional on the missingness probability. The idea is that under MAR the conditional distributions should be similar if the assumed model for creating multiple imputations has a good fit. An example is created as:

```
fit <- with(imp, glm(ici(imp) ~ age + bmi + hyp + chl,
                      family = binomial))
ps <- rep(rowMeans(sapply(fit$analyses, fitted.values)),
```

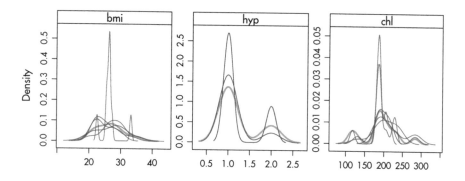

Figure 6.12: Kernel density estimates for the marginal distributions of the observed data (blue) and the $m = 5$ densities per variable calculated from the imputed data (thin red lines).

```
             imp$m + 1)
xyplot(imp, bmi ~ ps | as.factor(.imp),
       xlab = "Probability that record is incomplete",
       ylab = "BMI", pch = c(1, 19))
```

These statements first model the probability of each record being incomplete as a function of all variables in each imputed dataset. The probabilities (propensities) are then averaged over the imputed datasets to obtain stability. Figure 6.13 plots BMI against the propensity score in each dataset. Observe that the imputed data points are somewhat shifted to the right. In this case, the distributions of the blue and red points are quite similar, as expected under MAR.

Realize that the comparison is only as good as the propensity score is. If important predictors are omitted from the response model, then we may not be able to see the potential misfit. In addition, it could be useful to investigate the residuals of the regression of BMI on the propensity score. See Van Buuren and Groothuis-Oudshoorn (2011) on a technique for how to calculate and plot the relevant quantities.

Compared to conventional diagnostic methods, imputation comes with the advantage that we can directly compare the observed and imputed values. The marginal distributions of the observed and imputed data may differ because the missing data are MAR or MNAR. The diagnostics tell us in what way they differ, and hopefully also suggest whether these differences are expected and sensible in light of what we know about the data. Under MAR, any distributions that are conditional on the missing data process should be the same. If our diagnostics suggest otherwise (e.g., the blue and red points are very different), there might be something wrong with the imputations that we

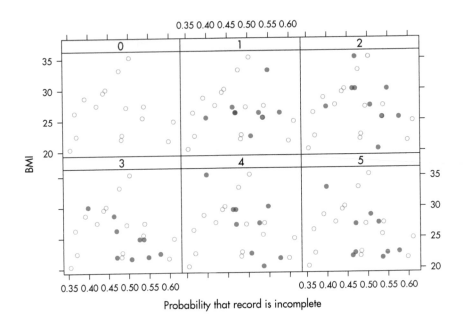

Figure 6.13: BMI against missingness probability for observed and imputed values.

created. Alternatively, it could be the case that the observed differences are justified, and that the missing data process is MNAR. The art of imputation is to distinguish between these two explanations.

6.7 Conclusion

Multiple imputation is not a quick automatic fix. Creating good imputations requires substantive knowledge about the data paired with some healthy statistical judgement. Impute close to the data. Real data are richer and more complex than the statistical models applied to them. Ideally, the imputed values should look like real data in every respect, especially if multiple models are to be fit on the imputed data. Keep the following points in mind:

- Plan time to create the imputed datasets. As a rule of thumb, reserve for imputation 5% of the time needed to collect the data.

- Check the modeling choices in Section 6.1. Though the software defaults are often reasonable, they may not work for the particular data.

- Use MAR as a starting point using the strategy outlined in Section 6.2.

- Choose the imputation methods and set the predictors using the strategies outlined in Section 6.3.

- If the data contain derived variables that are not needed for imputation, impute the originals and calculate the derived variables afterward.

- Use passive imputation if you need the derived variables during imputation. Carefully specify the predictor matrix to avoid feedback loops. See Section 6.4.

- Monitor convergence of the MICE algorithm for aberrant patterns, especially if the rate of missing data is high or if there are dependencies in the data. See Sections 4.5.6 and 6.5.2.

- Make liberal use of diagnostic graphs to compare the observed and the imputed data. Convince yourself that the imputed values could have been real data, had they not been missing. See Section 6.6.

Nguyen et al. (2017) present a concise overview of methodologies to check the quality of the imputation model. In addition to the points mentioned above, the authors advice the application of posterior predictive checks, the evaluation of the effect of the target analysis when making judgements about model adequacy, and the application of a wide range of methodologies to check imputation models.

6.8 Exercises

1. *Worm plot for normal model.* Repeat the imputations in Section 6.6.1 using the linear normal model for the numerical variables. Draw the worm plot.

   - Does the imputation model for `wgt` fit the observed data? If not, describe in which aspects they differ.

   - Does the imputation model for `wgt` fit the imputed data? If not, describe in which aspects they differ.

   - Are there striking differences between your worm plot and Figure 6.10? If so, describe.

   - Which imputation model do you prefer? Why?

2. *Defaults.* Select a real dataset that is familiar to you and that contains at least 20% missing data. Impute the data with `mice()` under all the default settings.

- Inspect the streams of the MICE algorithm. Does the sampler appear to converge?

- Extend the analysis with 20 extra iterations using `mice.mids()`. Does this affect your conclusion about convergence?

- Inspect the data with diagnostic plots for univariate data. Are the univariate distributions of the observed and imputed data similar? Do you have an explanation why they do (or do not) differ?

- Inspect the data with diagnostic plots for the most interesting bivariate relations. Are the relations similar in the observed and imputed data? Do you have an explanation why they do (or do not) differ?

- Consider each of the seven default choices in turn. Do you think the default is appropriate for your data? Explain why.

- Do you have particular suggestions for improved choices? Which?

- Implement one of your suggestions. Do the results now look more plausible or realistic? Explain.

Chapter 7

Multilevel multiple imputation

Multiple imputation is often a better tool for behavioral science data because it gives researchers the flexibility to tailor the missing data handling procedure to match a particular set of analysis goals.
Craig K. Enders

7.1 Introduction

Multiple imputation of multilevel data is one of the hot spots in statistical technology. Imputers and analysts now have a bewildering array of options for imputing missing values in multilevel data. This chapter summarizes the state of the art, and formulates advice and guidelines for practical application of multilevel imputation.

The structure of this chapter is as follows. We start with a concise overview of three ways to formulate the multilevel model. Section 7.3 reviews several non-imputation approaches for dealing with missing values in multilevel data. Sections 7.4 and 7.5 describe imputation using the joint modeling and fully conditional specification frameworks. Sections 7.6 and 7.7 review current procedures for imputation under multilevel models with continuous and discrete outcomes, respectively. Section 7.8 deals with missing data in the level-2 predictors, and Section 7.9 summarizes comparative work on the different approaches. Section 7.10 contains worked examples that illustrate how imputations can be generated in `mice`, provides guidelines on the practical application, written in the form of recipes for multilevel imputation. The chapter closes with an overview of unresolved issues and topics for further research.

7.2 Notation for multilevel models

Multilevel data have a hierarchical, or clustered, structure. The archetypical example is data from pupils who are nested within classes. Some of the data may relate to pupils (e.g., test scores), whereas other data concern the class level (e.g., class size). Another example arises in longitudinal studies, where the individual's responses over time are nested within individuals. Some of the data vary with time (e.g., disease history), whereas other data vary between individuals (e.g., sex). The term *multilevel analysis* refers to the methodology to analyze data with such multilevel structure, a methodology that can be traced back to the definition of the intra-class correlation (ICC) by Fisher (1925). Multilevel analysis is quite different from methods for single-level data. The analysis of multilevel data is a vast topic, and this is not the place to cover the model in detail. There are excellent introductions by Raudenbush and Bryk (2002), Gelman and Hill (2007), Snijders and Bosker (2012), Fitzmaurice et al. (2011), and Hox et al. (2018). This chapter assumes basic familiarity with these models.

A challenging aspect of multilevel analysis is the existence of a variety of notational systems and concepts. This section describes three different notations. In order to illustrate these notations, we use data on school performance of grade 8 pupils in Dutch schools. These data were collected by Brandsma and Knuver (1989), and were used as the primary examples in Chapters 4 and 5 of Snijders and Bosker (2012). The data are available as the `brandsma` object in the `mice` package. The data contain a mix of both pupil-level measurements and school-level measurements.

```
library(mice)
data("brandsma", package = "mice")
d <- brandsma[, c("sch", "lpo", "sex", "den")]
```

Let us concentrate on four variables, each representing a different role in the multilevel model:

sch School number, cluster variable;

lpo Language test post, outcome at pupil level;

sex Sex of pupil, predictor at pupil level;

den School denomination, predictor at school level.

The scientific interest is to create a model for predicting the outcome `lpo` from the level-1 predictor `sex` (coded as 0-1) and the level-2 predictor `den` (which takes values 1-4). Let the data be divided into C clusters (e.g., classes, schools), indexed by c ($c = 1, \ldots, C$). Each cluster holds n_c units, indexed by $i = 1, \ldots, n_c$. There are three ways to write the same model (Scott et al., 2013).

In *level notation*, introduced by Bryk and Raudenbush (1992), we formulate a multilevel model as a system of two equations, one at level-1, and two at level-2:

$$\mathbf{lpo}_{ic} = \beta_{0c} + \beta_{1c}\mathbf{sex}_{ic} + \epsilon_{ic} \qquad (7.1)$$
$$\beta_{0c} = \gamma_{00} + \gamma_{01}\mathbf{den}_c + u_{0c} \qquad (7.2)$$
$$\beta_{1c} = \gamma_{10} \qquad (7.3)$$

where $\mathbf{lpo}_{ic}$ is the test score of pupil i in school c, where $\mathbf{sex}_{ic}$ is the sex of pupil i in school c, and where $\mathbf{den}_c$ is the religious denomination of school c. Note that here the subscripts distinguish the level-1 and level-2 variables. In this notation, $\mathbf{sex}_{ic}$ only appears in the level-1 model 7.1, and $\mathbf{den}_c$ only appears in the level-2 model 7.2. The term β_{0c} is a random intercept that varies by cluster, while β_{1c} is a sex effect that is assumed to be the same across schools. The term ϵ_{ic} is the within-cluster random residual at the pupil level with a normal distribution $\epsilon_{ic} \sim N(0, \sigma_\epsilon^2)$. The first level-2 model describes the variation in the mean test score between schools as a function of the grand mean γ_{00}, a school-level effect γ_{01} of denomination and a school-level random residual with a normal distribution $u_{0c} \sim N(0, \sigma_{u_0}^2)$. The second level-2 model does not have a random residual, so this specifies that β_{1c} is a fixed effect equal in value to γ_{10}. The unknowns to be estimated are the fixed parameters γ_{00}, γ_{01} and γ_{10}, and the variance components σ_ϵ^2 and $\sigma_{u_0}^2$.

We may write the same model as a single predictive equation by substituting the level-2 models into the level-2 model:

$$\mathbf{lpo}_{ic} = \gamma_{00} + \gamma_{10}\mathbf{sex}_{ic} + \gamma_{01}\mathbf{den}_c + u_{0c} + \epsilon_{ic}, \qquad (7.4)$$

We do not need the double subscripts any more, so we write the model in *composite notation* as

$$\mathbf{lpo}_{ic} = \beta_0 + \beta_1\mathbf{sex}_{ic} + \beta_2\mathbf{den}_c + u_{0c} + \epsilon_{ic}, \qquad (7.5)$$

Note that these β's are fixed effects and the β's in the level-1 model 7.1 are random effects. They differ by the number of subscripts.

The same model written in *matrix notation* is widely known as the *linear mixed effects model* (Laird and Ware, 1982) and can be written as

$$y_c = X_c\beta + Z_c u_c + \epsilon_c \qquad (7.6)$$

where $y_c = \mathbf{lpo}_c$ is a column vector containing the scores in cluster c, where $X_c = (1, \mathbf{den}_c, \mathbf{sex}_c)$ is the $n_c \times 3$ design matrix in class c associated with the fixed effects, and where $Z_c = (1)$ is a column with n_c 1's associated with the random intercept u_{0c}. In the general model, u_c has length q and consists of both random intercepts and random slopes, which are normally distributed as $u_c \sim N(0, \Omega)$. The q random effects are typically a subset of the p fixed effects, so $q \leq p$.

The different formulations of the multilevel model reflect different scientific traditions. The matrix model formulation is favoured among statisticians because all multilevel models can be economically expressed by Equation 7.6, which eases estimation of parameters, statistical inference and prediction. The books by McCulloch and Searle (2001), Verbeke and Molenberghs (2000), De Leeuw and Meijer (2008) and Fitzmaurice et al. (2011) are representative for this tradition. The level formulation clearly separates the roles of level-1 and level-2 variables, which eases interpretation of the model. Also, the error structure in each equation is simpler than in the two other formulations. The level formulation was popularized by Bryk and Raudenbush (1992), and is common in the social sciences. The composite formulation covers the middle ground. It gives a balance between compactness and clarity, and is used by the introductions into multilevel analysis for applied researchers by Snijders and Bosker (2012) and Hox et al. (2018). Gelman and Hill (2007) provide a Bayesian approach to multilevel analysis using a related though slightly different terminology. See Fitzmaurice et al. (2011, pp. 203-208, Ch. 22) and Scott et al. (2013) for more detail on the relations between the terminologies.

Although the models are mathematically equivalent, the notation has implications on the ease with which imputation models can be specified. Equation 7.6 nicely separates the fixed X_c from random effects Z_c, but the same covariates may appear in both X_c and Z_c. This complicates imputation of those covariates in an FCS framework because the same variable appears two times in the model. We surely want to prevent the scenario where imputed versions of the same covariate would become different in X_c and Z_c. The level formulation distinguishes level-1 variables from level-2 predictors. No overlap occurs between the sets of variables, so level formulation is natural for developing imputation procedures for variables at different levels. Likewise, the composite notation makes it easier to see how the imputation models must be set up in FCS, and is natural for studying the effect of interactions.

7.3   Missing values in multilevel data

In single-level data, missing values may occur in the outcome, in the predictors, or in both. The situation for multilevel data is more complex. Missing values in the measured variables of the multilevel model can occur in

1. the outcome variable;

2. the level-1 predictors;

3. the level-2 predictors;

4. the class variable.

This chapter assumes that the class variable is always completely observed. In real life, this may not be the case and techniques detailed in Chapter 3 can be used to impute class membership. See Hill and Goldstein (1998) and Goldstein (2011b) for models to handle missing class identification.

7.3.1 Practical issues in multilevel imputation

In single-level models, the impact of the missing values on the analysis depends on where in the model they occur. This is also the case in multilevel analysis. In fact, the multilevel model is very well equipped to handle missing values in the outcomes. Missing values in the predictors are generally more difficult to handle. Some children may have missing values for the age of the child, occupational status of the father, ethnic background, and so on. In longitudinal applications, missing data may occur in time-varying covariates, like nutritional status and stage of pubertal development. Most mixed methods cannot handle such missing values, and will remove children with any missing values in the level-1 predictors prior to analysis.

Missing data in the level-2 predictors occur if, for example, it is not known whether a school is public or private. In a longitudinal setting, missing data in fixed person characteristics, like sex or education, lead to incomplete level-2 predictors. The consequences of such missing values can be even larger. The typical fix is to delete all records in the class. For example, suppose that the model contains the professional qualification of the teacher. If the qualification is missing, the data of all pupils in the class are disregarded.

Many multilevel models define *derived variables* as part of the analysis, like the cluster means of a level-1 predictor, the product of two level-1 predictors, the dummy-coded version of a categorical variable, the disaggregated version of a level-2 predictor, and so on. We can calculate such derived variables from the data and include them into the model as needed, but of course this is only possible when data are complete. Although the derived variables themselves need not be imputed (because we can always recalculate them from the imputed data), the imputation model needs to be aware of, and account for, the role that such derived variables play in the complete-data model.

In practice, complications may arise due to the nature of the data or model. Some of these are as follows:

1. For *small clusters* the within-cluster mean and variance are unreliable estimates, so the choice of the prior distribution becomes critical.

2. For a *small number of clusters*, it is difficult to estimate the between-cluster variance of the random effects.

3. In applications with *systematically missing data*, there are no observed values in the cluster, so the cluster location cannot be estimated.

4. The variation of the random slopes can be large, so the method used to deal with the missing data should account for this.

5. The error variance σ_ϵ^2 may differ across clusters (*heteroscedasticity*)

Table 7.1: Questions to gauge the complexity of a multilevel imputation task.

1.	Will the complete-data model include random slopes?
2.	Will the data contain systematically missing values?
3.	Will the distribution of the residuals be non-normal?
4.	Will the error variance differ over clusters?
5.	Will there be small clusters?
6.	Will there be a small number of clusters?
7.	Will the complete-data model have cross-level interactions?
8.	Will the dataset be very large?

6. The *residual error distributions* can be far from normal.

7. The model contains aggregates of the level-1 variables, such as cluster means, which need to be taken in account during imputation.

8. The model contains interactions, or other nonlinear terms.

9. The multilevel model may be very complex, it may not be possible to fit the model, or there are convergence problems.

There is not one super-method that will address all such issues. In practice, we may need to emphasize certain issues at the expense of others. In order to gauge the complexity of the imputation task for particular dataset and model, ask yourself the questions listed in Table 7.1. If your answer to all questions is "NO", then there are several methods for multilevel MI that are available in standard software. If many of your answers are "YES", the situation is less clear-cut, and you may need to think about the relative priority of the questions in light of the needs for the application.

7.3.2   Ad-hoc solutions for multilevel data

Missing values in the level-1 predictors or the level-2 predictors have long been treated by listwise deletion. This is easy to do, but may have severe adverse effects, especially for missing values in level-2 predictors. For example, we may not know whether a school is public or private. Ignoring all records pertaining to that school is not only wasteful, but may also lead to selection effects at cluster level. While listwise deletion could be useful when the variance of the slopes is large, it is not generally recommended (Grund et al., 2016a).

Another ad-hoc solution is to ignore the clustering and impute the data by a single-level method. It is known that this will underestimate the intra-class correlation (Taljaard et al., 2008; Van Buuren, 2011; Enders et al., 2016). Lüdtke et al. (2017) derived an expression for the asymptotic bias for the intra-class correlation under the random intercept model. The amount of underestimation grows with the ICC and with the missing data rate. Increasing the cluster size hardly aids in reducing this bias. In addition, the regression weights for the fixed effects will be biased. Grund et al. (2018b) conclude that single-level imputation should be avoided unless only a few cases contain missing data (e.g., less than 5%) and the intra-class correlation is low (e.g., less than .10). Conducting multiple imputation with the wrong model (e.g., single-level methods) can be more hazardous than listwise deletion.

Another ad-hoc technique is to add a dummy variable for each cluster, so that the model estimates a separate coefficient for each cluster. The coefficients are estimated by ordinary least squares, and the parameters are drawn from their posteriors. If the missing values are restricted to the outcome, this method will estimate the fixed effects quite well, but also artificially inflates the true variation between groups, and thus biases the ICC upwards (Andridge, 2011; Van Buuren, 2011; Graham, 2012). If there are also missing values in the predictors, the level-1 regression weights will be unbiased, but the level-2 weights are biased, in particular for small clusters and low ICC. See Lüdtke et al. (2017) for more detail, who also derive the asymptotic bias. If the primary interest is on the fixed effects, adding a cluster dummy is an easily implementable alternative, unless the missing rate is very large and/or the intra-class correlation is very low and the number of records in the cluster is small (Drechsler, 2015; Lüdtke et al., 2017). Since the bias in random slopes and variance components can be substantial, one should turn to multilevel imputation to obtain proper estimates of those parts of the multilevel model (Speidel et al., 2017).

Vink et al. (2015) described an application of Australian school data with over 2.8 million records, where a dummy variable per school was combined with predictive mean matching. Given the size and complexity of the imputation problem, this application would have been computationally infeasible with full multilevel imputation. Thus, for large databases, adding a dummy variable per cluster is a practical and useful technique for estimating the fixed effects.

7.3.3 Likelihood solutions

The multilevel model is actually "made to solve" the problem of missing values in the outcome. There is an extensive literature, especially for longitudinal data (Verbeke and Molenberghs, 2000; Molenberghs and Verbeke, 2005; Daniels and Hogan, 2008). For more details, see the encyclopaedic overview in Fitzmaurice et al. (2009). Multilevel models have the ability to handle models with varying time points, which is an advance over traditional repeated-measures ANOVA, where the usual treatment is to remove the entire case if one of the outcomes is missing. Multilevel models do not assume an equal number of occasions or fixed time points, so all cases can be used for analysis.

Missing outcome data are easily handled in modern likelihood-based methods. Snijders and Bosker (2012, p. 56) write that the model "can even be applied if some groups have sample size 1, as long as other groups have greater sizes." Of course, this statement will only go as far as the assumptions of the model are met: the data are missing at random and the model is correctly specified.

Mixed-effects models can be fit with maximum-likelihood methods, which take care of missing data in the dependent variable. This principle can be extended to address missing data in explanatory variables in (multilevel) software for structural equation modeling like M*plus* (Muthén et al., 2016) and `gllamm` (Rabe-Hesketh et al., 2002). Grund et al. (2018b) remarked that such extensions could alter the meaning and value of the parameters of interest.

7.4 Multilevel imputation by joint modeling

The joint model specifies a single model for all incomplete variables in data. Suppose that we have missing values in both outcome y_c and the level-1 predictors X_c in the linear mixed model 7.6 with random intercepts. A limitation of the standard model is that both X_c and Z_c need to be completely observed. Schafer and Yucel (2002) showed that it is possible to reformulate the model by placing all variables in the model as an outcome on the left-hand side of model 7.6, which gives rise to multilevel models with multivariate outcomes. Suppose that Y_{ic} is the $1 \times p$ matrix of all incomplete variables from unit i in class c. Under a random intercept model, the multivariate multilevel model decomposes the data into a between-group part and a within-group part as

$$Y_{ic} = \mu + Y_c^{\text{between}} + Y_{ic}^{\text{within}}. \tag{7.7}$$

The relations between the p variables at the group level are captured by the between component Y_c^{between}, while relations at the individual level are captured by Y_{ic}^{within}. Lüdtke et al. (2017) explain how this decomposition can be applied to draw imputations for the missing elements in Y_c.

The approach has been implemented in the **pan** package (Zhao and Schafer, 2016). Zhao and Yucel (2009) and Yucel (2011) discussed extensions for categorical data using the generalized linear mixed model. Goldstein et al. (2009) described a joint model for mixed continuous-categorical data with a multilevel structure. Carpenter and Kenward (2013) and Goldstein et al. (2014) proposed extensions for ordinal and unordered categorical data, which are implemented in the **REALCOM-IMPUTE** software (Carpenter et al., 2011). Quartagno and Carpenter (2016) extended this work by allowing for heteroscedasticity of the imputation model, combined with imputation of binary (and more generally categorical) variables, which is available through the **jomo** package (Quartagno and Carpenter, 2017). Related work has been done by Asparouhov and Muthén (2010) for M*plus*.

Because of the decomposition 7.7, imputation by joint modeling is very natural when the complete-data analysis focuses on different within- and between-cluster associations (Enders et al., 2016; Grund et al., 2016a). Multilevel imputation is not without problems. Except for `jomo`, most models assume a homoscedastic error structure in the level-1 residuals, which implies no random slope variation between Y_{ic} (Carpenter and Kenward, 2013; Enders et al., 2016). Imputations created by `jomo` reflect pairwise linear relationships in the data and ignore higher-orders interaction and non-linearities. Joint modeling may also experience difficulties with smaller samples and the default inverse Wishart prior (Grund et al., 2018b; Audigier et al., 2018).

Imputation models can also be formulated as latent class models (Vermunt et al., 2008; Vidotto et al., 2015). Vidotto (2018) proposed a Bayesian multilevel latent class model that is designed to capture heterogeneity in the data at both levels through local independence and conditional independence assumptions. This class of models is quite flexible. As the method is very recent, there is not yet much practical experience.

7.5 Multilevel imputation by fully conditional specification

Another possibility is to iterate univariate multilevel imputation over the variables (Van Buuren, 2011). For example, suppose there are missing values in `lpo` and `sex` in model 7.5. One way to draw imputations is to alternate the following two steps:

$$\dot{\text{lpo}}_{ic} \sim N(\beta_0 + \beta_1 \text{den}_c + \beta_2 \text{sex}_{ic} + u_{0c}, \sigma_\epsilon^2) \tag{7.8}$$

$$\dot{\text{sex}}_{ic} \sim N(\beta_0 + \beta_1 \text{den}_c + \beta_2 \text{lpo}_{ic} + u_{0c}, \sigma_\epsilon^2) \tag{7.9}$$

where all parameters are re-estimated at every iteration. Since the first equation corresponds to the complete-data model, there are no issues with this step. The second equation simply alternates the roles of `lpo` and `sex`, and uses the inverted mixed model to draw imputations. The above steps illustrate the key idea of multilevel imputation using FCS. It is not yet clear when and how the idea will work.

Resche-Rigon and White (2018) studied the consequences of model inversion, and found that the conditional expectation of the level-1 predictor in a multivariate multilevel model with random intercepts depends on the cluster mean of the predictor, and on the size of the cluster. In addition, the conditional variance depends on cluster size. These results hold for the random intercept model. Of course, including random slopes as well will only complicate matters. The naive FCS procedure in Equation 7.8 does not account for the cluster means or for the effects of cluster size, and hence might not provide good imputations. From their derivation, Resche-Rigon and White (2018) therefore hypothesized that the imputation model (1) should incorporate the cluster means, and (2) be heteroscedastic if cluster sizes vary. We now discuss these points in turn.

7.5.1   Add cluster means of predictors

Simulations done by Resche-Rigon and White (2018) showed little impact of adding the cluster means of the level-1 predictors to the imputation model, but did not hurt either. However, several other studies found substantial impact. Enders et al. (2016) described how the inclusion of cluster means preserves contextual effects. Adding a group mean of level-1 variables allows us to estimate the difference between within-group and between-group regressions. The aggregates are generally called contextual variables. Including both the individual and contextual variables into the same model is useful to find out whether the contextual variable would improve prediction of the outcome after the differences at the individual level have been taken into account. Enders et al. (2016) proposed to add contextual variables more generally to univariate multilevel imputation, requiring a change in the algorithm. In particular, cluster means need to be dynamically calculated from the currently imputed predictors, and depending on the model the original predictor needs to be replaced by its group-centered version, as in Kreft et al. (1995). In random slope models, the two specifications have different meanings, so the decision as to whether or not to use group-mean centered predictors at level 1 should be made in accordance with the analysis model. In random intercept models, the two specifications are equivalent and can be transformed into one another, so imputations should yield equivalent results regardless of centering. Mistler and Enders (2017) present a thorough and detailed analysis on the effects of this adaptation, both for a joint model and for an FCS model. Their paper demonstrated that in both cases, inclusion of the means of the clusters markedly improved performance. Grund et al. (2018b) present an extensive simulation study contrasting many state-of-the-art imputation techniques. In all scenarios involving multilevel FCS, including the cluster means in the imputation model was beneficial.

The difference between the results of Resche-Rigon and White (2018) on the one hand, and the other three studies is likely to be caused by a difference in complete-data models. The latter three studies used a contextual analysis model, which includes the cluster means into the substantive model, whereas Resche-Rigon and White (2018) were not interested in fitting such a model. Hence, it appears that these studies address separate issues. Resche-Rigon and White (2018) are interested in improving *compatibility* among the conditional models without reference to a particular analysis model. Their result indicates that trying to improve compatibility of conditionals seems to have little effect, a result that is in line with the existing literature on FCS. The other three studies address the problem of *congeniality*, a mismatch between the imputation model and the substantive model. Improving congeniality had a major effect, which is in line with the larger multiple imputation literature. Section 4.5.4 explains the confusion surrounding the term compatibility in some detail.

A problematic aspect of including cluster means is that the contextual variable may be an unreliable estimate in small clusters. It is known that the regression weight of the contextual variable is then biased (Lüdtke et al., 2008). A solution is to formulate the contextual variable as a latent variable, and use an estimator that essentially shrinks the weight towards zero. Most joint modeling approaches assume a multivariate mixed-effects model, where cluster means are latent.

It is not yet clear when the manifest cluster means can be regarded as "correct" in an FCS context. When clusters are large and of similar size, the manifest cluster means are likely to be valid and have little differential shrinkage. For smaller clusters or clusters of unequal size, including the cluster means in the imputation model also seems valid because proper imputation techniques will use draws from the posterior distribution of the group means rather than using the manifest means themselves. All in all, it appears preferable to include the cluster means into the imputation model.

7.5.2 Model cluster heterogeneity

Van Buuren (2011) considered the homoscedastic linear mixed model as invalid for imputing incomplete predictors, and investigated only the `2l.norm` method, which allows for heterogeneous error variances by employing an intricate Gibbs sampler. The `2l.norm` method is not designed to impute variables that are systematically missing for all cases in the cluster. Resche-Rigon and White (2018) developed a solution for this case using a heteroscedastic two-stage method, which also generalizes to binary and count data. Audigier et al. (2018) compared several univariate multilevel imputation methods, and concluded that "heteroscedastic imputation methods perform better than homoscedastic methods, which should be reserved with few individuals only." Apart from the last paper there is relatively little evidence on the benefits of allowing for heteroscedasticity. It could be very well that heteroscedasticity is a useful option to improve compatibility of the conditionals, but the last word has not yet been said.

7.6 Continuous outcome

This section discusses the problem of how to create multiple imputations under the multilevel model when missing values occur in the outcome y_c only.

7.6.1 General principle

Imputations under the linear mixed model in Equation 7.6 can be generated by taking draws from posterior distribution of the modeled parameters, followed by a step to draw the imputations. The sequence of steps is:

1. Fit model 7.6 to the observed data to obtain estimates $\hat{\sigma}^2$, $\hat{\beta}$, $\hat{\Omega}$ and $\hat{u}_c$;

2. Generate random draws $\dot{\sigma}^2$, $\dot{\beta}$, $\dot{\Omega}$ and $\dot{u}_c$ from their posteriors;

3. For each missing value, calculate its expected value under the drawn model, and add a randomly drawn error to it.

These steps form a template for any multilevel imputation method. For step 1 we may use standard multilevel software. Step 2 is needed to properly account for the uncertainty of the model, and this generally requires custom software that generates the draws. Alternatively, we may use bootstrapping to incorporate model uncertainty, although this is more complex than usual since resampling is needed at two levels (Goldstein, 2011a). Once we have obtained the parameter draws, generating the imputation is straightforward. It is also possible to use predictive mean matching in step 3.

We illustrate these steps by the method proposed by Jolani (2018) for imputing mixes of systematically and sporadically missing values. Step 1 of that method consists of calling the `lmer()` from the `lme4` package to fit the model. Step 2 draws successively $\dot{\sigma}^2$, $\dot{\beta}$, $\dot{\Omega}$ under a normal-inverse-Wishart prior for Ω, and $\dot{u}_c$ from the conditional normal model for sporadically missing data, and from an unconditional normal model for systematically missing values. See the paper for the exact specification of these steps. The expected value of the missing entry is then calculated, and a random draw from the residual error distribution is added to create the imputation. These steps are implemented as the `2l.lmer` method in `mice`.

Predictive mean matching works well for single-level continuous data, and is also an interesting option for imputing multilevel data. The idea is to calculate predicted values under the linear mixed model for all level-1 units. Level-1 units with observed outcomes are selected as potential donors depending on the distance in the predictive metric. It is up to the imputer to specify whether or not donors should be restricted to inherit from the same cluster. Drawing values inside the cluster may preserve heteroscedasticity better than taking donors from all clusters, which should strengthen the homoscedasticity assumption. So if preserving such heterogeneity is important, draws should be made locally. Intuitively, drawing from one's own cluster should be done only if the cluster is relatively large, so that the procedure can find enough good matches. If different clusters come from different reporting systems, i.e., using centimeters and converted inches, the imputer might wish to preserve such features by restricting draws to the local cluster. If clusters are geographically ordered, then one may try to preserve unmeasured local features by restricting donors to the neighboring clusters. Vink et al. (2015) presents an application that exploits this feature.

7.6.2   Methods

The `mice`, `miceadds`, `micemd` and `mitml` packages contain useful functions for multilevel imputation. The `mice` package implements two methods, `2l.lmer` and `2l.pan`. Method `2l.lmer` (Jolani, 2018) imputes both sporadically and systematically missing values. Under the appropriate model, the method is randomization-valid for the fixed effects, but the variance components were more difficult to estimate, especially for a small number of clusters. Method `2l.pan` uses the PAN method (Schafer and Yucel, 2002). Method `2l.continuous` from `miceadds` is similar to `2l.lmer` with some different options. The `2l.jomo` method from `micemd` is similar to `2l.pan`, but uses the `jomo` package as the computational engine. Method `2l.glm.norm` is similar to `2l.continuous` and `2l.lmer`.

Two functions for heteroscedastic errors are available. A method named `2l.2stage.norm` from `micemd` implements the two-stage method by Resche-Rigon and White (2018). The `2l.norm` method from `mice` implements the Gibbs sampler from Kasim and Raudenbush (1998). Method `2l.norm` can recover the intra-class correlation quite well, even for severe MAR cases and high amounts of missing data in the outcome or the predictor. However, it is fairly slow and fails to achieve nominal coverage for the fixed effects for small classes (Van Buuren, 2011).

The `2l.pmm` method in the `miceadds` package is a generalization of the default `pmm` method to data with two levels using linear mixed model fitted by `lmer` or `blmer` models. Method `2l.2stage.pmm` generalizes `pmm` by a two-stage method. The default in both methods is to obtain donors across all clusters, which is probably fine for most applications.

Table 7.2 presents an overview of `R` functions for univariate imputations according to a multilevel model for continuous outcomes. Each row represents a function. The functions belong to different packages, and there is overlap in functionality. All functions can be called from `mice()` as building blocks to form an iterative FCS algorithm.

2l.pan

7.6.3   Example

We use the `brandsma` data introduced in Section 7.2. Here we will analyze the full set of 4016 pupils. Apart from Chapter 9, Snijders and Bosker (2012) concentrated on the analysis of a reduced set of 3758 pupils. In order to keep things simple, this section restricts the analysis to just two variables.

```
d <- brandsma[, c("sch", "lpo")]
```

The cluster variable is `sch`. The variable `lpo` is the pupil's test score at grade 8. The cluster variable is complete, but `lpo` has 204 missing values.

Table 7.2: Overview of methods to perform univariate multilevel imputation of continuous data. Each of the methods is available as a function called mice.impute.[method] in the specified R package.

Package	Method	Description
Continuous		
mice	21.lmer	normal, lmer
mice	21.pan	normal, pan
miceadds	21.continuous	normal, lmer, blme
micemd	21.jomo	normal, jomo
micemd	21.glm.norm	normal, lmer
mice	21.norm	normal, heteroscedastic
micemd	21.2stage.norm	normal, heteroscedastic
Generic		
miceadds	21.pmm	pmm, homoscedastic, lmer
micemd	21.2stage.pmm	pmm, heteroscedastic, mvmeta

```
md.pattern(d, plot = FALSE)

     sch lpo
3902   1   1   0
204    1   0   1
       0 204 204
```

How do we impute the 204 missing values? Let's apply the following five methods:

1. sample: Find imputations by random sampling from the observed values in lpo. This method ignores sch;

2. pmm: Single-level predictive mean matching with the school indicator coded as a dummy variable;

3. 21.pan: Multilevel method using the linear mixed model to draw univariate imputations;

4. 21.norm: Multilevel method using the linear mixed model with heterogeneous error variances;

5. 21.pmm: Predictive mean matching based on predictions from the linear mixed model, with random draws from the regression coefficients and the random effects, using five donors.

The following code block will impute the data according to these five methods.

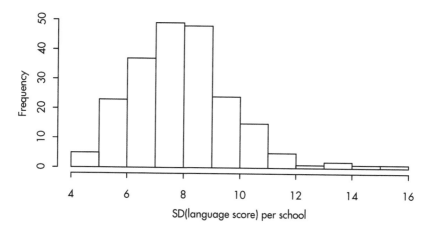

Figure 7.1: Distribution of standard deviations of language score per school.

```
library(miceadds)
methods <- c("sample", "pmm", "2l.pan", "2l.norm", "2l.pmm")
result <- vector("list", length(methods))
names(result) <- methods
for (meth in methods) {
  d <- brandsma[, c("sch", "lpo")]
  pred <- make.predictorMatrix(d)
  pred["lpo", "sch"] <- -2
  result[[meth]] <- mice(d, pred = pred, meth = meth,
                         m = 10, maxit = 1,
                         print = FALSE, seed = 82828)
}
```

The code -2 in the predictor matrix **pred** signals that **sch** is the cluster variable. There is only one variable with missing values here, so we do not need to iterate, and can set **maxit = 1**. The **miceadds** library is needed for the 2l.pmm method.

The 2l.pan and 2l.norm methods are the oldest multilevel methods. Method 2l.pan is very fast, while method 2l.norm is more flexible since the within-cluster error variances may differ. To see which of these methods should be preferred for these data, let us study the distribution of the standard deviation of **lpo** by schools. Figure 7.1 shows that the standard deviation per school varies between 4 and 16, a fairly large spread. This suggests that 2l.norm might be preferred here.

Figure 7.2 shows the box plot of the observed data (in blue) and the imputed data (in red) under each of the methods. Box plots are drawn for school with zero missing values, one missing value, two or three missing values and more than three missing values. Pupils in schools with one to three missing

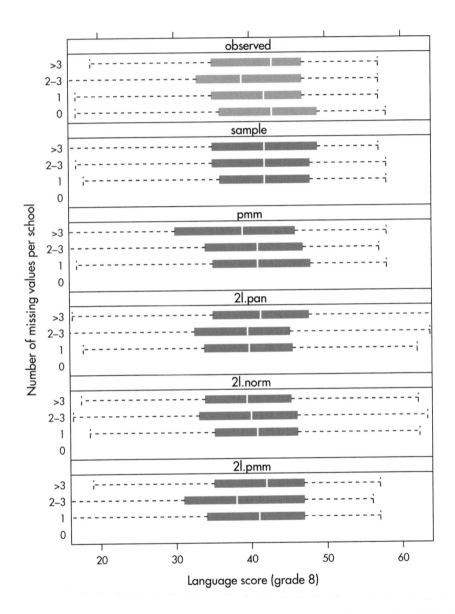

Figure 7.2: Box plots comparing the distribution of the observed data (blue), and the imputed data (red) under five methods, split according to the number of missing values per school.

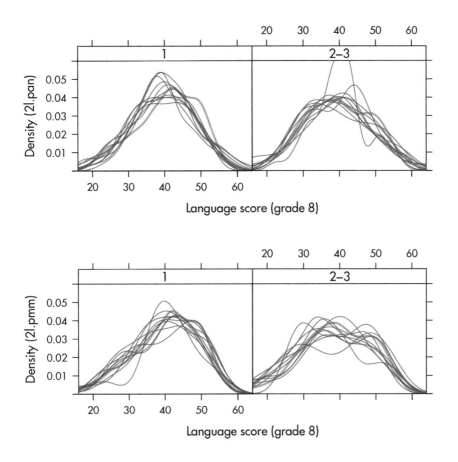

values have lower scores than pupils from a school with complete data. Pupils from schools with more than three missing values score similar to pupils from schools with complete data. It is interesting to study how well the different imputation methods preserve this feature in the data.

Method `sample` does not use any school information, and hence the imputations in all schools look alike. Methods `pmm`, `2l.pan`, `2l.norm` and `2l.pmm` preserve the pattern, though the differences are less outspoken than in the observed data. Note that the distribution of the two normal methods (`2l.pan` and `2l.norm`) have tails that extend beyond the range of the observed data (the maximum is 58). Hence, complete-data estimators based on the tails (e.g., finding the Top 10 Dutch schools) can be distorted by this use of the normal imputation.

Figure 7.3 shows the density plot of the 10 sets of imputed values (red) compared with the density plot of the observed values (blue). The top row corresponds to the 2l.pan method, and shows that some parts of the blue curve are not well represented by the imputed values. The method at the bottom row (2l.pmm) tracks the observed data distribution a little better.

Most research to date has concentrated on multilevel imputation using the normal model. In reality, normality is always an approximation, and it depends on the substantive question of how good this approximation should be. Two-level predictive mean matching is a promising alternative that can impute close to the data.

7.7 Discrete outcome

This section details how to create multiple imputations under the multilevel model when missing values occur in a discrete outcome only.

7.7.1 Methods

The generalized linear mixed model (GLMM) extends the mixed model for continuous data with link functions. For example, we can draw imputations for clustered binary data by positing a logit link with a binomial distribution. As before, all parameters need to be drawn from their respective posteriors in order to account for the sampling variation.

Jolani et al. (2015) developed a multilevel imputation method for binary data obtaining estimates of the parameters of model by the lme4::glmer() function in lme4 package (Bates et al., 2015), followed by a sequence of random draws from the parameter distributions. For meta-analysis of individual participant data, this method outperforms simpler methods that ignore the clustering, that assume MCAR or that split the data by cluster (Jolani et al., 2015). The method is available as method 2l.bin in mice. The miceadds package Robitzsch et al. (2017) contains a method 2l.binary that allows the user to choose between likelihood estimation with lme4::glmer() and penalized ML with blme::bglmer() (Chung et al., 2013). Related methods are available under sequential hierarchical regression imputation (SHRIMP) framework (Yucel, 2017).

Resche-Rigon and White (2018) proposed a two-stage estimator. At step 1, a linear regression model is fitted to each observed cluster. Any sporadically missing data are imputed, and the model per cluster ignores any systematically missing variables. At step 2, estimates obtained from each cluster are combined using meta-analysis. Systematically missing variables are modeled through a linear random effect model across clusters. A method for binary data is available as the method 2l.2stage.bin in the micemd package. The two-stage

Table 7.3: Methods to perform univariate multilevel imputation of missing discrete outcomes. Each of the methods is available as a function called `mice.impute.[method]` in the specified R package.

Package	Method	Description
Binary		
mice	2l.bin	logistic, glmer
miceadds	2l.binary	logistic, glmer
micemd	2l.2stage.bin	logistic, mvmeta
micemd	2l.glm.bin	logistic, glmer
Count		
micemd	2l.2stage.pois	Poisson, mvmeta
micemd	2l.glm.pois	Poisson, glmer
countimp	2l.poisson	Poisson, glmmPQL
countimp	2l.nb2	negative binomial, glmmadmb
countimp	2l.zihnb	zero-infl neg bin, glmmadmb

estimator is related to work done by Gelman et al. (1998) on data combinations of different surveys. These authors fitted a separate imputation for each survey using only the questions posed in the survey, and used hierarchical meta-analysis to combine the results from different surveys. Their term "not asked" translates into "systematically missing", whereas "not answered" translates into "sporadically missing".

Missing level-1 count outcomes can be imputed under the generalized linear mixed model using a Poisson or (zero-inflated) negative binomial distributions (Kleinke and Reinecke, 2015). Relevant functions can be found in the `micemd` and `countimp` packages. Table 7.3 presents an overview of R functions for univariate imputations for discrete outcomes. Discrete data can also be imputed by the predictive mean matching functions listed in Table 7.2.

7.7.2 Example

The `toenail` data were collected in a randomized parallel group trial comparing two treatments for a common toenail infection. A total of 294 patients were seen at seven visits, and severity of infection was dichotomized as "not severe" (0) and "severe" (1). The version of the data in the `DPpackage` is all numeric and easy to analyze. The following statements load the data, and expand the data to the full design with $7 \times 294 = 2058$ rows. There are in total 150 missed visits.

```
library(tidyr)
data("toenail", package = "DPpackage")
data <- tidyr::complete(toenail, ID, visit) %>%
  tidyr::fill(treatment) %>%
```

```
   dplyr::select(-month)
table(data$outcome, useNA = "always")
```

```
   0    1 <NA>
1500  408  150
```

Molenberghs and Verbeke (2005) described various analyses of these data. Here we impute the outcome of the missed visits. The next code block declares "ID" as the cluster variable, and creates $m = 5$ imputations for the missing outcomes by method 2l.bin.

```
pred <- make.predictorMatrix(data)
pred["outcome", "ID"] <- -2
imp <- mice(data, method = "2l.bin", pred = pred, seed = 12102,
            maxit = 1, m = 5, print = FALSE)
table(mice::complete(imp)$outcome, useNA = "always")
```

```
   0    1 <NA>
1635  423    0
```

Figure 7.4 visualizes the imputations. The plot shows the partially imputed profiles of 16 subjects in the toenail data. The general downward trend in the probability of infection severity with time is obvious, and was also found by Molenberghs and Verbeke (2005, p. 302). Subjects 9 (never severe) and 117 (always severe) have both complete data. They represent the extremes, and their random effect estimates are very similar in all five imputed datasets. They are close, but not identical — as you might have expected — because the multiple imputations will affect the random effects also for the fully observed subjects. Subjects 31, 41 and 309 are imputed such that their outcomes are equivalent to subject 9, and hence have similar random effect estimates. In contrast, subject 214 has the same observed data pattern as 31, but it is sometimes imputed as "severe". As a consequence, we see that there are now two random effect estimates for this subject that are quite different, which reflects the uncertainty due to the missing data. Subjects 48 and 99 even have three clearly different estimates. Imputation number 3 is colored green instead of grey, so the isolated lines in subjects 48 and 230 come from the same imputed dataset.

The complete-data model is a generalized linear mixed model for outcome given treatment status, time and a random intercept. This is similar to the models used by Molenberghs and Verbeke (2005), but here we use the visit instead of time (which is incomplete) as the timing variable. The estimates from the combined multiple imputation analysis are then obtained as

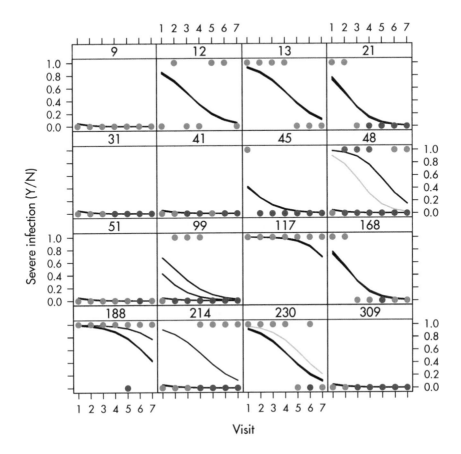

Figure 7.4: Plot of observed (blue) and imputed (red) infection (Yes/No) by visit for 16 selected persons in the toenail data (*m* = 5). The lines visualize the subject-wise infection probability predicted by the generalized linear mixed model given visit, treatment and their interaction per imputed dataset.

```
library(purrr)
mice::complete(imp, "all") %>%
  purrr::map(lme4::glmer,
             formula = outcome ~ treatment * visit + (1 | ID),
             family = binomial) %>%
  pool() %>%
  summary()

            estimate std.error statistic  df p.value
(Intercept)   -0.937    0.5778    -1.622 546  0.1052
```

treatment	0.152	0.6858	0.221 941	0.8250
visit	-0.770	0.0848	-9.079 154	0.0000
treatment:visit	-0.222	0.1219	-1.826 284	0.0682

As expected, these estimates are similar to the estimates obtained from the direct analysis of these data. The added value of multiple imputation here is that it produces a dataset with scores on all visits, which makes it easier to summarize. The added values of imputation increases if important covariates are available that are not present in the substantive model, or if missing values occur in the predictors. Section 7.10.2 contains an example of that problem.

7.8 Imputation of level-2 variable

The typical fix for missing values in a level-2 predictor is to delete all records in the cluster. Despite its potential impact on the analyses, the problem of incomplete level-2 predictors thus far received less attention than missingness in level-1 predictors.

Some authors studied the use of (inappropriate) single-level imputation methods that ignore the hierarchical group structure in multilevel data. Standard errors are underestimated, leading to confidence intervals that are too short. Early attempts to solve the problem with multiple imputation (Gibson and Olejnik, 2003; Cheung, 2007) were not successful.

Imputation methods for level-2 predictors should assign the same imputed value to all members within the same class. More recent attempts create two datasets, one with level-1 data, and one with level-2 data, and do separate imputations within each dataset while using the results from one in the other. Of course, these steps can be iterated (Gelman and Hill, 2007; Grund et al., 2018a).

The `mice` package contains several functions whose names start with `mice.impute.2lonly`. Method `2lonly.mean` fills in the class mean, and is primarily useful to repair errors in the data. Methods `2lonly.norm` and `2lonly.pmm` aggregate level-1 predictors, and impute the level-2 variables by the normal model and by predictive mean matching, respectively. The `miceadds` package contains two generic functions. The method `2lonly.function` allows the user to specify any univariate imputation function designed for level-1 data at level-2.

It is conceptually straightforward to extend imputations to higher levels (Yucel, 2008). If there are two levels, combine all level-2 predictors with an aggregate (e.g., the cluster means) of the level-1 predictors and the level-1 outcomes. Once we have this, we may choose suitable methods from Chapter 3 to impute the missing level-2 variables in the usual way. No new issues arise.

Method `ml.lmer` from `miceadds` implements a generalization to three or

Table 7.4: Overview of `mice.impute.[method]` functions to perform univariate multilevel imputation.

Package	Method	Description
Level-2		
mice	2lonly.mean	level-2 manifest class mean
miceadds	2l.groupmean	level-2 manifest class mean
miceadds	2l.latentgroupmean	level-2 latent class mean
mice	2lonly.norm	level-2 class normal
mice	2lonly.pmm	level-2 class pmm
miceadds	2lonly.function	level-2 class, generic
miceadds	ml.lmer	≥ 2 levels, generic

more levels. In addition, it also allows imputation at the lowest level (and any other level) with an arbitrary specification of (additive) random effects. This includes general nested models, cross-classified models, the ability to include cluster means at any level of clustering, and the specification of random slopes at any level of clustering. Table 7.4 lists the various methods.

7.9 Comparative work

Several comparisons on multilevel imputation methods are available. This section is a short summary of the main findings.

Enders et al. (2016) compared JM and FCS multilevel approaches, and found that both JM and FCS imputation are appropriate for random intercept analyses. The JM method was found to be superior for analyses that focus on different within- and between-cluster associations, whereas FCS provided a dramatic improvement over the JM in random slope models. Moreover, it turned out that the use of a latent variable for imputation of categorical variables worked well.

Mistler and Enders (2017) showed that more flexible and modern imputation methods for JM and FCS are preferable to older methods that assume homoscedastic distributions or multivariate normality. For random intercept models, JM and FCS are about equally good. The authors noted that JM does not preserve random slope variation, whereas FCS does.

Kunkel and Kaizar (2017) compared JM and FCS for models for random intercepts in the context of individual patient data. They found that, in spite of the theoretical differences, FCS and JM produced similar results. Moreover these authors highlighted that results were sensitive to the choice of the prior in high missingness scenarios.

Grund et al. (2018b) presents a detailed comparison between JM, FCS and FIML using current implementations. For random intercept models, they found JM and FCS equally effective, and better than ad-hoc approaches or FIML. A difference with Enders et al. (2016) was the addition of FCS methods that included cluster means. For models with random slopes and cross-level interactions, FCS was found almost unbiased for the main effects, but less reliable for higher-order terms. For categorical data, the conclusion was that both multilevel JM and FCS are suitable for creating multiple imputations. Incomplete level-2 variables were handled equally well by JM, FCS and FIML.

Audigier et al. (2018) found that JM, as implemented in `jomo`, worked well with large clusters and binary data, but had difficulty in modeling small (number of) clusters, tending to conservative inferences. The homogeneity assumption in the standard generalized linear mixed model was found to be limiting. The two-stage approach was found to perform well for systematically missing data, but was less reliable for small clusters.

The picture that emerges is that FIML is not inherently preferable for missing predictors or outcomes. Modern versions of JM and FCS are reliable ways of dealing with missing data in multilevel models with random intercepts. The FCS framework seems better suited to accommodate models with random slopes, but may have difficulty with higher-order terms.

7.10   Guidelines and advice

Many new multilevel methods have seen the light in the last five years. The comparative work as summarized above spawned a wealth of information. This section provides advice, guidelines and worked examples aimed to assist the applied statistician in solving practical multilevel imputation problems. The field moves rapidly, so the recommendations given here may change as more detailed comparative works become available in the future.

The advice given here builds upon the recommendations and code examples given in Table 6 in Grund et al. (2018b), supplemented by some of my personal biases.

There is not yet a fully satisfactory strategy for handling interactions with FCS. In this section, I will use passive imputation (Van Buuren and Groothuis-Oudshoorn, 2000), a technique that allows the user to specify deterministic relations between variables, which, amongst others, is useful for calculating interaction effects within the MICE algorithm. I will use passive imputation to enrich univariate imputation models with two-order interactions, in an attempt to preserve higher-order relations in the data. Passive imputation works reasonably well, and it is easy to apply in standard software, but it is only an approximate solution. In general, the joint distribution of the dependent and explanatory variables tends to become complex when the substantive model contains interactions (Seaman et al., 2012; Kim et al., 2015).

We revisit the **brandsma** data use in Chapters 4 and 5 of Snijders and

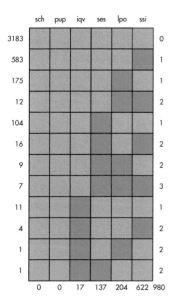

Figure 7.5: Missing data pattern of subset of brandsma data.

Bosker (2012). For reasons of clarity, the code examples are restricted to a subset of six variables.

```
data("brandsma", package = "mice")
dat <- brandsma[, c("sch", "pup", "lpo",
                    "iqv", "ses", "ssi")]
```

There is one cluster variable (`sch`), one administrative variable (`pup`), one outcome variable at the pupil level (`lpo`), two explanatory variables at the pupil level (`iqv`, `ses`) and one explanatory variable at the school level (`ssi`). The cluster variable and pupil number are complete, whereas the others contain missing values.

Figure 7.5, with the missing data patterns, reveals that there are 3183 (out of 4016) pupils without missing values. For the remaining sample, most have a missing value in just one variable: 583 pupils have only missing `ssi`, 175 pupils have only missing `lpo`, 104 pupils have only missing `ses` and 11 pupils have only missing `lpo`. The remaining 50 pupils have two or three missing values. The challenge is to perform the analyses from Snijders and Bosker (2012) using the full set with 4016 pupils.

7.10.1   Intercept-only model, missing outcomes

The intercept-only (or empty) model is the simplest multilevel model. We have already imputed the data according to this model in Section 7.6.3. Here we select the imputations according to the 2l.pmm method for further analysis.

```
d <- brandsma[, c("sch", "lpo")]
pred <- make.predictorMatrix(d)
pred["lpo", "sch"] <- -2
imp <- mice(d, pred = pred, meth = "2l.pmm", m = 10, maxit = 1,
            print = FALSE, seed = 152)
```

The empty model is fitted to the imputed datasets, and the estimates are pooled as

```
library(lme4)
fit <- with(imp, lmer(lpo ~ (1 | sch), REML = FALSE))
summary(pool(fit))
```

	estimate	std.error	statistic	df	p.value
(Intercept)	40.9	0.322	127	3368	0

We may obtain the variance components by the `testEstimates()` function from `mitml`:

```
library(mitml)
testEstimates(as.mitml.result(fit), var.comp = TRUE)$var.comp
```

	Estimate
Intercept~~Intercept|sch	18.021
Residual~~Residual	63.306
ICC|sch	0.222

See Example 4.1 in Snijders and Bosker (2012) for the interpretation of the estimates from this model.

7.10.2   Random intercepts, missing level-1 predictor

Let's now extend the model in order to quantify the impact of IQ on the language score. This random intercepts model with one explanatory variable is defined by

$$\texttt{lpo}_{ic} = \gamma_{00} + \gamma_{10}\texttt{iqv}_{ic} + u_{0c} + \epsilon_{ic}. \tag{7.10}$$

In level notation, the model reads as

$$\text{lpo}_{ic} = \beta_{0c} + \beta_{1c}\text{iqv}_{ic} + \epsilon_{ic} \tag{7.11}$$

$$\beta_{0c} = \gamma_{00} + u_{0c} \tag{7.12}$$

$$\beta_{1c} = \gamma_{10} \tag{7.13}$$

There are missing data in both `lpo` and `iqv`. Imputation can be done with both FCS and JM. For FCS my advice is to impute `lpo` and `iqv` by `2l.pmm` with added cluster means. Adding the cluster means is done here to improve compatibility among the conditional specified imputation models (cf. Section 7.5.1).

```
d <- brandsma[, c("sch", "lpo", "iqv")]
pred <- make.predictorMatrix(d)
pred["lpo", ] <- c(-2, 0, 3)
pred["iqv", ] <- c(-2, 3, 0)
imp <- mice(d, pred = pred, meth = "2l.pmm", seed = 919,
            m = 10, print = FALSE)
```

An entry of −2 in the predictor matrix signals the cluster variable, whereas an entry of 3 indicates that the cluster means of the covariates are added as a predictor to the imputation model. Thus, `lpo` is imputed from `iqv` *and* the cluster means of `iqv`, while `iqv` is imputed from `lpo` *and* the cluster means of `lpo`. If the residuals are close to normal and the within-cluster error variances are similar, then `2l.pan` is also a good choice.

Rescaling the variables as deviations from their mean often helps to improve stability of the estimates. We may rescale `lpo` to zero-mean by

```
d$lpo <- as.vector(scale(d$lpo, scale = FALSE))
```

The imputations will also adopt that scale, so we must back-transform the data if we want to analyze the data in the original scale. For the multilevel model with only random intercepts and fixed slopes, rescaling the data to the origin presents no issues since the model is invariant to linear transformations. This is not true when there are random slopes (Hox et al., 2018, p. 48). We return to this point in Section 7.10.6.

The JM can create multivariate imputations by the `jomoImpute` or `panImpute` methods. We use `panImpute` here.

```
fm1 <- lpo + iqv ~ 1 + (1 | sch)
mit <- mitml::panImpute(data = d, formula = fm1, m = 5,
                        silent = TRUE)
```

which returns as object of class `mitml`. The `panImputepanImpute` method can also be called from `mice` by creating one block for all variables as

```
blk <- make.blocks(d, "collect")
fm2 <- list(collect = fm1)
imp2 <- mice(d, meth = "panImpute", blocks = blk, form = fm2,
             print = FALSE, seed = 711)
```

This uses a new facility in `mice` that allows imputation of blocks of variables (cf. Section 4.7.2). The final estimates on the multiply imputed data under model 7.10 can be calculated (from the `21.pmm` method) as

```
fit <- with(imp, lmer(lpo ~ iqv + (1 | sch),
                      REML = FALSE))
summary(pool(fit))
```

```
            estimate std.error statistic    df p.value
(Intercept)    40.96    0.2381             172 3305       0
iqv             2.52    0.0525              48 2119       0
```

```
testEstimates(as.mitml.result(fit), var.comp = TRUE)$var.comp
```

```
                            Estimate
Intercept~~Intercept|sch      9.505
Residual~~Residual           40.819
ICC|sch                       0.189
```

which produces the estimates for the random intercept model with an effect for IQ with imputed IQ and language scores. See Example 4.2 in Snijders and Bosker (2012) for the interpretation of the parameters.

7.10.3   Random intercepts, contextual model

The ordinary least squares estimator does not distinguish between regressions within groups and between group. This section shows how we can allow for differences in the within- and between-group regressions. The models here parallel Example 4.3 and Table 4.4 in Snijders and Bosker (2012), and row 1 in Table 6 of Grund et al. (2018b).

We continue with the analysis of Section 7.10.2. We extend the complete-data multilevel model by an extra term, as follows:

$$\text{lpo}_{ic} = \gamma_{00} + \gamma_{01}\overline{\text{iqv}}_c + \gamma_{10}\text{iqv}_{ic} + u_{0c} + \epsilon_{ic}. \tag{7.14}$$

In level notation, we get

$$\begin{aligned}
\text{lpo}_{ic} &= \beta_{0c} + \beta_{1c}\text{iqv}_{ic} + \epsilon_{ic} & (7.15) \\
\beta_{0c} &= \gamma_{00} + \gamma_{01}\overline{\text{iqv}}_c + u_{0c} & (7.16) \\
\beta_{1c} &= \gamma_{10} & (7.17)
\end{aligned}$$

where the variable $\overline{iqv}_c$ stands for the cluster means of iqv. The model decomposes the contribution of IQ to the regression into a within-group component with parameter γ_{10}, and a between-group component with parameter γ_{01}. The interest in contextual analysis lies in testing the null hypothesis that $\gamma_{01} = 0$. Because of this decomposition we need to add the cluster means of lpo and iqv to the imputation model. Remember however that we just did that in the FCS imputation model of Section 7.10.2. Thus, we may use the same set of imputations to perform the within- and between-group regressions.

The following code block adds the cluster means to the imputed data, estimates the model parameters on each set, stores the results in a list, and pools the estimated parameters from the fitted models to get the combined results.

```
res <- mice::complete(imp, "long") %>%
  group_by(sch, .imp) %>%
  mutate(iqm = mean(iqv)) %>%
  group_by(.imp) %>%
  do(model = lmer(lpo ~ iqv + iqm + (1 | sch),
                  REML = FALSE, data = .)) %>%
  as.list() %>% .[[-1]]
summary(pool(res))
```

	estimate	std.error	statistic	df	p.value
(Intercept)	41.02	0.2279	180.04	3056	0.00000000
iqv	2.47	0.0535	46.20	2392	0.00000000
iqm	1.17	0.2571	4.55	667	0.00000562

```
testEstimates(res, var.comp = TRUE)$var.comp
```

	Estimate
Intercept~~Intercept|sch	8.430
Residual~~Residual	40.800
ICC|sch	0.171

An alternative could have been to use the imp2 object with the imputations under the joint imputation model.

Binary level-1 predictors can be imputed in the same way using one of the methods listed in Table 7.4. It is not yet clear which of the methods should be preferred. No univariate methods yet exist for multi-category variables, but 21.pmm may be a workable alternative. Categorical variables can be imputed by jomo (Quartagno and Carpenter, 2017), jomoImpute (Grund et al., 2018c), by latent class analysis (Vidotto, 2018), or by Blimp (Keller and Enders, 2017).

7.10.4 Random intercepts, missing level-2 predictor

The previous section extended the substantive model by the cluster means. Another extension is to add a measured level-2 predictor. For the sake of illustration we add another variable, religious denomination of the school, as a level-2 predictor. The corresponding complete-data model looks very similar:

$$\text{lpo}_{ic} = \gamma_{00} + \gamma_{01}\text{den}_c + \gamma_{10}\text{iqv}_{ic} + u_{0c} + \epsilon_{ic}. \tag{7.18}$$

In level notation, we get

$$\begin{aligned}
\text{lpo}_{ic} &= \beta_{0c} + \beta_{1c}\text{iqv}_{ic} + \epsilon_{ic} & (7.19) \\
\beta_{0c} &= \gamma_{00} + \gamma_{01}\text{den}_c + u_{0c} & (7.20) \\
\beta_{1c} &= \gamma_{10} & (7.21)
\end{aligned}$$

The missing values occur in lpo, iqv and den. The difference with model 7.14 is that den is a measured variable, so the value is identical for all members of the same cluster. If den is missing, it is missing for the entire cluster. Imputing a missing level-2 predictor is done by forming an imputation model at the cluster level.

Imputation can be done with both FCS and JM (cf. Table 6, row 3 in Grund et al. (2018b)). For FCS, the advice is to include aggregates of all level-1 variables into the cluster level imputation model. Methods 2lonly.norm and 2lonly.pmm add the means of all level-1 variables as predictors, and subsequently follow the rules for single-level imputation at level-2.

The following code block imputes missing values in the 2-level predictor den. For reasons of simplicity, I have used 2lonly.pmm, so imputations adhere to original four-point scale. This use of predictive mean matching rests on the assumption that the relative frequency of the denomination categories changes with a linear function. Alternatively, one might opt for a true categorical method to impute den, which would introduce additional parameters into the imputation model.

```
d <- brandsma[, c("sch", "lpo", "iqv", "den")]
meth <- make.method(d)
meth[c("lpo", "iqv", "den")] <- c("2l.pmm", "2l.pmm",
                                  "2lonly.pmm")
pred <- make.predictorMatrix(d)
pred["lpo", ] <- c(-2, 0, 3, 1)
pred["iqv", ] <- c(-2, 3, 0, 1)
pred["den", ] <- c(-2, 1, 1, 0)
imp <- mice(d, pred = pred, meth = meth, seed = 418,
            m = 10, print = FALSE)
```

The following statements address the same imputation task as a joint model by jomoImpute:

```
d$den <- as.factor(d$den)
fml <- list(lpo + iqv ~ 1 + (1 | sch), den ~ 1)
mit <- mitml::jomoImpute(data = d, formula = fml, m = 10,
                         silent = TRUE)
```

An alternative is to call `jomoImpute()` from `mice`, as follows:

```
blk <- make.blocks(d, "collect")
fm2 <- list(collect = fml)
imp2 <- mice(d, meth = "jomoImpute", blocks = blk, form = fm2,
             print = FALSE, seed = 418, maxit = 1,
             m = 10, n.burn = 100)
```

Because `mice` calls `jomoImpute` per replication, the latter method can be slow because the entire burn-in sequence is re-run for every call. Inspection of the trace lines revealed that autocorrelations were low and convergence was quick. To improve speed, the number of burn-in iterations was lowered from `n.burn` = 5000 (default) to `n.burn` = 100. The total number of iterations was set as `maxit` = 1, since all variables were members of the same block.

Figure 7.6 shows the density plots of the observed and imputed data after applying the joint mixed normal/categorical model, and after predictive mean matching. Both methods handle categorical data, so the figures for `den` have multiple modes. The imputations of `lpo` under JM and FCS are very similar, with `jomoImpute` slightly closer to normality. The complete-data analysis on the multiply imputed data can be fitted as

```
fit <- with(imp, lmer(lpo ~ 1 + iqv + as.factor(den)
                      + (1 | sch), REML = FALSE))
summary(pool(fit))
```

	estimate	std.error	statistic	df	p.value
(Intercept)	40.071	0.4549	88.09	187	0.000000
iqv	2.516	0.0532	47.34	1242	0.000000
as.factor(den)2	2.041	0.5925	3.45	430	0.000589
as.factor(den)3	0.234	0.6519	0.36	285	0.719226
as.factor(den)4	1.843	1.1642	1.58	1041	0.113706

```
testEstimates(as.mitml.result(fit), var.comp = TRUE)$var.comp
```

	Estimate
Intercept~~Intercept|sch	8.621
Residual~~Residual	40.761
ICC|sch	0.175

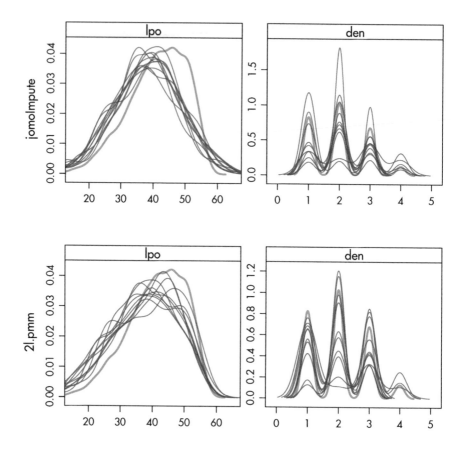

Figure 7.6: Density plots for language score and denomination after jomoImpute (top) and 2l.pmm (bottom).

7.10.5   Random intercepts, interactions

The random intercepts model may have predictors at level-1, at level-2, and possibly interactions within and/or across levels. A level-2 variable can be a level-1 aggregate (e.g., as in Section 7.10.3), or a measured level-2 variable (as in Section 7.10.4). Missing values in the measured variables will propagate through the interaction terms. This section suggests imputation methods for the model with random intercepts and interactions.

We continue with the `brandsma` data, and include three types of multiplicative interactions among the predictors into the model:

- a level-1 interaction, e.g., $\mathtt{iqv}_{ic} \times \mathtt{sex}_{ic}$;

- a cross-level interaction, e.g., $\mathtt{sex}_{ic} \times \mathtt{den}_c$;

- a level-2 interaction, e.g., $\overline{\text{iqv}}_c \times \text{den}_c$.

The extended model in composite notation is defined by:

$$
\begin{aligned}
\text{lpo}_{ic} \;=\; & \gamma_{00} + \gamma_{10}\text{iqv}_{ic} + \gamma_{20}\text{sex}_{ic} + \gamma_{30}\text{iqv}_{ic}\text{sex}_{ic} + \gamma_{40}\text{sex}_{ic}\text{den}_c + \\
& \gamma_{01}\overline{\text{iqv}}_c + \gamma_{02}\text{den}_c + \gamma_{03}\overline{\text{iqv}}_c\text{den}_c + u_{0c} + \epsilon_{ic}.
\end{aligned}
$$

In level notation, the model is

$$
\begin{aligned}
\text{lpo}_{ic} \;&=\; \beta_{0c} + \beta_{1c}\text{iqv}_{ic} + \beta_{2c}\text{sex}_{ic} + \beta_{3c}\text{iqv}_{ic}\text{sex}_{ic} + \beta_{4c}\text{sex}_{ic}\text{den}_c + \epsilon_{ic} \\
\beta_{0c} \;&=\; \gamma_{00} + \gamma_{01}\overline{\text{iqv}}_c + \gamma_{02}\text{den}_c + \gamma_{03}\overline{\text{iqv}}_c\overline{\text{den}}_c + u_{0c} \\
\beta_{1c} \;&=\; \gamma_{10} \\
\beta_{2c} \;&=\; \gamma_{20} \\
\beta_{3c} \;&=\; \gamma_{30} \\
\beta_{4c} \;&=\; \gamma_{40}
\end{aligned}
$$

How should we impute the missing values in `lpo`, `iqv`, `sex` and `den`, and obtain valid estimates for the interaction term? Grund et al. (2018b) recommend FCS with passive imputation of the interaction terms. As a first step, we initialize a number of derived variables.

```
d <- brandsma[, c("sch", "lpo", "iqv", "sex", "den")]
d <- data.frame(d, lpm = NA, iqm = NA, sxm = NA,
                iqd = NA, lpd = NA,
                iqd.sex = NA, lpd.sex = NA, iqd.lpd = NA,
                iqd.den = NA, sex.den = NA, lpd.den = NA,
                iqm.den = NA, sxm.den = NA, lpm.den = NA)
```

The new variables `lpm`, `iqm` and `sxm` will hold the cluster means of `lpo`, `iqv` and `sex`, respectively. Variables `iqd` and `lpd` will hold the values of `iqv` and `lpo` in deviations from their cluster means. Variables `iqd.sex`, `lpd.sex` and `iqd.lpd` are two-way interactions of level-1 variables scaled as deviations from the cluster means. Variables `iqd.den`, `sex.den` and `lpd.den` are cross-level interactions. Finally, `iqm.den`, `sxm.den` and `lpm.den` are interactions at level-2. For simplicity, we ignore further level-2 interactions between `iqm`, `sxm` and `lpm`.

The idea is that we impute `lpo`, `iqv`, `sex` and `den`, and update the other variables accordingly. Level-1 variables are imputed by two-level predictive mean matching, and include as predictor the other level-1 variables, the two-way interactions between the other level-1 variables (in deviations from their group means), level-2 variables, and cross-level interactions.

```
# level-1 variables
meth <- make.method(d)
meth[c("lpo", "iqv", "sex")] <- "2l.pmm"

pred <- make.predictorMatrix(d)
pred[,] <- 0
pred[, "sch"] <- -2
codes <- c(3, 3, rep(1, 6))
pred["lpo", c("iqv", "sex", "iqd.sex", "sex.den", "iqd.den",
              "den", "iqm.den", "sxm.den")] <- codes
pred["iqv", c("lpo", "sex", "lpd.sex", "sex.den", "lpd.den",
              "den", "lpm.den", "sxm.den")] <- codes
pred["sex", c("lpo", "iqv", "iqd.lpd", "lpd.den", "iqd.den",
              "den", "iqm.den", "lpm.den")] <- codes
```

Level-2 variables are imputed by predictive mean matching on level 2, using as predictors the aggregated level-1 variables, and the aggregated two-way interactions of the level-1 variables, and - if available - other level-2 variables and their two-way interactions.

```
# level-2 variables
meth["den"] <- "2lonly.pmm"
pred["den", c("lpo", "iqv", "sex",
              "iqd.sex", "lpd.sex", "iqd.lpd")] <- 1
```

The *transpose* of the relevant entries of the predictor matrix illustrates the symmetric structure of the imputation model.

```
          lpo iqv sex den
sch        -2  -2  -2  -2
lpo         0   3   3   1
iqv         3   0   3   1
sex         3   3   0   1
den         1   1   1   0
iqd.sex     1   0   0   1
lpd.sex     0   1   0   1
iqd.lpd     0   0   1   1
iqd.den     1   0   1   0
sex.den     1   1   0   0
lpd.den     0   1   1   0
iqm.den     1   0   1   0
sxm.den     1   1   0   0
lpm.den     0   1   1   0
```

The entries corresponding to the level-1 predictors are coded with a 3, indicating that both the original values as well as the cluster means of the predictor are included into the imputation model. Interactions are coded with a 1. One could also code these with a 3, in order to improve compatibility, but this is not done here because the imputation model becomes too heavy. Because we cannot have the same variable appearing at both sides of the equation, any interaction terms involving the target have been deleted from the conditional imputation models.

The specification above defines the imputation model for the variables in the data. All other variables (e.g., cluster means, interactions) are calculated on-the-fly by passive imputation. The code below centers `iqm` and `lpo` relative to their cluster means.

```
# derive group means
meth[c("iqm", "sxm", "lpm")] <- "21.groupmean"
pred[c("iqm", "sxm", "lpm"), c("iqv", "sex", "lpo")] <- diag(3)

# derive deviations from cluster mean
meth["iqd"] <- "~ I(iqv - iqm)"
meth["lpd"] <- "~ I(lpo - lpm)"
```

The `21.groupmean` method from the `miceadds` package returns the cluster mean pertaining to each observation. Centering on the cluster means is widely practiced, but significantly alters the multilevel model. In the context of imputation, centering on the cluster means often enhances stability and robustness of models to generate imputations, especially if interactions are involved. When the complete-data model uses cluster centering, then the imputation model should also do so. See Section 7.5.1 for more details.

The next block of code specifies the interaction effects, by means of passive imputation.

```
# derive interactions
meth["iqd.sex"] <- "~ I(iqd * sex)"
meth["lpd.sex"] <- "~ I(lpd * sex)"
meth["iqd.lpd"] <- "~ I(iqd * lpd)"
meth["iqd.den"] <- "~ I(iqd * den)"
meth["sex.den"] <- "~ I(sex * den)"
meth["lpd.den"] <- "~ I(lpd * den)"
meth["iqm.den"] <- "~ I(iqm * den)"
meth["sxm.den"] <- "~ I(sxm * den)"
meth["lpm.den"] <- "~ I(lpm * den)"
```

The visit sequence specified below updates the relevant derived variables after any of the measured variables is imputed, so that interactions are always in sync. The specification of the imputation model is now complete, so it can be run with `mice()`.

```
visit <- c("lpo", "lpm", "lpd",
           "lpd.sex", "iqd.lpd", "lpd.den", "lpm.den",
           "iqv", "iqm", "iqd",
           "iqd.sex", "iqd.lpd", "iqd.den", "iqm.den",
           "sex", "sxm",
           "iqd.sex", "lpd.sex", "sex.den", "sxm.den",
           "den", "iqd.den", "sex.den", "lpd.den",
           "iqm.den", "sxm.den", "lpm.den")

imp <- mice(d, pred = pred, meth = meth, seed = 188,
            visit = visit, m = 10, print = FALSE,
            allow.na = TRUE)
```

The analysis of the imputed data according to the specified model first transforms den into a categorical variable, and then fits and pools the mixed model.

```
long <- mice::complete(imp, "long", include = TRUE)
long$den <- as.factor(long$den)
imp2 <- as.mids(long)
fit <- with(imp2, lmer(lpo ~ 1 + iqv*sex + iqm*den + sex*den
                       + (1 | sch), REML = FALSE))
summary(pool(fit))
```

	estimate	std.error	statistic	df	p.value
(Intercept)	39.371	0.4620	85.228	660	0.00e+00
iqv	2.540	0.0742	34.222	391	0.00e+00
sex	2.503	0.3785	6.613	873	4.95e-11
iqm	1.497	0.4336	3.453	271	5.68e-04
den2	1.795	0.6245	2.875	444	4.09e-03
den3	-0.613	0.6742	-0.909	391	3.63e-01
den4	1.935	1.4749	1.312	823	1.90e-01
iqv:sex	-0.139	0.1084	-1.284	192	1.99e-01
iqm:den2	-0.400	0.6503	-0.615	304	5.39e-01
iqm:den3	-0.757	0.5757	-1.315	968	1.89e-01
iqm:den4	-1.841	1.4083	-1.307	1403	1.91e-01
sex:den2	-0.653	0.5014	-1.302	1372	1.93e-01
sex:den3	0.787	0.5742	1.371	433	1.71e-01
sex:den4	-0.370	1.0052	-0.368	1811	7.13e-01

7.10.6 Random slopes, missing outcomes and predictors

So far our examples were restricted to models with random intercepts. We continue here with the contextual model that includes random slopes for

IQ (cf. Example 5.1 in Snijders and Bosker (2012)). Section 7.10.3 showed how to impute the contextual model. Including random slopes extends the complete-data model as

$$\texttt{lpo}_{ic} = \gamma_{00} + \gamma_{01}\overline{\texttt{iqv}}_c + \gamma_{10}\texttt{iqv}_{ic} + u_{0c} + u_{1c}\texttt{iqv}_{ic} + \epsilon_{ic} \tag{7.22}$$

When expressed in level notation, the model is

$$
\begin{aligned}
\texttt{lpo}_{ic} &= \beta_{0c} + \beta_{1c}\texttt{iqv}_{ic} + \epsilon_{ic} & (7.23) \\
\beta_{0c} &= \gamma_{00} + \gamma_{01}\overline{\texttt{iqv}}_c + u_{0c} & (7.24) \\
\beta_{1c} &= \gamma_{10} + u_{1c} & (7.25)
\end{aligned}
$$

The addition of the term u_{1c} to the equation for β_{1c} allows for β_{1c} to vary over clusters, hence the name "random slopes".

Missing data may occur in `lpo` and `iqv`. Enders et al. (2016) and Grund et al. (2018b) recommend FCS for this problem. The procedure is almost identical to that in Section 7.10.2, but now including both the cluster means and random slopes into the imputation model.

```
d <- brandsma[, c("sch", "lpo", "iqv")]
d$lpo <- as.vector(scale(d$lpo, scale = FALSE))
pred <- make.predictorMatrix(d)
pred["lpo", ] <- c(-2, 0, 4)
pred["iqv", ] <- c(-2, 4, 0)
pred

    sch lpo iqv
sch   0   1   1
lpo  -2   0   4
iqv  -2   4   0

imp <- mice(d, pred = pred, meth = "21.pmm", seed = 441,
            m = 10, print = FALSE, maxit = 20)
```

The entry of 4 at cell (`lpo`, `iqv`) in the predictor matrix adds three variables to the imputation model for `lpo`: the value of `iqv`, the cluster means of `iqv` and the random slopes of `iqv`. Conversely, imputing `iqv` adds the three covariates: the values of `lpo`, the cluster means of `lpo` and the random slopes of `lpo`.

The `iqv` variable had zero mean in the data, so this could be imputed right away, but `lpo` needs to be centered around the grand mean in order to reduce the large number of warnings about unstable estimates. It is known that the random slopes model is not invariant to a shift in origin in the predictors (Hox et al., 2018), so we may wonder what the effect of centering on the grand mean will be on the quality of the imputations. See Kreft et al. (1995) and Enders and Tofighi (2007) for discussions on the effects of centering in multilevel models. In imputation, we generally have no desire to attach a meaning to the parameters of the imputation model, so centering on the grand mean is often beneficial. Grand-mean centering implies a little extra work because we must back-transform the data if we want the values in the original scale. What remains is that rescaling improves speed and stability, so for the purpose of imputation I recommend to scale level-1 variables in deviations from their means.

The following code block unfolds the `mids` object, adds the IQ cluster means, restores the rescaling of `lpo`, and estimates and combines the parameters of the random slopes model.

```
imp2 <- mice::complete(imp, "long", include = TRUE) %>%
  group_by(sch) %>%
    mutate(iqm = mean(iqv, na.rm = TRUE),
           lpo = lpo + mean(brandsma$lpo, na.rm = TRUE)) %>%
  as.mids()
fit <- with(imp2, lmer(lpo ~ iqv + iqm + (1 + iqv | sch),
                       REML = FALSE))
summary(pool(fit))
```

	estimate	std.error	statistic	df	p.value
(Intercept)	41.061	0.228	179.85	3508	0.000000
iqv	2.495	0.063	39.63	1511	0.000000
iqm	0.975	0.261	3.74	620	0.000185

```
testEstimates(as.mitml.result(fit), var.comp = TRUE)$var.comp
```

	Estimate
Intercept~~Intercept|sch	8.591
Intercept~~iqv|sch	-0.781
iqv~~iqv|sch	0.188
Residual~~Residual	39.791
ICC|sch	0.178

See Example 5.1 in Snijders and Bosker (2012) for the interpretation of these model parameters. Interestingly, if we don't restore the mean of `lpo`, the estimated intercept represents the average difference between the observed and imputed language scores. Its value here is -0.271 (not shown), so on average pupils without a language test score a little lower than pupils with a score. The difference is not statistically significant ($p = 0.23$).

7.10.7 Random slopes, interactions

Random slopes models may also include interactions among level-1 predictors, among level-2 predictors, and between level-1 and level-2 predictor (cross-level interactions). This section concentrates on imputation under the model described in Example 5.3 of Snijders and Bosker (2012). This is a fairly elaborate model that can best be understood in level notation:

$$
\begin{aligned}
\texttt{lpo}_{ic} &= \beta_{0c} + \beta_{1c}\texttt{iqv}_{ic} + \beta_{2c}\texttt{ses}_{ic} + \beta_{3c}\texttt{iqv}_{ic}\texttt{ses}_{ic} + \epsilon_{ic} & (7.26)\\
\beta_{0c} &= \gamma_{00} + \gamma_{01}\overline{\texttt{iqv}}_c + \gamma_{02}\overline{\texttt{ses}}_c + \gamma_{03}\overline{\texttt{iqv}}_c\overline{\texttt{ses}}_c + u_{0c} & (7.27)\\
\beta_{1c} &= \gamma_{10} + \gamma_{11}\overline{\texttt{iqv}}_c + \gamma_{12}\overline{\texttt{ses}}_c + u_{1c} & (7.28)\\
\beta_{2c} &= \gamma_{20} + \gamma_{21}\overline{\texttt{iqv}}_c + \gamma_{22}\overline{\texttt{ses}}_c + u_{2c} & (7.29)\\
\beta_{3c} &= \gamma_{30} & (7.30)
\end{aligned}
$$

which can be reorganized into composite notation as:

$$
\begin{aligned}
\texttt{lpo}_{ic} = \; & \gamma_{00} + \gamma_{10}\texttt{iqv}_{ic} + \gamma_{20}\texttt{ses}_{ic} + \gamma_{30}\texttt{iqv}_{ic}\texttt{ses}_{ic} + \\
& \gamma_{01}\overline{\texttt{iqv}}_c + \gamma_{02}\overline{\texttt{ses}}_c + \\
& \gamma_{11}\texttt{iqv}_{ic}\overline{\texttt{iqv}}_c + \gamma_{12}\texttt{iqv}_{ic}\overline{\texttt{ses}}_c + \gamma_{21}\texttt{ses}_{ic}\overline{\texttt{iqv}}_c + \gamma_{22}\texttt{ses}_{ic}\overline{\texttt{ses}}_c + \\
& \gamma_{03}\overline{\texttt{iqv}}_c\overline{\texttt{ses}}_c + \\
& u_{0c} + u_{1c}\texttt{iqv}_{ic} + u_{2c}\texttt{ses}_{ic} + \\
& \epsilon_{ic}.
\end{aligned}
$$

Although this expression may look somewhat horrible, it clarifies that the expected value of `lpo` depends on the following terms:

- the level-1 variables $\texttt{iqv}_{ic}$ and $\texttt{ses}_{ic}$;
- the level-1 interaction $\texttt{iqv}_{ic}\texttt{ses}_{ic}$;
- the cluster means $\overline{\texttt{iqv}}_c$ and $\overline{\texttt{ses}}_c$;
- the within-variable cross-level interactions $\texttt{iqv}_{ic}\overline{\texttt{iqv}}_c$ and $\texttt{ses}_{ic}\overline{\texttt{ses}}_c$;
- the between-variable cross-level interactions $\texttt{iqv}_{ic}\overline{\texttt{ses}}_c$ and $\texttt{ses}_{ic}\overline{\texttt{iqv}}_c$;
- the level-2 interaction $\overline{\texttt{iqv}}_c\overline{\texttt{ses}}_c$;
- the random intercepts;
- the random slopes for `iqv` and `ses`.

All terms need to be included into the imputation model for `lpo`. Univariate imputation models for `iqv` and `ses` can be specified along the same principles by reversing the roles of outcome and predictor. As a first step, let us pad the data with the set of all relevant interactions from model 7.26.

```
d <- brandsma[, c("sch", "lpo", "iqv", "ses")]
d$lpo <- as.vector(scale(d$lpo, scale = FALSE))
d <- data.frame(d,
                iqv.ses = NA, ses.lpo = NA, iqv.lpo = NA,
```

```
lpm = NA, iqm = NA, sem = NA,
iqv.iqm = NA, ses.sem = NA, lpo.lpm = NA,
iqv.sem = NA, iqv.lpm = NA,
ses.iqm = NA, ses.lpm = NA,
lpo.iqm = NA, lpo.sem = NA,
iqm.sem = NA, lpm.sem = NA, iqm.lpm = NA)
```

Here `iqv.ses` represents the multiplicative interaction term for `iqv` and `ses`, and `lpm` represents the cluster means of `lpo`, and so on. Imputation models for `lpo`, `iqv` and `ses` are specified by setting the relevant entries in the *transformed* predictor matrix as follows:

```
        lpo iqv ses
sch     -2  -2  -2
lpo      0   3   3
iqv      3   0   3
ses      3   3   0
iqv.ses  1   0   0
ses.lpo  0   1   0
iqv.lpo  0   0   1
iqv.iqm  1   0   1
ses.sem  1   1   0
lpo.lpm  0   1   1
iqv.sem  1   0   0
iqv.lpm  0   0   1
ses.iqm  1   0   0
ses.lpm  0   1   0
lpo.iqm  0   0   1
lpo.sem  0   1   0
iqm.sem  1   0   0
lpm.sem  0   1   0
iqm.lpm  0   0   1
```

The model for `lpo` is almost equivalent to model 7.26. According to the model, both cluster means and random effects should be included, thus values `pred["lpo", c("iqv", "ses")]` should be coded as a 4, and not as a 3. However, the cluster means and random effects are almost linearly dependent, which causes slow convergence and unstable estimates in the imputation model. These problems disappear when only the cluster means are included as covariates. An alternative is to scale the predictors in deviations from the cluster means, as was done in Section 7.10.5. This circumvents many of the computational issues of raw-scored variables, and the parameters are easier to interpret.

The specifications for `iqv` and `ses` correspond to the inverted models. Inverting the random slope model produces reasonable estimates for the fixed effect and the intercept variance, but estimates of the slope variance can be unstable and biased, especially in small samples (Grund et al., 2016a). Unless the interest is in the slope variance (for which listwise deletion appears to be better), using FCS by inverting the random slope model is the currently preferred method to account for differences in slopes between clusters.

Next, we need to specify the derived variables. The cluster means are updated by the 2l.groupmean method.

```
meth[c("iqm", "sem", "lpm")] <- "2l.groupmean"
pred[c("iqm", "sem", "lpm"), ] <- 0
pred["iqm", c("sch", "iqv")] <- c(-2, 1)
pred["sem", c("sch", "ses")] <- c(-2, 1)
pred["lpm", c("sch", "lpo")] <- c(-2, 1)
```

The level-1 interactions are updated by passive imputation.

```
meth["iqv.ses"] <- "~ I(iqv * ses)"
meth["iqv.lpo"] <- "~ I(iqv * lpo)"
meth["ses.lpo"] <- "~ I(ses * lpo)"
```

The remaining interactions are updated by passive imputation in an analogous way (code not shown).

The visit sequence updates the derived variables that depend on the target variable.

```
visit <- c("lpo", "iqv.lpo", "ses.lpo",
            "lpm", "lpo.lpm", "iqv.lpm", "ses.lpm",
            "lpo.iqm", "lpo.sem", "iqm.lpm", "lpm.sem",
            "iqv", "iqv.ses", "iqv.lpo",
            "iqm", "iqv.iqm", "iqv.sem", "iqv.lpm",
            "ses.iqm", "lpo.iqm", "iqm.sem", "iqm.lpm",
            "ses", "iqv.ses", "ses.lpo",
            "sem", "ses.sem", "iqv.sem", "ses.iqm",
            "ses.lpm", "lpo.sem", "iqm.sem", "lpm.sem")
```

```
imp <- mice(d, pred = pred, meth = meth, seed = 211,
            visit = visit, m = 10, print = FALSE, maxit = 10,
            allow.na = TRUE)
```

The model can now be fitted to the full data as

```
fit <- with(imp, lmer(lpo ~ iqv * ses + iqm * sem +
                iqv * iqm + iqv * sem +
                ses * iqm + ses * sem + (1 + ses + iqv | sch),
                REML = FALSE))
```

```
summary(pool(fit))
```

	estimate	std.error	statistic	df	p.value
(Intercept)	0.1801828	0.25242	0.7138	2584	0.47538
iqv	2.2421838	0.06205	36.1369	2619	0.00000
ses	0.1709524	0.01238	13.8100	695	0.00000
iqm	0.7675273	0.30994	2.4763	502	0.01332
sem	-0.0921057	0.04372	-2.1066	2756	0.03522
iqv:ses	-0.0172118	0.00631	-2.7261	370	0.00644
iqm:sem	-0.1167091	0.03758	-3.1060	807	0.00191
iqv:iqm	-0.0631837	0.07480	-0.8446	3675	0.39836
iqv:sem	0.0045330	0.01371	0.3307	487	0.74086
ses:iqm	0.0171123	0.01882	0.9095	268	0.36317
ses:sem	0.0000898	0.00235	0.0382	470	0.96953

```
testEstimates(as.mitml.result(fit), var.comp = TRUE)$var.comp
```

	Estimate
Intercept~~Intercept|sch	7.93524
Intercept~~ses|sch	-0.00920
Intercept~~iqv|sch	-0.75078
ses~~ses|sch	0.00114
ses~~iqv|sch	-0.00830
iqv~~iqv|sch	0.16489
Residual~~Residual	37.78840
ICC|sch	0.17355

The estimates are quite close to Table 5.3 in Snijders and Bosker (2012). These authors continue with simplifying the model. The same set of imputations can be used for these simpler models since the imputation model is more general than the substantive models.

7.10.8   Recipes

The term "cookbook statistics" is sometimes used to refer to thoughtless and rigid applications of statistical procedures. Minute execution of a sequence of steps won't earn you a Nobel Prize, but a good recipe will enable you to produce a decent meal from ingredients that you may not have seen before. The recipes given here are intended to assist you to create a decent set of imputations for multilevel data.

Table 7.5 contains two recipes for imputing multilevel data. There are separate recipes for level-1 and level-2 data. The recipes follow the inclusive strategy advocated by Collins et al. (2001), and extend the predictor specification strategy in Section 6.3.2 to multilevel data. Including all two-way (or higher-order) interactions may quickly inflate the number of parameters in

Table 7.5: Recipes for imputing multilevel data for models with random intercepts and random slopes. There are different procedures for level-1 and level-2 variables.

	Recipe for a level-1 target
1.	Define the most general analytic model to be applied to imputed data
2.	Select a 21 method that imputes close to the data
3.	Include all level-1 variables
4.	Include the disaggregated cluster means of all level-1 variables
5.	Include all level-1 interactions implied by the analytic model
6.	Include all level-2 predictors
7.	Include all level-2 interactions implied by the analytic model
8.	Include all cross-level interactions implied by the analytic model
9.	Include predictors related to the missingness and the target
10.	Exclude any terms involving the target
	Recipe for a level-2 target
1.	Define the most general analytic model to be applied to imputed data
2.	Select a 21only method that imputes close to the data
3.	Include the cluster means of all level-1 variables
4.	Include the cluster means of all level-1 interactions
5.	Include all level-2 predictors
6.	Include all interactions of level-2 variables
7.	Include predictors related to the missingness and the target
8.	Exclude any terms involving the target

the model, especially for categorical data, so some care is needed in selecting the interactions that seem most important to the application at hand.

Sections 7.10.1 to 7.10.7 demonstrated applications of these recipes for a variety of multilevel models. One very important source of information was not yet included. For clarity, all procedures were restricted to the subset of data that was actually used in the model. This strategy is not optimal in general because it fails to include potentially auxiliary information that is not modeled. For example, the **brandsma** data also contains the test scores from the same pupils taken one year before the outcome was measured. This score is highly correlated to the outcome, but it was not part of the model and hence not used for imputation. Of course, one could extend the substantive model (e.g., include the pre-test score as a covariate), but this affects the interpretation and may not correspond to the question of scientific interest. A better way is to include these variables only into the imputation model. This will decrease the between-imputation variability and hence lead to sharper statistical inferences. Including extra predictive variables is left as an exercise for the reader.

The procedure in Section 7.10.7 may be a daunting task when the number of variables grows, especially keeping track of all required interaction effects. The whole process can be automated, but currently there is no software that will perform these steps behind the screen. This may be a matter of time. In general, it is good to be aware of the steps taken, so specification by hand could also be considered an advantage.

Monitoring convergence is especially important for models with many random slopes. Warnings from the underlying multilevel routines may indicate over-specification of the model, for example, with a too large number of parameters. The imputer should be attentive to such messages by reducing the complexity of imputation model in the light of the analytic model. In multilevel modeling, overparameterization occurs almost always in the variance part of the model. Reducing the number of random slopes, or simplifying the level-2 model structure could help to reduce computational complexity.

7.11 Future research

The first edition of this book featured only three pages on multilevel imputation, and concluded: *"Imputation of multilevel data is an area where work still remains to be done"* (Van Buuren, 2012, p. 87). The progress over the last few years has been tremendous, and we can now see the contours of an emerging methodology. There are still open issues, and we may expect to see further advances in the near future.

The multilevel model does not assume the regressions to be identical in different subsets of the data. This allows for more general and interesting patterns in the data to be studied, but the added flexibility comes at the price of increased modeling effort. The current software needs to become more robust and forgiving, so that application of multilevel imputation eventually becomes a routine component of multilevel analysis. We need faster imputation algorithms, automatic model specification, and good defaults that will work across a wide variety of practical data types and models. We also need more experience with imputation in three-level data, and beyond, e.g., as supported by `Blimp` and `ml.lmer`, as well as more experience in handling of categorical data with many categories. We need better insight into the convergence properties, and more generally into the strengths and limitations of the procedures.

There is little consensus about the optimal way to handle interaction effects in multiple imputation. I used passive imputation because it is easy to apply in standard software, and has been found to work reasonably well. In the future we may see model-based imputation procedures that enhance the handling of interactions by combining the imputation and analysis models into larger Bayesian models. See Section 4.5.5 for some pointers into the literature.

Chapter 8

Individual causal effects

> *I believe that the notion of cause that operates in an experiment and in an observational study is the same.*
> Paul W. Holland

People differ widely in how they react to events. Most scientific studies express the effect of a treatment as an average over a group of persons. This is informative if the effect is thought to be similar for all persons, but is less useful if the effect is expected to differ. This chapter uses multiple imputation to estimate the *individual causal effect* (ICE), or the unit-level causal effect, for one or more units in the data. The hope is that this allows us to develop a deeper understanding of how and why people differ in their reactions to an intervention.

8.1   Need for individual causal effects

John had a stroke and was brought into the emergency department at the local hospital. After his initial rescue his cardiologist told him that his medical condition made him eligible for two types of surgery, a standard surgery and a new surgery. Both are known to prolong life, but the effect varies across patients. How do John and his doctor determine which of these two interventions would be best?

In order to answer this question, we would ideally like to know John's survival under both options, and choose the option that gives him the longest life. Table 8.1 contains the hypothetical number of years lived for eight patients under a new surgery, labeled $Y(1)$, and the number for years under standard treatment, labeled $Y(0)$.

Let us define the individual causal effect τ_i for individual i as the difference between the two outcomes

$$\tau_i = Y_i(1) - Y_i(0) \tag{8.1}$$

so John gains one year because of surgery. We see that the new surgery is beneficial for John, Peter and Torey, but harmful to the others.

Table 8.1: Number of years lived for eight patients under a new surgery $Y(1)$ and under standard treatment $Y(0)$. Hypothetical data.

Patient	Age	$Y(1)$	$Y(0)$	τ_i
John	68	14	13	+1
Caren	76	0	6	−6
Joyce	66	1	4	−3
Robert	81	2	5	−3
Ruth	70	3	6	−3
Nick	72	1	6	−5
Peter	81	10	8	+2
Torey	72	9	8	+1
Average causal effect τ				−2

In addition, let the *average causal effect*, or ACE, be the mean ICE over all units, i.e.,

$$\tau = \frac{1}{n} \sum_{i=1}^{n} Y_i(1) - Y_i(0) \tag{8.2}$$

In Table 8.1 it is equal to $\tau = (1 - 6 - 3 - 3 - 3 - 5 + 2 + 1)/8 = -2$, so applying the new surgery will reduce average life expectancy in these patients by two years. Knowing this, we would be inclined to conclude that the new surgery is harmful, and should thus not be performed. However, that would also take away valuable life years from John, Peter and Torey.

What would the *perfect doctor* do instead? The perfect doctor would assign the best treatment to each patient, so that only John, Peter and Torey would get the new surgery. Under that assignment of treatments, these three persons live for another $(14 + 10 + 9)/3 = 11$ years, whereas the others live for another 5.4 years. Seeing these two numbers only, we might be tempted to conclude that surgery increases life expectancy by $11 - 5.4 = 5.6$ years, but that conclusion would be far off. If, because of this apparent benefit, we were to provide surgery to everyone, we would actually be shortening their lives by an average of two years, a decision worse than withholding surgery for everybody. Evidently the best policy is to treat some, but not others. But how do we know whom to treat? The answer is that we need to know the ICE for every patient.

The ICE is of genuine interest in many practical settings. In clinical practice, we treat an individual, not a group, so we need an estimate of the effect for that individual. Distinguishing the ICE from the ACE allows for a more precise and clear understanding of causal inference. The ICE is more fundamental than the ACE. We can calculate ACE from a set of ICE estimates, but cannot go the other way around. Thus, knowing the ICE allows for easy estimation of every other causal estimand. It is true that estimates of the ICE might turn out to be more variable than group-wise causal estimates. But, paraphrasing Tukey, it might be better to have an approximate answer to the right question than a precise answer to the wrong one.

The case of the perfect doctor above is an example of a phenomenon known as *heterogeneity in treatment effect* (HTE). There is no consensus about the importance of HTE in practice. For example, Rothwell (2005) contends that genuine HTE is very rare, especially in those cases where treatment effects reverse in different groups. Brand and Xie (2010) however argued that HTE is "the norm, not an exception." In an informal search of the scientific literature, I had no difficulty in locating examples of HTE in a wide variety of disciplines. Here are some:

- The effect of financial deterrents on giving birth to a third child depends on whether the first two children have the same sex (Angrist, 2004);

- The effect of job training programs on earnings depends on age, sex, race and level of education (Imai and Ratkovic, 2013);

- Coronary artery bypass grafting (CABG) reduces total mortality in medium- and high-risk patients, while low-risk patients showed a non-significant trend toward increased mortality (Yusuf et al., 1994);

- Estrogen replacement therapy increased HDL cholesterol, but the increase was twice as high in women with the ER-α IVS1-401 C/C genotype (Herrington et al., 2002);

- Social skills training programs had no effect in reconviction rates of criminal offenders, but did unintentionally increase reconviction rates of psychopaths (Hare et al., 2000);

- Individuals who are least likely to obtain a college education benefit most from college (Brand and Xie, 2010).

In all of these cases treatment heterogeneity was partly explained by covariates. Of course, in practice heterogeneity may be present but we may be unable to explain it. Thus, here we see only a (possible tiny) subset of forms of HTE.

8.2 Problem of causal inference

In reality we will only be able to observe part of the values in Table 8.1. This is the *fundamental problem of causal inference* (Rubin, 1974; Holland, 1986). If Joyce gets the standard treatment, we will observe that she lives for another 4 years, but we will not know that she would have died after one year had she been given the new surgery. We can observe only one of the two outcomes, and hence these outcomes are now known as *potential outcomes*. Of course, we observe no outcome at all if the patient is not yet treated.

At least 50% of the information needed to calculate the ICE is missing, so the quantification of ICE may not be a particularly easy task. Classic linear statistical methods rely on the (mostly implicitly made) simplifying assumption of *unit treatment additivity*, which implies that the treatment has exactly the same effect on each experimental unit. When the assumption is dropped, things become complicated. Neyman (1923) defined individual causal effects for the first time, and he was well aware that these could vary over units. But he also knew that he needed assumptions beyond the data to estimate them. This might have led him to believe that these effects are not interesting or relevant (Neyman, 1935, p. 126):

> So long as the *average* yields of any treat are identical, the question as to whether these treats affect *separate* yields on a *single* plot seems to be uninteresting and academic.

Rubin (1974, p. 690) said:

> ... we assume that the average causal effect is the desired typical effect...

and Imbens and Rubin (2015, p. 18) wrote:

> There are many such unit-level causal effects, and we often wish to summarize them for the finite sample or for subpopulations.

These authors also note that estimating the ICE is difficult because the estimates are sensitive to choices for the prior distribution of the dependence structure between the two potential outcomes. Morgan and Harding (2006) wrote

> Because it is usually impossible to effectively estimate individual-level causal effects, we typically shift attention to aggregated causal effects.

Weisberg (2010, p. 36) observed that

> Mainstream statistical theory has almost nothing to say about individual causal effects.

For the better or worse, mainstream statistical methodology silently accepted the unit treatment additivity assumption. The assumption is at the heart of the Neyman-Fisher controversy, and curiously Neyman's argument in 1935 as quoted above may actually have upheld wider use of his own invention. See Sabbaghi and Rubin (2014) for additional historic background.

8.3 Framework

Let us explore the use of multiple imputation of the missing potential outcomes, with the objective of estimating τ_i for some target person i. We use the potential outcomes framework using the notation of Imbens and Rubin (2015). Let the individual causal effect for unit i be defined as $\tau_i = Y_i(1) - Y_i(0)$. Let $W_i = 0$ if unit i received the control treatment, and let $W_i = 1$ if i received the active treatment. We assume that assignment to treatments is unconfounded by the unobserved outcomes Y_{mis}, so $P(W|Y(0), Y(1), X) = P(W|Y_{\text{obs}}, X)$ specifies ignorable treatment assignment mechanism where each unit has a non-zero probability for each treatment (Imbens and Rubin, 2015, p. 39). Optionally, we may assume a joint distribution $P(Y(0), Y(1), X)$ of potential outcomes $Y(0)$ and $Y(1)$ and covariates X. This is not strictly needed for creating valid inferences under known randomized treatment assignments, but it is beneficial in more complex situations.

Imbens and Rubin (2015) specified a series of joint normal models to generate multiple imputations of the missing values in the potential outcomes. Here we will use the FCS framework to create multiple imputations of the missing potential outcomes. The idea is that we alternate two univariate imputations:

$$\dot{Y}_1 \sim P(Y_1^{\text{mis}}|Y_1^{\text{obs}}, Y_0, X, \dot{\phi}_1) \tag{8.3}$$
$$\dot{Y}_0 \sim P(Y_0^{\text{mis}}|Y_0^{\text{obs}}, Y_1, X, \dot{\phi}_0) \tag{8.4}$$

where $\dot{\phi}_1$ and $\dot{\phi}_0$ are draws from the parameters of the imputation model. Let $\dot{Y}_{i\ell}(W_i)$ denote an independent draw from the posterior predictive distributions of Y for unit i, imputation ℓ, and treatment W_i. The replicated individual causal effect $\dot{\tau}_{i\ell}$ in the ℓth imputed dataset is equal to

$$\dot{\tau}_{i\ell} = \dot{Y}_{i\ell}(1) - \dot{Y}_{i\ell}(0) \tag{8.5}$$

so the individual causal effect τ_i is estimated by

$$\hat{\tau}_i = \frac{1}{m} \sum_{\ell=1}^{m} \dot{\tau}_{i\ell} \tag{8.6}$$

The variance of $\hat{\tau}_i$ is equal to the within-unit between-replication spread

$$\hat{\sigma}_i^2 = \frac{m+1}{m^2 - m} \sum_{\ell=1}^{m} (\dot{\tau}_{i\ell} - \hat{\tau}_i)^2 \tag{8.7}$$

Note that both $\dot{Y}_{i\ell}(1)$ and $\dot{Y}_{i\ell}(0)$ vary over ℓ in Equation 8.5, but this is only needed if both outcomes are missing for unit i. In general, we may equate $Y_{i\ell}(W_i) = Y_i(W_i)$ for the observed outcomes. If unit i was allocated to the experimental treatment and if the outcome was observed, the replicated causal effect 8.5 simplifies to

$$\dot{\tau}_{i\ell} = Y_i(1) - \dot{Y}_{i\ell}(0) \tag{8.8}$$

Likewise, if unit i was measured under the control condition, we find

$$\dot{\tau}_{i\ell} = \dot{Y}_{i\ell}(1) - Y_i(0) \tag{8.9}$$

8.4   Generating imputations by FCS

We return to the data in Table 8.1. Suppose that the assignment mechanism depends on age only, where patients up to age 70 years are allocated to the new surgery with a probability of 0.75, and older patients are assigned the new surgery with a probability of 0.5. The next code creates the data that we might see after the study. John, Caren and Joyce (mean age 70) were assigned to the new surgery, whereas the other five (mean age 75) obtained standard surgery.

```
ideal <- data.frame(
  x = c(68, 76, 66, 81, 70, 72, 81, 72),
  y1 = c(14, 0, 1, 2, 3, 1, 10, 9),
  y0 = c(13, 6, 4, 5, 6, 6, 8, 8),
  row.names = c("John", "Caren", "Joyce", "Robert",
                "Ruth", "Nick", "Peter", "Torey")
)

# assign first three units to trt
data <- ideal
data[1:3, "y0"] <- NA
data[4:8, "y1"] <- NA
```

8.4.1   Naive FCS

We are interested in obtaining estimates of τ_i from the data. We generate imputations by assuming a normal distribution for the potential outcomes y1 and y0. Let us for the moment ignore the impact of age on assignment. The next code block represents a naive first try to impute the missing values.

```
library(mice)
data2 <- data[, -1]
imp <- mice(data2, method = "norm", seed = 188, print = FALSE)
Warning: Number of logged events: 7
```

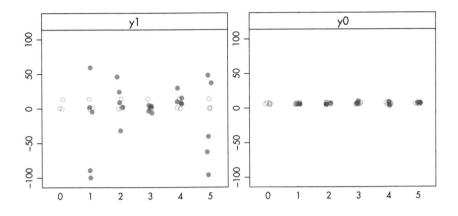

Figure 8.1 shows the values of the observed and imputed data of the potential outcomes. The imputations look very bad, especially for y1. The spread is much larger than in the data, resulting in illogical negative values and implausible high values. The problem is that there are no persons for which y1 and y0 are jointly observed, so the relation between y1 and y0 is undefined. We may see this clearly from the correlations $\rho(Y(0), Y(1))$ between the two potential outcomes in each imputed dataset.

```
sapply(mice::complete(imp, "all"), function(x) cor(x$y1, x$y0))
```

```
     1      2      3      4      5
-0.994 -0.552  0.952  0.594 -0.558
```

The ρ's are all over the place, signaling that the correlation $\rho(Y(0), Y(1))$ is not identified from the data in y1 and y0, a typical finding for the *file matching* missing data pattern.

8.4.2 FCS with a prior for ρ

We stabilize the solution by specifying a prior for $\rho(Y(0), Y(1))$. The data are agnostic to the specification, in the sense that the data will not contradict or support a given value. However, some ρ's will be more plausible than others in the given scientific context. A high value for ρ implies that the variation between the different τ_i is relatively small. The extreme case $\rho = 1$ corresponds to the assumption of homogeneous treatment effects. Setting lower values (e.g., $\rho = 0$) allows for substantial heterogeneity in treatment effects. If we set the extreme $\rho = -1$, we expect that the treatment would entirely reverse the order of units, so the unit with the maximum outcome under treatment will have the minimum outcome under control, and vice versa. It is hard to imagine interventions for which that would be realistic.

The ρ parameter can act as a *tuning knob* regulating the amount of heterogeneity in the imputation. In my experience, ρ has to be set at fairly high value, say in the range $0.9 - 0.99$. The correlation in Table 8.1 is 0.9, which allows for fairly large differences in τ_i, here from -6 years to $+2$ years.

The specification ρ in `mice` can be a little tricky, and is most easily achieved by appending hypothetical extra cases to the data with both `y1` and `y0` observed given the specified correlation. Following Imbens and Rubin (2015, p. 165) we assume a bivariate normal distribution for the potential outcomes:

$$\begin{pmatrix} Y_i(0) \\ Y_i(1) \end{pmatrix} \Big| \theta \sim N\left(\begin{pmatrix} \mu_0 \\ \mu_1 \end{pmatrix}, \begin{pmatrix} \sigma_0^2 & \rho\sigma_0\sigma_1 \\ \rho\sigma_0\sigma_1 & \sigma_1^2 \end{pmatrix} \right) \tag{8.10}$$

where $\theta = (\mu_0, \mu_1, \sigma_0^2, \sigma_1^2)$ are informed by the available data, and where ρ is set by the user. The corresponding sample estimates are $\hat{\mu}_0 = 6.6$, $\hat{\mu}_1 = 5.0$, $\hat{\sigma}_0^2 = 1.8$ and $\hat{\sigma}_1^2 = 61$. However, we do not use these estimates right away in Equation 8.10. Rather we equate $\mu_0 = \mu_1$ because we wish to avoid using the group difference twice. The location is arbitrary, but a convenient choice is grand mean $\mu = 6$, which gives quicker convergence than, say, $\mu = 0$. Also, we equate $\sigma_0^2 = \sigma_1^2$ because the scale units of the potential outcomes must be the same in order to calculate meaningful differences. A convenient choice is the variance of the observed outcome data $\hat{\sigma}^2 = 19.1$. For very high ρ, we found that setting $\sigma_0^2 = \sigma_1^2 = 1$ made the imputation algorithm more stable. Finally, we need to account for the difference in means between the data and the prior. Define $D_i = 1$ if unit i belongs to the data, and $D_i = 0$ otherwise. The bivariate normal model for drawing the imputation is

$$\begin{pmatrix} Y_i(0) \\ Y_i(1) \end{pmatrix} \Big| \theta \sim N\left(\begin{pmatrix} 6 + D_i\dot{\alpha}_0 \\ 6 + D_i\dot{\alpha}_1 \end{pmatrix}, \begin{pmatrix} 19.1 & 19.1\rho \\ 19.1\rho & 19.1 \end{pmatrix} \right) \tag{8.11}$$

where $\dot{\alpha}_0$ and $\dot{\alpha}_1$ are drawn as usual. The number of cases used for the prior is arbitrary, and will give essentially the same result. We have set it here to 100, so that the empirical correlation in the extra data will be reasonably close to the specified value. The following code block generates the extra cases.

```
set.seed(84409)
rho <- 0.9
mu <- mean(unlist(data2[, c("y1", "y0")]), na.rm = TRUE)
sigma2 <- var(unlist(data2), na.rm = TRUE)
# sigma2 <- 1
cv <- rho * sigma2
s2 <- matrix(c(sigma2, cv, cv, sigma2), nrow = 2)
prior <- data.frame(MASS::mvrnorm(n = 100, mu = rep(mu, 2),
```

```
                                    Sigma = s2))
names(prior) <- c("y1", "y0")
```

The next statements combine the observed data and the prior, and calculate two variables. The binary indicator d separates the intercepts of the observed data unit and prior units.

```
# combine data and prior
stacked <- dplyr::bind_rows(prior, data2, .id = "d")
stacked$d <- as.numeric(stacked$d) - 1
```

The tau variable is included to ease monitoring and analysis. It is calculated by passive imputation during the imputations. We need to remove tau from the predictor matrix in order to evade circularities.

```
stacked$tau <- stacked$y1 - stacked$y0
pred <- make.predictorMatrix(stacked)
pred[, "tau"] <- 0
meth <- c("", "norm", "norm", "~ I(y1 - y0)")
imp <- mice(stacked, maxit = 100, pred = pred,
            meth = meth, print = FALSE)
```

The trace lines in Figure 8.2, produced by plot(imp), look well-behaved. Imputations of y1 and y0 hover in the same range, but imputations for y0 have more spread. Note that tau (the average over the individual causal effects) is mostly negative.

Figure 8.3 shows the imputed data (red) against a backdrop of the prior of 100 data points. The left-hand plot shows the imputations for the five patients who received the standard surgery, while the right-hand plot shows imputations for the three patients who received the new surgery. Although there are a few illogical negative values, most imputations are in a plausible range. The correlations between y1 and y0 vary around the expected value of 0.9:

```
sapply(mice::complete(imp, "all"), function(x) {
  x <- x[x$d == 1, ]; cor(x$y0, x$y1)})

    1     2     3     4     5
0.952 0.976 0.960 0.953 0.840
```

Figure 8.4 is created as

```
xyplot(imp, y1 ~ y0 | as.factor(.imp), layout = c(3, 2),
       groups =   d, col = c("grey70", mdc(2)), pch = c(1, 19))
```

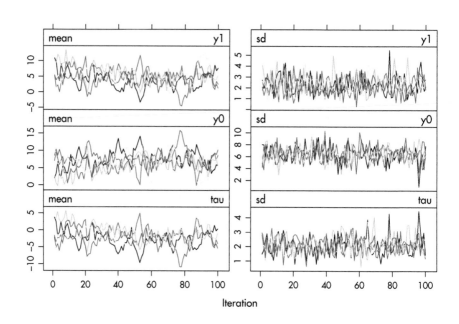

Figure 8.2: Trace lines of FCS with a $\rho = 0.9$ and $m = 5$.

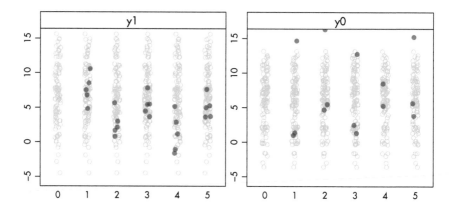

Figure 8.3: FCS with a ρ prior. Stripplots of the prior data (gray) and the imputed data (red) for the potential outcomes y1 and y0.

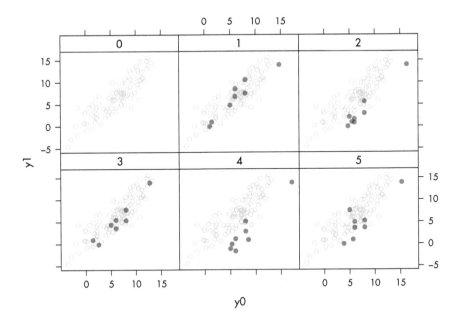

Figure 8.4: Multiple imputation ($m = 5$) of the potential outcomes (in red) plotted against the hypothetical prior data (in gray) with $\rho = 0.9$.

It visualizes the eight pairs of imputed potential outcomes against a backdrop of the hypothetical prior data. The imputed values are now reasonably behaved in the sense that they look like the prior data. Note that the red points may move as a whole in the horizontal or vertical directions (e.g., as in imputation 4) as the `norm` method draws the regression parameters from their respective posteriors to account for the sampling uncertainty of the imputation model.

Figure 8.5 displays imputations by patients for three different values of ρ. Each panel contains the observed outcome for the patient, and $m = 100$ imputed values for the missing outcome. We used $\sigma_0^2 = \sigma_1^2 = 1$ here. John, Caren and Joyce had the new surgery, so in their panel the observed value is at the left-hand side (labeled 1), and the imputed 100 values are on the right-hand side (labeled 0). The direction is reversed for patients that had standard surgery. The values are connected, resulting in a fan-shaped image.

For $\rho = 0$, there is substantial regression to the global mean in the imputed values. For example, John's imputed outcomes are nearly all lower than his observed value, whereas the opposite occurs for Caren. John would benefit from the new surgery, whereas Caren would benefit from the standard treatment. Thus, for $\rho = 0$ there is a strong tendency to pull the imputations towards

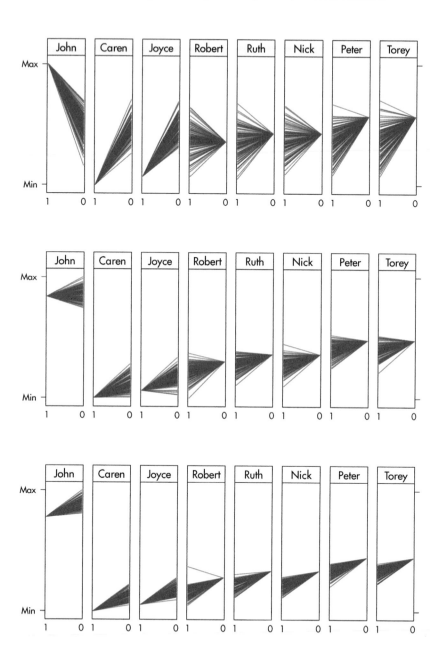

Figure 8.5: Fan plot. Observed and imputed ($m = 100$) outcomes under new (1) and standard (0) surgery. John, Caren and Joyce had the new surgery. The three rows correspond to $\rho = 0.00$ (top), $\rho = 0.90$ (middle) and $\rho = 0.99$ (bottom). Data from Table 8.1.

the mean. The effects are heterogeneous. Convergence of this condition is very quick.

The pattern for the $\rho = 0.99$ condition (bottom row) is different. All patients would benefit from the standard surgery, but the magnitude of the benefit is smaller than those under $\rho = 0$. Observe that all effects, except for Caren and Joyce, go against the direction of the regression to the mean. The effects are almost identical, and the between-imputation variance is small. The solution in the middle row ($\rho = 0.9$) is a compromise between the two. Intuitively, this setting is perhaps the most realistic and reasonable.

8.4.3 Extensions

Imbens and Rubin (2015) observe that the inclusion of covariates does not fundamentally change the underlying method for imputing the missing outcomes. This generalizes Neyman's method to covariates, and has the advantage that covariates can improve the imputations by providing additional information on the outcomes. A clear causal interpretation is only warranted if the covariates are not influenced by the treatment. This includes pre-treatment factors that led up to the decision to treat or not (e.g., age, disease history), pre-treatment factors that are predictive of the later outcome, such as baseline outcomes, and post-treatment factors that are not affected by treatment. Covariates that may have changed as a result of the experimental treatment should not be included in a causal model.

We may distinguish various types of covariates. A covariate can be part of the assignment mechanism, related to the potential outcomes $Y_i(0)$ and $Y_i(1)$, or related to τ_i. The first covariate type is often included into the imputation model in order to account for design effects, in particular to achieve comparability of the experimental units. A causal effect must be a comparison of the ordered sets $\{Y_i(1), i \in S\}$ and $\{Y_i(0), i \in S\}$ (Rubin, 2005), which can always be satisfied once the potential outcomes have been imputed for units $i \in S$. Hence, we have no need to stratify on design variables to achieve comparability. However, we still need to include design factors into the imputation model in order to satisfy the condition of ignorable treatment assignment. Including the second covariate type will make the imputations more precise, and so this is generally beneficial. The third covariate type directly explains heterogeneity of causal effects over the units. Because τ_i is an entirely missing (latent) variable, it is difficult to impute τ_i directly in `mice`. The method in Section 8.4.2 explains heterogeneity of τ_i indirectly via the imputation models for $Y_i(0)$ and $Y_i(1)$. Any covariates of type 3 should be included on the imputation models, and their regression weights should be allowed to differ between models for $Y_i(0)$ and $Y_i(1)$.

Suppose we wish to obtain the average causal effect for $n_S > 1$ units $i \in S$. Calculate the within-replication average causal effect τ_ℓ over the units in set

S as

$$\dot{\tau}_\ell = \frac{1}{n_S} \sum_{i \in S} \dot{\tau}_{i\ell} \tag{8.12}$$

and its variance as

$$\dot{\sigma}_\ell^2 = \frac{1}{n_S - 1} \sum_{i \in S} (\dot{\tau}_{i\ell} - \dot{\tau}_\ell)^2 \tag{8.13}$$

and then combine the results over the replications by means of Rubin's rules. We can use the same principles for categorical outcomes. Mortality is a widely used outcome, and can be imputed by logistic regression. The ICE can then takes on one of four possible values: (alive, alive), (alive, dead), (dead, alive) and (dead, dead). An still open question is how the dependency between the potential outcomes is best specified.

We may add a potential outcome for every additional treatment, so extension to three or more experimental conditions does not present new conceptual issues. However the imputation problem becomes more difficult. With four treatments, 75 percent of each outcome will need to be imputed, and the number of outcomes to impute will double. There are practical limits to what can be done, but I have done analyses with seven treatment arms. Careful monitoring of convergence is needed, as well as a reasonably size dataset in each experimental group.

After imputation, the individual causal effect estimates can be analyzed for patterns that explain the heterogeneity. The simplest approach takes $\hat{\tau}_i$ as the outcome for a regression at the unit level, and ignores the often substantial variation around $\hat{\tau}_i$. This is primarily useful for exploratory analysis. Alternatively, we may utilize the full multiple imputation cycle, so perform the regression on $\dot{\tau}_{i\ell}$ within each imputed dataset, and then pool the results by Rubin's rules.

8.5   Bibliographic notes

Multiple imputation of the potential outcomes was suggested by Rubin (2004b). Several authors experimented with multiple imputation of potential outcomes, all with the goal of estimating the ACE. Piesse et al. (2010) empirically demonstrated that, with proper adjustments, multiple imputation of potential outcomes in non-randomized experiments can approximate the results of randomized experiments. Bondarenko and Raghunathan (2010) augmented the data matrix with prior information, and showed the sensitivity of the results due to different modelling assumptions. For the randomized design, Aarts et al. (2010) found that multiple imputation of potential outcomes is more efficient than the t-test and on par with ANCOVA when all usual linear assumptions are met, and better if assumptions were violated.

Lam (2013) found that predictive mean matching performed well for imputing potential outcomes. Gutman and Rubin (2015) described a spline-based imputation method for binary data with good statistical properties. Imbens and Rubin (2015) show how the ACE and ρ are independent, discuss various options of setting ρ and derive estimates of the ACE. Smink (2016) found that the quality of the ICE estimate depends on the quantile of the realized outcome, and concluded that proper modeling of the correlation between the potential outcomes is needed.

There is a vast class of methods that relate the observed scores Y_i to covariates X by least-squares or machine learning methods. These methods are conceptually and analytically distinct from the methods presented in this chapter. Some methods are advertised as estimating individual causal effects, but actually target a different estimand. The relevant literature typically defines individual causal effect as something like

$$\tilde{\tau}_i = \mathrm{E}[Y|X = x_i, W_i = 1] - \mathrm{E}[Y|X = x_i, W_i = 0] \tag{8.14}$$

which is the difference between the predicted value under treatment and predicted value under control for each individual. In order to quantify $\tilde{\tau}_i$, one needs to estimate the components $\mathrm{E}[Y|X = x_i, W_i = 1]$ and $\mathrm{E}[Y|X = x_i, W_i = 0]$ from the data. Now in practice, the set of units $i \in S_1$ for estimating the first component differs from the set of units $i \in S_0$ for estimating the second. In that case, $\tilde{\tau}_i$ takes the expectation over different sets of units, so $\tilde{\tau}_i$ reflects not only the treatment effect, but also any effects that arise because the units in S_1 and S_0 are different, and even mutually exclusive. This violates the critical requirement for causal inference that "the comparison must be a comparison of $Y_i(1)$ and $Y_i(0)$ for a common set of units" (Rubin, 2005, p. 323). If we aspire taking expectations over the *same* set of units, we will need to make additional assumptions. Depending on such assumptions about the treatment assignment mechanism and about ρ, there will be circumstances where τ_i and $\tilde{\tau}_i$ lead to the same estimates, but without such assumptions, the estimands τ_i and $\tilde{\tau}_i$ are generally different.

I realize that the methods presented in this chapter only scratch the surface of a tremendous, yet unexplored field. The methodology is in a nascent state, and I hope that the materials in this chapter will stimulate further research in the area.

Part III

Case studies

Chapter 9

Measurement issues

Measurement is the contact of reason with nature.
Henry Margenau

This chapter contains three case studies using real data. The common theme is that all have "problems with the columns." Section 9.1 illustrates a number of useful steps to take when confronted with a dataset that has an overwhelming number of variables. Section 9.2 continues with the same data, and shows how a simple sensitivity analysis can be done. Section 9.3 illustrates how multiple imputation can be used to estimate overweight prevalence from self-reported data. Section 9.4 shows a way to do a sensible analysis on data that are incomparable.

9.1 Too many columns

Suppose that your colleague has become enthusiastic about multiple imputation. She asked you to create a multiply imputed version of her data, and forwarded you her entire database. As a first step, you use R to read it into a data frame called `data`. After this is done, you type in the following commands:

```
library(mice)
## DO NOT DO THIS
imp <- mice(data)      # not recommended
```

If you are lucky, the program may run and impute, but after a few minutes it becomes clear that it takes a long time to finish. And after the wait is over, the imputations turn out to be surprisingly bad. What happened?

Some exploration of the data reveals that your colleague sent you a dataset with 351 columns, essentially all the information that was sampled in the study. By default, the `mice()` function uses all other variables as predictors, so `mice()` will try to calculate regression analyses with 350 explanatory variables, and repeat that for every incomplete variable. Categorical variables are internally represented as dummy variables, so the actual number of predictors could easily double. This makes the algorithm extremely slow, if it runs at all.

Some further exploration reveals some variables are free text fields, and that some of the missing values were not marked as such in the data. As a consequence, `mice()` treats impossible values such as "999" or "−1" as real data. Just one forgotten missing data mark may introduce large errors into the imputations.

In order to evade such practical issues, it is necessary to spend some time exploring the data first. Furthermore, it is helpful if you understand for which scientific question the data are used. Both will help in creating sensible imputations.

This section concentrates on what can be done based on the data values themselves. In practice, it is far more productive and preferable to work together with someone who knows the data really well, and who knows the questions of scientific interest that one could ask from the data. Sometimes the possibilities for cooperation are limited. This may occur, for example, if the data have come from several external sources (as in meta analysis), or if the dataset is so diverse that no one person can cover all of its contents. It will be clear that this situation calls for a careful assessment of the data quality, well before attempting imputation.

9.1.1   Scientific question

There is a paradoxical inverse relation between blood pressure (BP) and mortality in persons over 85 years of age (Boshuizen et al., 1998; Van Bemmel et al., 2006). Normally, people with a lower BP live longer, but the oldest old with lower BP live a shorter time.

The goal of the study was to determine if the relation between BP and mortality in the very old is due to frailty. A second goal was to know whether high BP was a still risk factor for mortality after the effects of poor health had been taken into account.

The study compared two models:

1. The relation between mortality and BP adjusted for age, sex and type of residence.

2. The relation between mortality and BP adjusted for age, sex, type of residence and health.

Health was measured by 28 different variables, including mental state, handicaps, being dependent in activities of daily living, history of cancer and others. Including health as a set of covariates in model 2 might explain the relation between mortality and BP, which, in turn, has implications for the treatment of hypertension in the very old.

9.1.2   Leiden 85+ Cohort

The data come from the 1236 citizens of Leiden who were 85 years or older on December 1, 1986 (Lagaay et al., 1992; Izaks et al., 1997). These individuals were visited by a physician between January 1987 and May 1989. A full medical history, information on current use of drugs, a venous blood sample, and other health-related data were obtained. BP was routinely measured during the visit. Apart from some individuals who were bedridden, BP was measured while seated. An Hg manometer was used and BP was rounded to the nearest 5 mmHg. Measurements were usually taken near the end of the interview. The mortality status of each individual on March 1, 1994 was retrieved from administrative sources.

Of the original cohort, a total of 218 persons died before they could be visited, 59 persons did not want to participate (some because of health problems), 2 emigrated and 1 was erroneously not interviewed, so 956 individuals were visited. Effects due to subsampling the visited persons from the entire cohort were taken into account by defining the date of the home visit as the start (Boshuizen et al., 1998). This type of selection will not be considered further.

9.1.3   Data exploration

The data are stored as a **SAS** export file. The **read.xport()** function from the **foreign** package can read the data.

```
library(foreign)
file.sas <- file.path(dataproject, "original/master85.xport")
## xport.info <- lookup.xport(file.sas)
original.sas <- read.xport(file.sas)
names(original.sas) <- tolower(names(original.sas))
dim(original.sas)
```

```
[1] 1236   351
```

The dataset contains 1236 rows and 351 columns. When I tracked down the origin of the data, the former investigators informed me that the file was composed during the early 1990's from several parts. The basic component consisted of a **Dbase** file with many free text fields. A dedicated **Fortran** program was used to separate free text fields. All fields with medical and drug-related information were hand-checked against the original forms. The information not needed for analysis was not cleaned. All information was kept, so the file contains several versions of the same variable.

A first scan of the data makes clear that some variables are free text fields, person codes and so on. Since these fields cannot be sensibly imputed, they are removed from the data. In addition, only the 956 cases that were initially visited are selected, as follows:

```
# remove 15 columns (text, administrative)
all <- names(original.sas)
```

```
drop <- c(3, 22, 58, 162:170, 206:208)
keep <- !(1:length(all) %in% drop)
leiden85 <- original.sas[original.sas$abr == "1", keep]
data <- leiden85
```

The frequency distribution of the missing cases per variable can be obtained as:

```
ini <- mice(data, maxit = 0)    # recommended

Warning: Number of logged events: 28

table(ini$nmis)
```

```
   0   2   3   5   7  14  15  28  29  32  33  34  35  36  40  42
  87   2   1   1   1   1   2   1   3   2  34  15  25   4   1   1
  43  44  45  46  47  48  49  50  51  54  64  72  85 103 121 126
   2   1   4   2   3  24   4   1  20   2   1   4   1   1   1   1
 137 155 157 168 169 201 202 228 229 230 231 232 233 238 333 350
   1   1   1   2   1   7   3   5   4   2   4   1   1   1   3   1
 501 606 635 636 639 642 722 752 753 812 827 831 880 891 911 913
   3   1   2   1   1   2   1   5   3   1   1   3   3   3   3   1
 919 928 953 954 955
   1   1   3   3   3
```

Ignoring the warning for a moment, we see that there are 87 variables that are complete. The set includes administrative variables (e.g., person number), design factors, date of measurement, survival indicators, selection variables and so on. The set also included some variables for which the missing data were inadvertently not marked, containing values such as "999" or "−1." For example, the frequency distribution of the complete variable "beroep1" (occupation) is

```
table(data$beroep1, useNA = "always")
```

```
  -1    0    1    2    3    4    5    6 <NA>
  42    1  576  125  104   47   44   17    0
```

There are no missing values, but a variable with just categories "−1" and "0" is suspect. The category "−1" likely indicates that the information was missing (this was the case indeed). One option is to leave this "as is," so that mice() treats it as complete information. All cases with a missing occupation are then seen as a homogeneous group.

Two other variables without missing data markers are `syst` and `diast`, i.e., systolic and diastolic BP classified into six groups. The correlation (using the observed pairs) between `syst` and `rrsyst`, the variable of primary interest, is 0.97. Including `syst` into the imputation model for `rrsyst` will ruin the imputations. The "as is" option is dangerous, and shares some of the same perils of the indicator method (cf. Section 1.3.7). The message is that variables that are 100% complete deserve appropriate attention.

After a first round of screening, I found that 57 of the 87 complete variables were uninteresting or problematic in some sense. Their names were placed on a list named `outlist1` as follows:

```
v1 <- names(ini$nmis[ini$nmis == 0])
outlist1 <- v1[c(1, 3:5, 7:10, 16:47, 51:60, 62, 64:65, 69:72)]
length(outlist1)
```

```
[1] 57
```

9.1.4   Outflux

We should also scrutinize the variables at the other end. Variables with high proportions of missing data generally create more problems than they solve. Unless some of these variables are of genuine interest to the investigator, it is best to leave them out. Virtually every dataset contains some parts that could better be removed before imputation. This includes, but is not limited to, uninteresting variables with a high proportion of missing data, variables without a code for the missing data, administrative variables, constant variables, duplicated, recoded or standardized variables, and aggregates and indices of other information.

Figure 9.1 is the influx-outflux pattern of Leiden 85+ Cohort data. The influx of a variable quantifies how well its missing data connect to the observed data on other variables. The outflux of a variable quantifies how well its observed data connect to the missing data on other variables. See Section 4.1.3 for more details. Though the display could obviously benefit from a better label-placing strategy, we can see three groups. All points are relatively close to the diagonal, which indicates that influx and outflux are balanced.

The group at the left-upper corner has (almost) complete information, so the number of missing data problems for this group is relatively small. The intermediate group has an outflux between 0.5 and 0.8, which is small. Missing data problems are more severe, but potentially this group could contain important variables. The third group has an outflux with 0.5 and lower, so its predictive power is limited. Also, this group has a high influx, and is thus highly dependent on the imputation model.

Note that there are two variables (`hypert1` and `aovar`) in the third group that are located above the diagonal. Closer inspection reveals that the missing

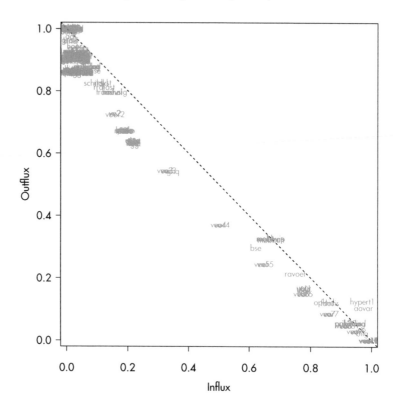

Figure 9.1: Global influx-outflux pattern of the Leiden 85+ Cohort data. Variables with higher outflux are (potentially) the more powerful predictors. Variables with higher influx depend strongly on the imputation model.

data mark had not been set for these two variables. Variables that might cause problems later on in the imputations are located in the lower-right corner. Under the assumption that this group does not contain variables of scientific interest, I transferred 45 variables with an outflux < 0.5 to outlist2:

```
outlist2 <- row.names(fx)[fx$outflux < 0.5]
length(outlist2)
```

```
[1] 45
```

In these data, the set of selected variables is identical to the group with more than 500 missing values, but this need not always be the case. I removed the 45 variables, recalculated influx and outflux on the smaller dataset and selected 32 new variables with outflux < 0.5.

```
data2 <- data[, !names(data) %in% outlist2]
fx2 <- flux(data2)
outlist3 <- row.names(fx2)[fx2$outflux < 0.5]
```

Variable `outlist3` contains 32 variable names, among which are many laboratory measurements. I prefer to keep these for imputation since they may correlate well with BP and survival. Note that the outflux changed considerably as I removed the 45 least observed variables. Influx remained nearly the same.

9.1.5   Finding problems: `loggedEvents`

Another source of information is the list of logged events produced by `mice()`. The warning we ignored previously indicates that `mice` found some peculiarities in the data that need the user's attention. The logged events form a structured report that identify problems with the data, and details which corrective actions were taken by `mice()`. It is a component called `loggedEvents` of the `mids` object.

```
head(ini$loggedEvents, 2)
```

```
  it im dep      meth out
1 0  0      constant abr
2 0  0      constant vo7
```

```
tail(ini$loggedEvents, 2)
```

```
   it im dep       meth    out
27 0  0     collinear voor10
28 0  0     collinear voor11
```

At initialization, a log entry is made for the following actions:

- A constant variable is removed from the imputation model, unless the `remove.constant = FALSE` argument is specified;

- A variable that is collinear with another variable is removed from the imputation model, unless the `remove.collinear = FALSE` argument is specified.

A variable is removed from the model by internal edits of the `predictorMatrix`, `method`, `visitSequence` and `post` components of the model. The data are kept intact. Note that setting `remove.constant = FALSE` or `remove.collinear = FALSE` bypasses usual safety measures in `mice`, and could cause problems further down the road. If a variable has only `NA`'s, it is considered a constant variable, and will not be imputed. Setting `remove.constant = FALSE` will cause numerical problems since there are no observed cases to estimate the imputation model, but such variables can be imputed by passive imputation by specifying the `allow.na = TRUE` argument.

During execution of the main algorithm, the entries in `loggedEvents` can signal the following actions:

- A predictor that is constant or correlates higher than 0.999 with the target variable is removed from the univariate imputation model. The cut-off value can be specified by the `threshold` argument;

- If all predictors are removed, this is noted in `loggedEvents`, and the imputation model becomes an intercept-only model;

- The degrees of freedom can become negative, usually because there are too many predictors relative to the number of observed values for the target. In that case, the degrees of freedom are set to 1, and a note is written to `loggedEvents`.

A few events may happen just by chance, in which case they are benign. However, if there are many entries, it is likely that the imputation model is overparametrized, causing sluggish behavior and unstable estimates. In that case, the imputation model needs to be simplified.

The `loggedEvents` component of the `mids` object is a data frame with five columns. The columns `it`, `im` stand for iteration and imputation number. The column `dep` contains the name of the target variable, and is left blank at initialization. Column `meth` entry signals the type of problem, e.g. `constant`, `df set to 1`, and so on. Finally, the column `out` contains the names of the removed variables. The `loggedEvents` component contains vital hints about possible problems with the imputation model. Closer examination of these logs could provide insight into the nature of the problem. In general, strive for zero entries, in which case the `loggedEvent` component is equal to `NULL`.

Unfortunately, `loggedEvents` is not available if `mice` crashes. If that happens, inspect the console output to see what the last variable was, and think of reasons that might have caused the breakdown, e.g., using a categorical predictor with many categories as a predictor. Then remove this from the model. Alternatively, lowering `maxit`, setting `ridge` to a high value (`ridge = 0.01`), or using a more robust imputation method (e.g., `pmm`) may get you beyond the point where the program broke down. Then, obtain `loggedEvents` to detect any problems.

Continuing with the analysis, based on the initial output by `mice()`, I placed the names of all constant and collinear variables on `outlist4` by

```
outlist4 <- as.character(ini$loggedEvents[, "out"])
```

This outlist contains 28 variables.

9.1.6 Quick predictor selection: quickpred

The `mice` package contains the function `quickpred()` that implements the predictor selection strategy of Section 6.3.2. In order to apply this strategy to the Leiden 85+ Cohort data, I first deleted the variables on three of the four outlists created in the previous sections.

```
outlist <- unique(c(outlist1, outlist2, outlist4))
length(outlist)
```

```
[1] 108
```

There are 108 unique variables to be removed. Thus, before doing any imputations, I cleaned out about one third of the data that are likely to cause problems. The downsized data are

```
data2 <- data[, !names(data) %in% outlist]
```

The next step is to build the imputation model according to the strategy outlined above. The function `quickpred()` is applied as follows:

```
inlist <- c("sex", "lftanam", "rrsyst", "rrdiast")
pred <- quickpred(data2, minpuc = 0.5, include = inlist)
```

There are 198 incomplete variables in `data2`. The character vector `inlist` specifies the names of the variables that should be included as covariates in every imputation model. Here I specified age, sex and blood pressure. Blood pressure is the variable of central interest, so I included it in all models. This list could be longer if there are more outcome variables. The `inlist` could also include design factors.

The `quickpred()` function creates a binary predictor matrix of 198 rows and 198 columns. The rows correspond to the incomplete variables and the columns report the same variables in their role as predictor. The number of predictors varies per row. We can display the distribution of the number of predictors by

```
table(rowSums(pred))
```

```
 0   7  11  12  13  14  15  16  17  18  19  20  21  22  23  24  25  26  27  28  29
30   1   2   1   1   2   5   2  13   8  16   9  13   7   5   6  10   6   3   6   4
30  31  32  33  34  35  36  37  38  39  40  41  42  44  45  46  49  50  57  59  60
 8   3   6   9   2   4   6   2   5   2   4   2   3   4   3   3   3   1   1   1   1
61  68  79  83  85
 1   1   1   1   1
```

The variability in model sizes is substantial. The 30 rows with no predictors are complete. The mean number of predictors is equal to 24.8. It is possible to influence the number of predictors by altering the values of `mincor` and `minpuc` in `quickpred()`. A number of predictors of 15–25 is about right (cf. Section 6.3.2), so I decided to accept this predictor matrix. The number of predictors for systolic and diastolic BP are

```
rowSums(pred[c("rrsyst", "rrdiast"),])
```

```
rrsyst rrdiast
    41      36
```

The names of the predictors for `rrsyst` can be obtained by

```
names(data2)[pred["rrsyst", ] == 1]
```

It is sometimes useful the inspect the correlations of the predictors selected by `quickpred()`. Table 3 in Van Buuren et al. (1999) provides an example. For a given variable, the correlations can be tabulated by

```
vname <- "rrsyst"
y <- cbind(data2[vname], r =! is.na(data2[, vname]))
vdata <- data2[,pred[vname,] == 1]
round(cor(y = y, x = vdata, use = "pair"), 2)
```

9.1.7　Generating the imputations

Everything is now ready to impute the data as

```
imp.qp <- mice(data2, pred = pred, seed = 29725)
```

Thanks to the smaller dataset and the more compact imputation model, this code runs about 50 times faster than "blind imputation" as practiced in Section 9.1. More importantly, the new solution is much better. To illustrate the latter, take a look at Figure 9.2.

The figure is the scatterplot of `rrsyst` and `rrdiast` of the first imputed dataset. The left-hand figure shows what can happen if the data are not properly screened. In this particular instance, a forgotten missing data mark of "−1" was counted as a valid blood pressure value, and produced imputation that are far off. In contrast, the imputations created with the help of `quickpred()` look reasonable.

The plot was created by the following code:

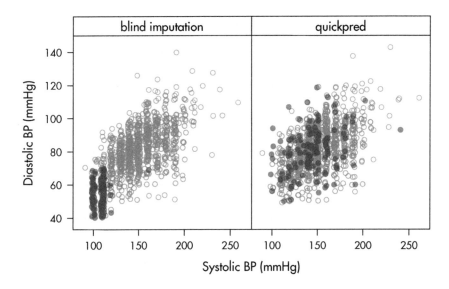

Figure 9.2: Scatterplot of systolic and diastolic blood pressure from the first imputation. The left-hand-side plot was obtained after just running `mice()` on the data without any data screening. The right-hand-side plot is the result after cleaning the data and setting up the predictor matrix with `quickpred()`. Leiden 85+ Cohort data.

```
vnames <- c("rrsyst", "rrdiast")
cd1 <- mice::complete(imp)[, vnames]
cd2 <- mice::complete(imp.qp)[, vnames]
typ <- factor(rep(c("blind imputation", "quickpred"),
                  each = nrow(cd1)))
mis <- ici(data2[, vnames])
mis <- is.na(imp$data$rrsyst) | is.na(imp$data$rrdiast)
cd <- data.frame(typ = typ, mis = mis, rbind(cd1, cd2))
xyplot(jitter(rrdiast, 10) ~ jitter(rrsyst, 10) | typ,
       data = cd, groups = mis,
       col = c(mdc(1), mdc(2)),
       xlab = "Systolic BP (mmHg)",
       type = c("g","p"), ylab = "Diastolic BP (mmHg)",
       pch = c(1, 19),
       strip = strip.custom(bg = "grey95"),
       scales = list(alternating = 1, tck = c(1, 0)))
```

Table 9.1: Pearson correlations between the cumulative death hazard $H_0(T)$, survival time T, $\log(T)$, systolic and diastolic blood pressure.

	$H_0(T)$	T	$\log(T)$	SBP	DBP
$H_0(T)$	1.000	0.997	0.830	0.169	0.137
T	0.997	1.000	0.862	0.176	0.141
$\log(T)$	0.830	0.862	1.000	0.205	0.151
SBP	0.169	0.176	0.205	1.000	0.592
DBP	0.137	0.141	0.151	0.592	1.000

9.1.8 A further improvement: Survival as predictor variable

If the complete-data model is a survival model, incorporating the cumulative hazard to the survival time, $H_0(T)$, as one of the predictors provide slightly better imputations (White and Royston, 2009). In addition, the event indicator should be included into the model. The Nelson-Aalen estimate of $H_0(T)$ in the Leiden 85+ Cohort can be calculated as

```
dat <- cbind(data2, dead = 1 - data2$dwa)
hazard <- nelsonaalen(dat, survda, dead)
```

where `dead` is coded such that "1" means death. The `nelsonaalen()` function is part of `mice`. Table 9.1 lists the correlations beween several key variables. The correlation between $H_0(T)$ and T is almost equal to 1, so for these data it matters little whether we take $H_0(T)$ or T as the predictor. The high correlation may be caused by the fact that nearly everyone in this cohort has died, so the percentage of censoring is low. The correlation between $H_0(T)$ and T could be lower in other epidemiological studies, and thus it might matter whether we take $H_0(T)$ or T. Observe that the correlation between $\log(T)$ and blood pressure is higher than for $H_0(T)$ or T, so it makes sense to add $\log(T)$ as an additional predictor. This strong relation may have been a consequence of the design, as the frail people were measured first.

9.1.9 Some guidance

Imputing data with many columns is challenging. Even the most carefully designed and well-maintained data may contain information or errors that can send the imputations awry. I conclude this section by summarizing advice for imputation of data with "too many columns."

1. Inspect all complete variables for forgotten missing data marks. Repair or remove these variables. Even one forgotten mark may ruin the imputation model. Remove outliers with improbable values.

2. Obtain insight into the strong and weak parts of the data by studying the influx-outflux pattern. Unless they are scientifically important, remove variables with low outflux, or with high fractions of missing data.

3. Perform a dry run with `maxit=0` and inspect the logged events produced by `mice()`. Remove any constant and collinear variables before imputation.

4. Find out what will happen after the data have been imputed. Determine a set of variables that are important in subsequent analyses, and include these as predictors in all models. Transform variables to improve predictability and coherence in the complete-data model.

5. Run `quickpred()`, and determine values of `mincor` and `minpuc` such that the average number of predictors is around 25.

6. After imputation, determine whether the generated imputations are sensible by comparing them to the observed information, and to knowledge external to the data. Revise the model where needed.

7. Document your actions and decisions, and obtain feedback from the owner of the data.

It is most helpful to try out these techniques on data gathered within your own institute. Some of these steps may not be relevant for other data. Determine where you need to adapt the procedure to suit your needs.

9.2 Sensitivity analysis

The imputations created in Section 9.1 are based on the assumption that the data are MAR (cf. Sections 1.2 and 2.2.4). While this is often a good starting assumption, it may not be realistic for the data at hand. When the data are not MAR, we can follow two strategies to obtain plausible imputations. The first strategy is to make the data "more MAR." In particular, this strategy requires us to identify additional information that explains differences in the probability to be missing. This information is then used to generate imputations conditional on that information. The second strategy is to perform a sensitivity analysis. The goal of the sensitivity analysis is to explore the result of the analysis under alternative scenarios for the missing data. See Section 6.2 for a more elaborate discussion of these strategies.

This section explores sensitivity analysis for the Leiden 85+ Cohort data. In sensitivity analysis, imputations are generated according to one or more scenarios. The number of possible scenarios is infinite, but these are not equally likely. A scenario could be very simple, like assuming that everyone with a missing value had scored a "yes," or assuming that those with missing blood pressures have the minimum possible value. While easy to interpret, such extreme scenarios are highly unlikely. Preferably, we should attempt to make an educated guess about both the direction and the magnitude of the missing

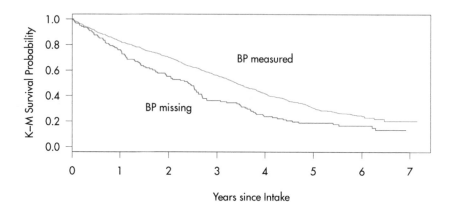

data had they been observed. By definition, this guess needs to be based on external information beyond the data.

9.2.1   Causes and consequences of missing data

We continue with the Leiden 85+ Cohort data described in Section 9.1. The objective is to estimate the effect of blood pressure (BP) on mortality. BP was not measured for 126 individuals (121 systolic, 126 diastolic).

The missingness is strongly related to survival. Figure 9.3 displays the Kaplan-Meier survival curves for those with ($n = 835$) and without ($n = 121$) a measurement of systolic BP (SBP). BP measurement was missing for a variety of reasons. Sometimes there was a time constraint. In other cases the investigator did not want to place an additional burden on the respondent. Some subjects were too ill to be measured.

Table 9.2 indicates that BP was measured less frequently for very old persons and for persons with health problems. Also, BP was measured more often if the BP was too high, for example if the respondent indicated a previous diagnosis of hypertension, or if the respondent used any medication against hypertension. The missing data rate of BP also varied during the period of data collection. The rate gradually increases during the first seven months of the sampling period from 5 to 40 percent of the cases, and then suddenly drops to a fairly constant level of 10–15 percent. A complicating factor here is

Table 9.2: Some variables that have different distributions in the response ($n = 835$) and nonresponse groups ($n = 121$). Shown are rounded percentages. Significance levels correspond to the χ^2-test.

Variable		Observed BP	Missing BP
Age (year)	$p < 0.0001$		
85–89		63	48
90–94		32	34
95+		6	18
Type of residence	$p < 0.0001$		
Independent		52	35
Home for elderly		35	54
Nursing home		13	12
Activities of daily living (ADL)	$p < 0.001$		
Independent		73	54
Dependent on help		27	46
History of hypertension	$p = 0.06$		
No		77	85
Yes		23	15

that the sequence in which the respondents were interviewed was not random. High-risk groups, that is, elderly in hospitals and nursing homes and those over 95, were visited first.

Table 9.3 contains the proportion of persons for which BP was not measured, cross-classified by three-year survival and history of hypertension as measured during anamnesis. Of all persons who die within three years and that have no history of hypertension, more than 19% have no BP score. The rate for other categories is about 9%. This suggests that a relatively large group of individuals without hypertension and with high mortality risk is missing from the sample for which BP is known.

Table 9.3: Proportion of persons for which no BP was measured, cross-classified by three-year survival and previous hypertension history. Shown are proportions per cell (number of cases with missing BP/total cell count).

Survived	History of Previous Hypertension	
	No	Yes
Yes	8.7% (34/390)	8.1% (10/124)
No	19.2% (69/360)	9.8% (8/82)

Using only the complete cases could lead to confounding by selection. The complete-case analysis might underestimate the mortality of the lower and normal BP groups, thereby yielding a distorted impression of the influence of BP on survival. This reasoning is somewhat tentative as it relies on the use of hypertension history as a proxy for BP. If true, however, we would expect more missing data from the lower BP measures. It is known that BP and mortality are inversely related in this age group, that is, lower BP is associated with higher mortality. If there are more missing data for those with low BP and high mortality (as in Table 9.3), selection of the complete cases could blur the effect of BP on mortality.

9.2.2   Scenarios

The previous section presented evidence that there might be more missing data for the lower blood pressures. Imputing the data under MAR can only account for nonresponse that is related to the observed data. However, the missing data may also be caused by factors that have not been observed. In order to study the influence of such factors on the final inferences, let us conduct a sensitivity analysis.

Section 3.8 advocated the use of simple adjustments to the imputed data as a way to perform sensitivity analysis. Table 3.6 lists possible values for an offset δ, together with an interpretation whether the value would be (too) small or (too) large. The next section uses the following range for δ: 0 mmHg (MCAR, too small), -5 mmHg (small), -10 mmHg (large), -15 mmHg (extreme) and -20 mmHg (too extreme). The last value is unrealistically low, and is primarily included to study the stability of the analysis in the extreme.

9.2.3   Generating imputations under the δ-adjustment

Subtracting a fixed amount from the imputed values is easily achieved by the `post` processing facility in `mice()`. The following code first imputes under $\delta = 0$ mmHg (MAR), then under $\delta = -5$ mmHg, and so on.

```
delta <- c(0, -5, -10, -15, -20)
post <- imp.qp$post
imp.all.undamped <- vector("list", length(delta))

for (i in 1:length(delta)) {
  d <- delta[i]
  cmd <- paste("imp[[j]][,i] <- imp[[j]][,i] +", d)
  post["rrsyst"] <- cmd
  imp <- mice(data2, pred = pred, post = post, maxit = 10,
              seed = i * 22)
  imp.all.undamped[[i]] <- imp
}
```

Table 9.4: Realized difference in means of the observed and imputed SBP (mmHg) data under various δ-adjustments. The number of multiple imputations is $m = 5$.

δ	Difference
0	-8.2
-5	-12.3
-10	-20.7
-15	-26.1
-20	-31.5

Note that we specify an adjustment in SBP only. Since imputed SBP is used to impute other incomplete variables, δ will also affect the imputations in those. The strength of the effect depends on the correlation between SBP and the variable. Thus, using a δ-adjustment for just one variable will affect many.

The mean of the observed systolic blood pressures is equal to 152.9 mmHg. Table 9.4 provides the differences in means between the imputed and observed data as a function of δ. For $\delta = 0$, i.e., under MAR, we find that the imputations are on average 8.2 mmHg lower than the observed blood pressure, which is in line with the expectations. As intended, the gap between observed and imputed increases as δ decreases.

Note that for $\delta = -10$ mmHg, the magnitude of the difference with the MAR case $(-20.7 + 8.2 = -12.5$ mmHg$)$ is somewhat larger in size than δ. The same holds for $\delta = -15$ mmHg and $\delta = -20$ mmHg. This is due to feedback of the δ-adjustment itself via third variables. It is possible to correct for this, for example by multiplying δ by a damping factor $\sqrt{1 - r^2}$, with r^2 the proportion of explained variance of the imputation model for SBP. In R this can be done by changing the expression for cmd as

```
cmd <- paste("fit <- lm(y ~ as.matrix(x));
             damp <- sqrt(1 - summary(fit)$r.squared);
             imp[[j]][, i] <- imp[[j]][, i] + damp * ", d)
```

As the estimates of the complete-data model turned out to be very similar to the "raw" δ, this route is not further explored.

9.2.4 Complete-data model

The complete-data model is a Cox regression with survival since intake as the outcome, and with blood pressure groups as the main explanatory variable. The analysis is stratified by sex and age group. The preliminary data transformations needed for this analysis were performed as follows:

δ	<125 mmHg		125–140 mmHg		>200 mmHg	
0	1.76	(1.36–2.28)	1.43	(1.16–1.77)	0.86	(0.44–1.67)
-5	1.81	(1.42–2.30)	1.45	(1.18–1.79)	0.88	(0.50–1.55)
-10	1.89	(1.47–2.44)	1.50	(1.21–1.86)	0.90	(0.51–1.59)
-15	1.82	(1.39–2.40)	1.45	(1.14–1.83)	0.88	(0.49–1.57)
-20	1.80	(1.39–2.35)	1.46	(1.17–1.83)	0.85	(0.48–1.50)
CCA	1.76	(1.36–2.28)	1.48	(1.19–1.84)	0.89	(0.51–1.57)

```
cda <- expression(
    sbpgp <- cut(rrsyst, breaks = c(50, 124, 144, 164, 184, 200,
                                    500)),
    agegp <- cut(lftanam, breaks = c(85, 90, 95, 110)),
    dead  <- 1 - dwa,
    coxph(Surv(survda, dead)
            ~ C(sbpgp, contr.treatment(6, base = 3))
          + strata(sexe, agegp)))
imp <- imp.all.damped[[1]]
fit <- with(imp, cda)
```

The cda object is an expression vector containing several statements needed for the complete-data model. The cda object will be evaluated within the environment of the imputed data, so (imputed) variables like rrsyst and survda are available during execution. Derived variables like sbpgp and agegp are temporary and disappear automatically. When evaluated, the expression vector returns the value of the last expression, in this case the object produced by coxph(). The expression vector provides a flexible way to apply R code to the imputed data. Do not forget to include commas to separate the individual expressions. The pooled hazard ratio per SBP group can be calculated by

```
as.vector(exp(summary(pool(fit))[, 1]))
```

```
[1] 1.758 1.433 1.065 1.108 0.861
```

Table 9.5 provides the hazard ratio estimates under the different scenarios for three SBP groups. A risk ratio of 1.76 means that the mortality risk (after correction for sex and age) in the group "<125 mmHg" is 1.76 times the risk of the reference group "145–160 mmHg." The inverse relation relation between mortality and blood pressure in this age group is consistent, where even the group with the highest blood pressures have (nonsignificant) lower risks.

Though the imputations differ dramatically under the various scenarios, the hazard ratio estimates for different δ are close. Thus, the results are essentially the same under all specified MNAR mechanisms. Also observe that the results are close to those from the analysis of the complete cases.

9.2.5 Conclusion

Sensitivity analysis is an important tool for investigating the plausibility of the MAR assumption. This section explored the use of an informal, simple and direct method to create imputations under nonignorable models by simply deducting some amount from the imputations.

Section 3.8.1 discussed shift, scale and shape parameters for nonignorable models. We only used a shift parameter here, which suited our purposes in the light of what we knew about the causes of the missing data. In other applications, scale or shape parameters could be more natural. The calculations are easily adapted to such cases.

9.3 Correct prevalence estimates from self-reported data

9.3.1 Description of the problem

Prevalence estimates for overweight and obesity are preferably based on standardized measured data of height and weight. However, obtaining such measures is logistically challenging and costly. An alternative is to ask persons to report their own height and weight. It is well known that such measures are subject to systematic biases. People tend to overestimate their height and underestimate their weight. A recent overview covering 64 studies can be found in Gorber et al. (2007).

Body Mass Index (BMI) is calculated from height and weight as kg/m^2. For BMI both biases operate in the same direction, so any self-reporting biases are amplified in BMI. Figure 9.4 is drawn from data of Krul et al. (2010). Self-reported BMI is on average $1-2\,kg/m^2$ lower than measured BMI.

BMI values can be categorized into underweight (BMI $<$ 18.5), normal ($18.5 \leq$ BMI $<$ 25), overweight ($25 \leq$ BMI $<$ 30), and obese (BMI $\geq$ 30). Self-reported BMI may assign subjects to a category that is too low. In Figure 9.4 persons in the white area labeled "1" are obese according to both self-reported and measured BMI. Persons in the white area labeled "3" are non-obese. The shaded areas represent disagreement between measured and self-reported obesity. The shaded area "4" are obese according to measured BMI, but not to self-report. The reverse holds for the shaded area "2." Due to self-reporting

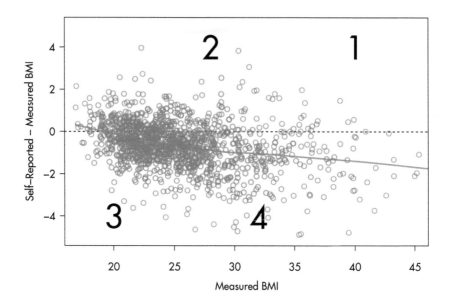

Figure 9.4: Underestimation of obesity prevalence in self-reported data. Self-reported BMI is on average 1–2 kg/m² too low. Lines are fitted by LOWESS.

bias, the number of persons located in area "4" is generally larger than in area "2," leading to underestimation.

There have been many attempts to correct measured height and weight for bias using predictive equations. These attempts have generally not been successful. The estimated prevalences were often still found to be too low after correction. Moreover, there is substantial heterogeneity in the proposed predictive formulae, resulting in widely varying prevalence estimates. See Visscher et al. (2006) for a summary of these issues. The current consensus is that it is not possible to estimate overweight and obesity prevalence from self-reported data. Dauphinot et al. (2008) even suggested to lower cut-off values for obesity based on self-reported data.

The goal is to estimate obesity prevalence in the population from self-reported data. This estimate should be unbiased in the sense that, on average, it should be equal to the estimate that would have been obtained had data been truly measured. Moreover, the estimate must be accompanied by a standard error or a confidence interval.

9.3.2   Don't count on predictions

Table 4 in Visscher et al. (2006) lists 36 predictive equations that have been proposed over the years. Visscher et al. (2006) observed that these equations predict too low. This section explains why this happens.

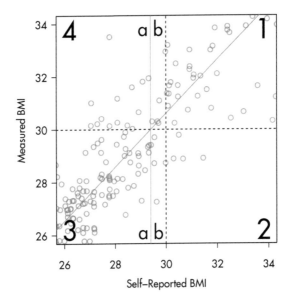

Figure 9.5: Illustration of the bias of predictive equations. In general, the combined region 2 + 3b will have fewer cases than region 4a. This causes a downward bias in the prevalence estimate.

Figure 9.5 plots the data of Figure 9.4 in a different way. The figure is centered around the BMI of $30 \, \text{kg/m}^2$. The two dashed lines divide the area into four quadrants. Quadrant 1 contains the cases that are obese according to both BMI values. Quadrant 3 contains the cases that are classified as non-obese according to both. Quadrant 2 holds the subjects that are classified as obese according to self-report, but not according to measured BMI. Quadrant 4 has the opposite interpretation. The area and quadrant numbers used in Figures 9.4 and 9.5 correspond to identical subdivisions in the data.

The "true obese" in Figure 9.5 lie in quadrants 1 and 4. The obese according to self-report are located in quadrants 1 and 2. Observe that the number of cases in quadrant 2 is smaller than in quadrant 4, a result of the systematic bias that is observed in humans. Using uncorrected self-report thus leads to an underestimate of the true prevalence.

The regression line that predicts measured BMI from self-reported BMI is added to the display. This line intersects the horizontal line that separates quadrant 3 from quadrant 4 at a (self-reported) BMI value of $29.4 \, \text{kg/m}^2$. Note that using the regression line to predict obese versus non-obese is in fact equivalent to classifying all cases with a self-report of $29.4 \, \text{kg/m}^2$ or higher as obese. Thus, the use of the regression line as a predictive equation effectively shifts the vertical dashed line from $30 \, \text{kg/m}^2$ to $29.4 \, \text{kg/m}^2$. Now we can make the same type of comparison as before. We count the number of cases in

Table 9.6: Basic variables needed to correct overweight/obesity prevalence for self-reporting.

Table 9.6: Basic variables needed to correct overweight/obesity prevalence for self-reporting.

Name	Description
age	Age (years)
sex	Sex (M/F)
hm	Height measured (cm)
hr	Height reported (cm)
wm	Weight measured (kg)
wr	Weight reported (kg)

Note: The survey data are representative for the population of interest, possibly after correction for design factors.

quadrant 2 + section 3b (n_1), and compare it to the count in region 4a (n_2). The difference $n_2 - n_1$ is now much smaller, thanks to the correction by the predictive equation.

However, there is still bias remaining. This comes from the fact that the distribution on the left side is more dense. The number of subjects with a BMI of $28\,\mathrm{kg/m^2}$ is typically larger than the number of subjects with a BMI of $32\,\mathrm{kg/m^2}$. Thus, even if a symmetric normal distribution around the regression line is correct, n_2 is on average larger than n_1. This yields bias in the predictive equation.

Observe that this effect will be stronger if the regression line becomes more shallow, or equivalently, if the spread around the regression line increases. Both are manifestation of less-than-perfect predictability. Thus, predictive equations only work well if the predictability is very high, but they are systematically biased in general.

9.3.3   The main idea

Table 9.6 lists the six variable names needed in this application. Let us assume that we have two data sources available:

- The *calibration dataset* contains n_c subjects for which both self-reported and measured data are available;

- The *survey dataset* contains n_s subjects with only the self-reported data.

We assume that the common variables in these two datasets are comparable.

The idea is to stack the datasets, multiply impute the missing values for hm and wm in the survey data and estimate the overweight and obesity prevalence (and their standard errors) from the imputed survey data. See Schenker et al. (2010) for more background.

9.3.4 Data

The calibration sample is taken from Krul et al. (2010). The dataset contains of $n_c = 1257$ Dutch subjects with both measured and self-reported data. The survey sample consists of $n_s = 803$ subjects of a representative sample of Dutch adults aged 18–75 years. These data were collected in November 2007 either online or using paper-and-pencil methods. The missing data pattern in the combined data is summarized as

```
data <- selfreport[, c("age", "sex", "hm", "hr", "wm", "wr")]
md.pattern(data, plot = FALSE)
```

```
     age sex hr wr  hm   wm
1257   1   1  1  1   1    1    0
803    1   1  1  1   0    0    2
       0   0  0  0 803  803 1606
```

The row containing all ones corresponds to the 1257 observations from the calibration sample with complete data, whereas the rows with a zero on `hm` and `wm` correspond to 803 observations from the survey sample (where `hm` and `wm` were not measured).

We apply predictive mean matching (cf. Section 3.4) to impute `hm` and `wm` in the 803 records from the survey data. The number of imputations $m = 10$. The complete-data estimates are calculated on each imputed dataset and combined using Rubin's pooling rules to obtain prevalence rates and the associated confidence intervals as in Sections 2.3.2 and 2.4.

9.3.5 Application

The `mice()` function can be used to create $m = 10$ multiply imputed datasets. We imputed measured height, measured weight and measured BMI using the following code:

```
bmi <- function(h, w) w / (h / 100)^2
meth <- make.method(selfreport)
meth[c("prg", "edu", "etn")] <- ""
meth["bm"] <- "~ bmi(hm, wm)"
pred <- make.predictorMatrix(selfreport)
pred[, c("src", "id", "pop", "prg", "edu", "etn",
         "web", "bm", "br")] <- 0
imp <- mice(selfreport, pred = pred, meth = meth, m = 10,
            seed = 66573, maxit = 20, print = FALSE)
```

The code defines a `bmi()` function for use in passive imputation to calculate `bmi`. The predictor matrix is set up so that only `age`, `sex`, `hr` and `wr` are permitted to impute `hm` and `wm`.

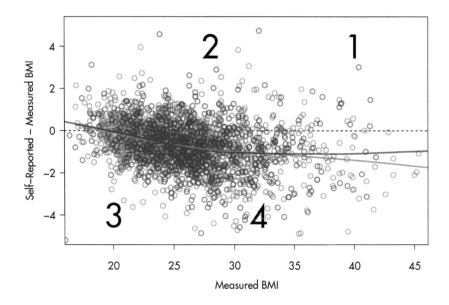

Figure 9.6: Relation between measured BMI and self-reported BMI in the calibration (blue) and survey (red) data in the first imputed dataset.

Figure 9.6 is a diagnostic plot to check whether the imputations maintain the relation between the measured and the self-reported data. The plot is identical to Figure 9.4, except that the imputed data from the survey data (in red) have been added. Imputations have been taken from the first imputed dataset. The figure shows that the red and blue dots are similar in terms of location and spread. Observe that BMI in the survey data is slightly higher. The very small difference between the smoothed lines across all measured BMI values confirms this notion. We conclude that the relation between self-reported and measured BMI as observed in the calibration data successfully "migrated" to the survey data.

Table 9.7 contains the prevalence estimates based on the survey data given for self-report and corrected for self-reporting bias. The estimates themselves are variable and have large standard errors. It is easy to infer that the size of the correction depends on age. Note that the standard errors of the corrected estimates are always larger than for the self-report. This reflects the information lost due to the correction. To obtain an equally precise estimate, the sample size of the study with only self-reports needs to be larger than the sample size of the study with direct measures.

Table 9.7: Obesity prevalence estimate (%) and standard error (se) in the survey data ($n = 803$), reported (observed data) and corrected (imputed).

Sex	Age	n	Reported %	se	Corrected %	se
Male	18–29	69	8.7	3.4	9.4	3.9
	30–39	73	11.0	3.7	15.7	5.0
	40–49	66	9.1	3.6	12.5	4.8
	50–59	91	20.9	4.3	25.4	5.2
	60–75	101	7.9	2.7	15.6	4.2
	18–75	400	11.7	1.6	16.0	2.0
Female	18–29	68	14.7	4.3	16.3	5.7
	30–39	69	26.1	5.3	28.4	6.6
	40–49	68	19.1	4.8	25.4	6.1
	50–59	81	25.9	4.9	32.8	6.0
	60–75	117	11.1	2.9	17.1	4.6
	18–75	403	18.6	1.9	23.0	2.4
M & F	18–75	803	15.2	1.3	19.5	1.5

9.3.6 Conclusion

Predictive equations to correct for self-reporting bias will only work if the percentage of explained variance is very high. In the general case, they have a systematic downward bias, which makes them unsuitable as correction methods. The remedy is to explicitly account for the residual distribution. We have done so by applying multiple imputation to impute measured height and weight. In addition, multiple imputation produces the correct standard errors of the prevalence estimates.

9.4 Enhancing comparability

9.4.1 Description of the problem

Comparability of data is a key problem in international comparisons and meta analysis. The problem of comparability has many sides. An overview of the issues and methodologies can be found in Van Deth (1998), Harkness et al. (2002), Salomon et al. (2004), King et al. (2004), Matsumoto and Van de Vijver (2010) and Chevalier and Fielding (2011).

This section addresses just one aspect, incomparability of the data obtained on survey items with different questions or response categories. This is a very common problem that hampers many comparisons.

One of the tasks of the European Commission is to provide insight into the level of disability of the populations in each of the 27 member states of the European Union. Many member states conduct health surveys, but the precise way in which disability is measured are very different. For example, The U.K. Health Survey contains a question *How far can you walk without stopping/experiencing severe discomfort, on your own, with aid if normally used?* with response categories "can't walk," "a few steps only," "more than a few steps but less than 200 yards" and "200 yards or more." The Dutch Health Interview Survey contains the question *Can you walk 400 metres without resting (with walking stick if necessary)?* with response categories "yes, no difficulty," "yes, with minor difficulty," "yes, with major difficulty" and "no." Both items obviously intend to measure the ability to walk, but it is far from clear how an answer on the U.K. item can be compared with one on the Dutch item.

Response conversion (Van Buuren et al., 2005) is a way to solve this problem. The technique transforms responses obtained on different questions onto a common scale. Where this can be done, comparisons can be made using the common scale. The actual data transformation can be repeatedly done on a routine basis as new information arrives. The construction of *conversion keys* is only possible if enough overlapping information can be identified. Keys have been constructed for dressing disability (Van Buuren et al., 2003), personal care disability, sensory functioning and communication, physical well-being (Van Buuren and Tennant, 2004), walking disability (Van Buuren et al., 2005) and physical activity (Hopman-Rock et al., 2012).

This section presents an extension based on multiple imputation. The approach is more flexible and more general than response conversion. Multiple imputation does not require a common items into each other, whereas response conversion scales the data on a common scale. Multiple imputation does not require a common unidimensional latent scale, thereby increasing the range of applications.

9.4.2 Full dependence: Simple equating

In principle, the comparability problem is easy to solve if all sources would collect the same data. In practice, setting up and maintaining a centralized, harmonized data collection is easier said than done. Moreover, even where such efforts are successful, comparability is certainly not guaranteed (Harkness et al., 2002). Many factors contribute to the incomparability of data, but we will not go into details here.

In the remainder, we take an example of two bureaus that each collect health data on its own population. The bureaus use survey items that are similar, but not the same. The survey used by bureau A contains an item for measuring walking disability (item A):

Are you able to walk outdoors on flat ground?

0: Without any difficulty
1: With some difficulty
2: With much difficulty
3: Unable to do

The frequencies observed in sample A are 242, 43, 15 and 0. There are six missing values. Bureau A produces a yearly report containing an estimate of the mean of the distribution of population A on item A. Assuming MCAR, a simple random sample and equal inter-category distances, we find $\hat{\theta}_{AA} = (242*0 + 43*1 + 15*2)/300 = 0.243$, the disability estimate for population A using the method of bureau A.

The survey of bureau B contains item B:

Can you, fully independently, walk outdoors (if necessary, with a cane)?

0: Yes, no difficulty
1: Yes, with some difficulty
2: Yes, with much difficulty
3: No, only with help from others

The frequencies observed in sample B are 145, 110, 29 and 8. There were no missing values reported by bureau B. Bureau B publishes the proportion of cases in category 0 as a yearly health measure. Assuming a simple random sample, $P(Y_B = 0)$ is estimated by $\hat{\theta}_{BB} = 145/292 = 0.497$, the health estimate for population B using the method of bureau B.

Note that $\hat{\theta}_{AA}$ and $\hat{\theta}_{BB}$ are different statistics calculated on different samples, and hence cannot be compared. On the surface, the problem is trivial and can be solved by just equating the four categories. After that is done, and we can apply the methods of bureau A or B, and compare the results. Such recoding to "make data comparable" is widely practiced.

Let us calculate the result using simple equating. To estimate walking disability in population B using the method of bureau A we obtain $\hat{\theta}_{BA} = (145*0 + 110*1 + 29*2 + 8*3)/292 = 0.658$. Remember that the mean disability estimate for population A was equal to 0.243, so population B appears to have substantially more walking disability. The difference equals $\hat{\theta}_{BA} - \hat{\theta}_{AA} = 0.658 - 0.243 = 0.414$ on a scale from 0 to 3.

Likewise, we may estimate bureau's B health measure θ_{AB} in population A as $\hat{\theta}_{AB} = 242/300 = 0.807$. Thus, over 80% of population A scores in category 0. This is substantially more than in population B, which was $\hat{\theta}_{BB} = 145/292 = 0.497$.

So by equating categories both bureaus conclude that the healthier population is A, and by a fairly large margin. As we will see, this result is however highly dependent on assumptions that may not be realistic for these data.

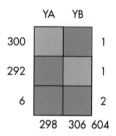

YA YB

Figure 9.7: Missing data pattern for walking data without a bridge study.

9.4.3 Independence: Imputation without a bridge study

Let Y_A be the item of bureau A, and let Y_B be the item of bureau B. The comparability problem can be seen as a missing data problem, where Y_A is missing for population B, and where Y_B is missing for population A. This formulation suggest that we can use imputation to solve the problem, and calculate $\hat{\theta}_{AB}$ and $\hat{\theta}_{BA}$ from the imputed data.

Let's see what happens if we put mice() to work to solve the problem. We first create the dataset:

```
fA <- c(242, 43, 15, 0, 6)
fB <- c(145, 110, 29, 8)
YA <- rep(ordered(c(0:3, NA)), fA)
YB <- rep(ordered(c(0:3)), fB)
Y <- rbind(data.frame(YA, YB = ordered(NA)),
           data.frame(YB, YA = ordered(NA)))
```

The data Y is a data frame with 604 rows and 2 columns: YA and YB. Figure 9.7 shows that the missing data pattern is unconnected (cf. Section 4.1.1), with no observations linking YA to YB. There are six records that contain no data at all.

For this problem, we monitor the behavior of a rank-order correlation, Kendall's τ, between YA and YB. This is not a standard facility in mice(), but we can easily write a small function micemill() that calculates Kendall's τ after each iteration as follows.

```
ra <- function(x, simplify = FALSE) {
  if (!is.mira(x)) return(NULL)
  ra <- x$analyses
  if (simplify) ra <- unlist(ra)
  return(ra)
}
```

```
micemill <- function(n) {
  for (i in 1:n) {
    imp <<- mice.mids(imp, print = FALSE)
    cors <- with(imp, cor(as.numeric(YA), as.numeric(YB),
                          method = "kendall"))
    tau <<- rbind(tau, ra(cors, simplify = TRUE))
  }
}
```

This function calls `mice.mids()` to perform just one iteration, calculates Kendall's τ, and stores the result. Note that the function contains two double assignment operators. This allows the function to overwrite the current `imp` and `tau` object in the global environment. This is a dangerous operation, and not really an example of good programming in general. However, we may now write

```
tau <- NULL
imp <- mice(Y, maxit = 0, m = 10, seed = 32662, print = FALSE)
micemill(50)
```

This code executes 50 iterations of the MICE algorithm. After any number of iterations, we may plot the trace lines of the MICE algorithm by

```
plotit <- function()
  matplot(x = 1:nrow(tau), y = tau,
          ylab = expression(paste("Kendall's ", tau)),
          xlab = "Iteration", type = "l", las = 1)
plotit()
```

Figure 9.8 contains the trace plot of 50 iterations. The traces start near zero, but then freely wander off over a substantial range of the correlation. In principle, the traces could hit values close to $+1$ or -1, but that is an extremely unlikely event. The MICE algorithm obviously does not know where to go, and wanders pointlessly through parameter space. The reason that this occurs is that the data contain no information about the relation between Y_A and Y_B.

Despite the absence of any information about the relation between Y_A and Y_B, we can calculate $\hat{\theta}_{AB}$ and $\hat{\theta}_{BA}$ without a problem from the imputed data. We find $\hat{\theta}_{AB} = 0.500$ (SD: 0.031), which is very close to $\hat{\theta}_{BB}$ (0.497), and far from the estimate under simple equating (0.807). Likewise, we find $\hat{\theta}_{BA} = 0.253$ (SD: 0.034), very close to $\hat{\theta}_{AA}$ (0.243) and far from the estimate under equating (0.658). Thus, if we perform the analysis without any information that links the items, we consistently find no difference between the estimates for populations A and B, despite the huge variation in Kendall's τ.

We have now two estimates of $\hat{\theta}_{AB}$ and $\hat{\theta}_{BA}$. In particular, in Section 9.4.2

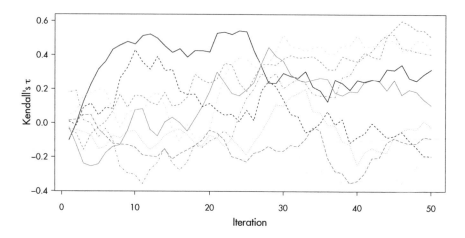

Figure 9.8: The trace plot of Kendall's τ for Y_A and Y_B using $m = 10$ multiple imputations and 50 iterations. The data contain no cases that have observations on both Y_A and Y_B.

Table 9.8: Contingency table of responses on Y_A and Y_B in an external sample E ($n = 292$).

		Y_B			
Y_A	0	1	2	3	Total
0	128	45	3	2	178
1	13	45	10	0	68
2	3	20	14	5	42
3	0	0	1	1	2
NA	1	0	1	0	2
Total	145	110	29	8	292

we calculated $\hat{\theta}_{BA} = 0.658$ and $\hat{\theta}_{AB} = 0.807$, whereas in the present section the results are $\hat{\theta}_{BA} = 0.253$ and $\hat{\theta}_{AB} = 0.500$, respectively. Thus, both health measures are very dissimilar due to the assumptions made. The question is which method yields results that are closer to the truth.

9.4.4　Fully dependent or independent?

Equating categories is equivalent to assuming that the pairs are 100% concordant. In that case Kendall's τ is equal to 1. Figure 9.8 illustrates that it is extremely unlikely that $\tau = 1$ will happen by chance. On the other hand, the two items look very similar, so Kendall's τ could be high on that basis. In order to make progress, we need to look at the data, and estimate τ.

Suppose that item Y_A and Y_A had both been administered to an external sample, called sample E. Table 9.8 contains the contingency table of Y_A and

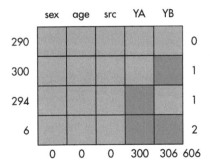

sex   age   src   YA   YB

290						0
300						1
294						1
6						2

0   0   0   300   306   606

Figure 9.9: Missing data pattern for walking data with a bridge study.

Y_B in sample E, taken from Van Buuren et al. (2005). Although there is a strong relation between Y_A and Y_B, the contingency table is far from diagonal. For example, category 1 of Y_B has 110 observations, whereas category 1 of Y_A contains only 68 persons. The table is also not symmetric, and suggests that Y_A is more difficult than Y_B. In other words, a given score on Y_A corresponds to more walking disability compare to the same score on Y_B. Kendall's τ is equal to 0.57, so about 57% of the pairs are concordant. This is far better than chance (0%), but also far worse than 100% concordance implied by simple equating. Thus even though the four response categories of Y_A and Y_B look similar, the information from sample E suggests that there are large and systematic differences in the way the items work. Given these data, the assumption of equal categories is in fact untenable. Likewise, the solution that assumes independence is also unlikely.

The implication is that both estimates of θ_{AB} and θ_{BA} presented thus far are doubtful. At this stage, we cannot yet tell which of the estimates is the better one.

9.4.5   Imputation using a bridge study

We will now rerun the imputation, but with sample E appended to the data from the sample for populations A and B. Sample E acts as a bridge study that connects the missing data patterns from samples A and B. The combined data are available in `mice` as the dataset `walking`. Figure 9.9 shows the missing data pattern of the combined data. Observe that YA and YB are now connected by 290 records from the bridge study on sample E. We assume that the data are missing at random. More specifically, the conditional distributions of Y_A and Y_B given the other item is equivalent across the three sources. Let S be an administrative variable taking on values A, B and E for the three sources.

The assumptions are

$$P(Y_A|Y_B, X, S = B) = P(Y_A|Y_B, X, S = E) \qquad (9.1)$$
$$P(Y_B|Y_A, X, S = A) = P(Y_B|Y_A, X, S = E) \qquad (9.2)$$

where X contains any relevant covariates, like age and sex, and/or interaction terms. In other words, the way in which Y_A depends on Y_B and X is the same in sources B and E. Likewise, the way in which Y_B depends on Y_A and X is the same in sources A and E. The inclusion of such covariates allows for various forms of differential item functioning (Holland and Wainer, 1993).

The two assumptions need critical evaluation. For example, if the respondents in source $S = E$ answered the items in a different language than the respondents in sources A or B, then the assumption may not be sensible unless one has great faith in the translation. It is perhaps better then to search for a bridge study that is more comparable.

Note that it is only required that the conditional distributions are identical. The imputations remain valid when the samples have different marginal distributions. For efficiency reasons and stability, it is generally advisable to have match samples with similar distribution, but it is not a requirement. The design is known as the *common-item nonequivalent groups* design (Kolen and Brennan, 1995) or the *non-equivalent group anchor test (NEAT)* design (Dorans, 2007).

Multiple imputation on the dataset `walking` is straightforward.

```
tau <- NULL
pred <- make.predictorMatrix(walking)
pred[, c("src", "age", "sex")] <- 0
imp <- mice(walking, maxit = 0, m = 10, pred = pred,
            seed = 92786, print = FALSE)
micemill(20)
```

The behavior of the trace plot is very different now (cf. Figure 9.10). After the first few iterations, the trace lines consistently move around a value of approximately 0.53, with a fairly small range. Thus, after five iterations, the conditional distributions defined by sample E have percolated into the imputations for item A (in sample B) and item B (in sample A).

The behavior of the samplers is dependent on the relative size of the bridge study. In these data, the bridge study is about one third of the total data. If the bridge study is small relative to the other two data sources, the sampler may be slow to converge. As a rule of the thumb, the bridge study should be at least 10% of the total sample size. Also, carefully monitor convergence of the most critical linkages using association measures.

Note that we can also monitor the behavior of $\hat{\theta}_{AB}$ and $\hat{\theta}_{BA}$. In order to calculate $\hat{\theta}_{AB}$ after each iteration we add two statements to the `micemill()` function:

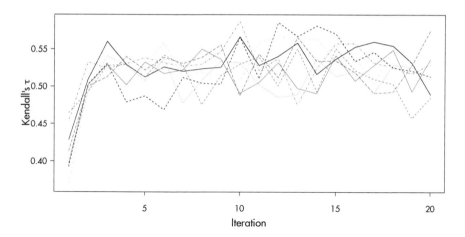

Figure 9.10: The trace plot of Kendall's τ for Y_A and Y_B using $m = 10$ multiple imputations and 20 iterations. The data are linked by the bridge study.

```
props <- with(imp, mean(YB[src == "A"] == '0'))
thetaAB <<- rbind(thetaAB, ra(props, simplify = TRUE))
```

The results are assembled in the variable **thetaAB** in the working directory. This variable should be initialized as **thetaAB <- NULL** before milling.

It is possible that the relation between Y_A and Y_B depends on covariates, like age and sex. If so, including covariates into the imputation model allows for differential item functioning across the covariates. It is perfectly possible to change the imputation model between iterations. For example, after the first 20 iterations (where we impute Y_A from Y_B and vice versa) we add age and sex as covariates, and do another 20 iterations. This goes as follows:

```
tau <- NULL
thetaAB <- NULL
pred2 <- pred1 <- make.predictorMatrix(walking)
pred1[, c("src", "age", "sex")] <- 0
pred2[, "src"] <- 0
imp <- mice(walking, maxit = 0, m = 10, pred = pred1,
            seed = 99786)
micemill(20)
imp <- mice(walking, maxit = 0, m = 10, pred = pred2)
micemill(20)
```

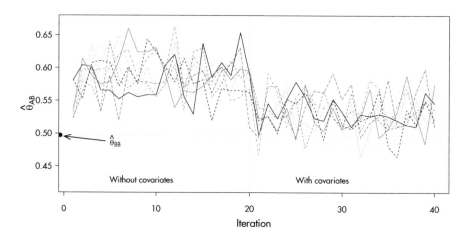

Figure 9.11: Trace plot of θ_{AB} (proportion of sample A that scores in category 0 of item B) after multiple imputation ($m = 10$), without covariates (iteration 1–20), and with covariates age and sex as part of the imputation model (iterations 21–40).

Table 9.9: Disability and health estimates for populations A and B under four assumptions. θ_{AA} and θ_{BA} are the item means on item A for samples A and B, respectively. θ_{AB} and θ_{BB} are the proportions of cases into category 0 of item B for samples A and B, respectively. MI=multiple imputation.

Assumption	$\hat{\theta}_{AA}$	$\hat{\theta}_{BA}$	$\hat{\theta}_{AB}$	$\hat{\theta}_{BB}$
Simple equating	0.243	0.658	0.807	0.497
Independence	0.243	0.253	0.500	0.497
MI (no covariate)	0.243	0.450	0.587	0.497
MI (covariate)	0.243	0.451	0.534	0.497

9.4.6   Interpretation

Figure 9.11 plots the traces of MICE algorithm, where we calculated θ_{AB}, the proportion of sample A in category 0 of item B. Without covariates, the proportion is approximately 0.587. Under equating, this proportion was found to be equal to 0.807 (cf. Section 9.4.2). The difference between the old (0.807) and the new (0.587) estimate is dramatic. After adding age and sex to the imputation model, θ_{AB} drops further to about 0.534, close to θ_{BB}, the estimate for population B (0.497).

Table 9.9 summarizes the estimates from the four analyses. Large differences are found between population A and B when we simply assume that the four categories of both items are identical (simple equating). In this case, population A appears much healthier by both measures. In constrast, if we assume independence between Y_A and Y_B, all differences vanish, so now it appears that the populations A and B are equally healthy. The solutions based on multiple imputation strike a balance between these extremes. Population A is considerably healthier than B on the item mean statistic (0.243 versus 0.451). However, the difference is much smaller on the proportion in category 0, especially after taking age and sex into account. The solutions based on multiple imputation are preferable over the first two because they have taken the relation between items A and B into account.

Which of the four estimates is best? The method of choice is multiple imputation including the covariates. This method not only accounts for the relation between Y_A and Y_B, but also incorporates the effects of age and sex. Consequently, the method provides estimates with the lowest bias in θ_{AB} and θ_{BA}.

9.4.7 Conclusion

Incomparability of data is a key problem in many fields. It is natural for scientists to adapt, refine and tweak measurement procedures in the hope of obtaining better data. Frequent changes, however, will hamper comparisons.

Equating categories is widely practiced to "make the data comparable." It is often not realized that recoding and equating data amplify differences. The degree of exaggeration is inversely related to Kendall's τ. For the item mean statistic, the difference in mean walking disability after equating is about twice the size of that under multiple imputation. Also, the estimate of 0.807 after simple equating is a gross overestimate. Overstated differences between populations may spur inappropriate interventions, sometimes with substantial financial consequences. Unless backed up by appropriate data, equating categories is not a solution.

The section used multiple imputation as a natural and attractive alternative. The first major application of multiple imputation addressed issues of comparability (Clogg et al., 1991). The advantage is that bureau A can interpret the information of bureau B using the scale of bureau A, and vice versa. The method provides possible contingency tables of items A and B that could have been observed if both had been measured.

Dorans (2007) describes techniques for creating valid equating tables. Such tables convert the score of instrument A into that of instrument B, and vice versa. The requirements for constructing such tables are extremely high: the measured constructs should be equal, the reliability should be equal, the conversion of B to A should be the inverse of that from B to A (symmetry), it should not matter whether A or B is measured and the table should be independent of the population. Holland (2007) presents a logical sequence of linking methods that progressively moves toward higher forms of equating. Multiple imputation in general fails on the symmetry requirement, as it produces m scores on B for one score of A, and thus cannot be invertible. The method as presented here can be seen as a first step toward obtaining formal equating of test items. It can be improved by correcting for the reliabilities of both items. This is an area of future research.

For simplicity, the statistical analyses used only one bridge item. In general, better strategies are possible. It is wise to include as many bridge items as there are. Also, linking and equating at the sub-scale and scale levels could be done (Dorans, 2007). The double-coded data could also comprise a series of vignettes (Salomon et al., 2004). The use of such strategies in combination with multiple imputation has yet to be explored.

9.5   Exercises

1. *Contingency table.* Adapt the `micemill()` function for the `walking` data so that it prints out the contingency table of Y_A and Y_B of the first imputation at each iteration. How many statements do you need?

2. *Pool* τ. Find out what the variance of Kendall's τ is, and construct its 95% confidence intervals under multiple imputation. Use the auxiliary function `pool.scalar()` for pooling.

3. *Covariates.* Calculate the correlation between age and the items A and B under two imputation models: one without covariates, and one with covariates. Which of the correlations is higher? Which solution do you prefer? Why?

4. *Heterogeneity.* Kendall's τ in the source E is 0.57 (cf. Section 9.4.4). The average of the sampler is slightly lower (Figure 9.10). Adapt the `micemill()` function to calculate the τ-values separately for the three sources. Which population has the lowest τ-values?

5. *Sample size.* Repeat the previous exercise, but with the samples for A and B taken 10 times as large. Does the sample size have an effect on convergence? If so, can you come up with an explanation? (Hint: Think of how τ is calculated.)

6. *True values.* For sample B, we do actually have the data on Item A from sample E. Calculate the "true" value θ_{BA}, and compare it with the simulated values. How do these values compare? Should these values be the same? If they are different, what could be the explanations? How could you reorganize the `walking` data so that no iteration is needed?

Chapter 10

Selection issues

There are known knowns. These are things we know that we know.
There are known unknowns. That is to say, there are things that
we know we don't know. But there are also unknown unknowns.
There are things we don't know we don't know.
Donald Rumsfeld

This chapter changes the perspective to the rows of the data matrix. An important consequence of nonresponse and missing data is that the remaining sample may not be representative any more. Multiple imputation of entire blocks of variables (Section 10.1) can be useful to adjust for selective loss of cases in panel and cohort studies. Section 10.2 takes this idea a step further by appended and imputing new synthetic records to the data. This can also work for cross-sectional studies.

10.1 Correcting for selective drop-out

Panel attrition is a problem that plagues all studies in which the same people are followed over time. People who leave the study are called *drop-outs*. The persons who drop out may be systematically different from those who remain, thus providing an opportunity for bias. This section assumes that the drop-out mechanism is MAR and that the parameters of the complete-data model and the response mechanism are distinct (cf. Section 2.2.5). Techniques for nonignorable drop-outs are described by Little (1995), Diggle et al. (2002), Daniels and Hogan (2008) and Wu (2010).

10.1.1 POPS study: 19 years follow-up

The Project on Preterm and Small for Gestational Age Infants (POPS) is an ongoing collaborative study in the Netherlands on the long-term effect of prematurity and dysmaturity on medical, psychological and social outcomes. The cohort was started in 1983 and enrolled 1338 infants with a gestational age below 32 weeks or with a birth weight of below 1500 grams (Verloove-Vanhorick et al., 1986). Of this cohort, 312 infants died in the first 28 days, and another 67 children died between the ages of 28 days and 19 years, leaving 959 survivors at the age of 19 years. Intermediate outcome measures from earlier follow-ups were available for 89% of the survivors at age 14 ($n = 854$), 77% at age 10 ($n = 712$), 84% at age 9($n = 813$), 96% at age 5 ($n = 927$) and 97% at age 2($n = 946$).

To study the effect of drop-out, Hille et al. (2005) divided the 959 survivors into three response groups:

1. *Full responders* were examined at an outpatient clinic and completed the questionnaires ($n = 596$);

2. *Postal responders* only completed the mailed questionnaires ($n = 109$);

3. *Nonresponders* did not respond to any of the mailed requests or telephone calls, or could not be traced ($n = 254$).

10.1.2   Characterization of the drop-out

Of the 254 nonresponders, 38 children (15%) did not comply because they were "physically or mentally unable to participate in the assessment." About half of the children (132, 52%) refused to participate. No reason for drop-out was known for 84 children (33%).

Table 10.1 lists some of the major differences between the three response groups. Compared to the postal and nonresponders, the full response group consists of more girls, contains more Dutch children, has higher educational and social economic levels and has fewer handicaps. Clearly, the responders form a highly selective subgroup in the total cohort.

Differential drop-out from the less healthy children leads to an obvious underestimate of disease prevalence. For example, the incidence of handicaps would be severely underestimated if based on data from the full responders only. In addition, selective drop-out could bias regression parameters in predictive models if the reason for drop-out is related to the outcome of interest. This may happen, for example, if we try to predict handicaps at the age of 19 years from the full responders only. Thus, statistical parameters may be difficult to interpret in the presence of selective drop-out.

10.1.3   Imputation model

The primary interest of the investigators focused on 14 different outcomes at 19 years: cognition, hearing, vision, neuromotor functioning, ADHD, respiratory symptoms, height, BMI, health status (Health Utilities Index Mark 3), perceived health (London Handicap Scale), coping, self-efficacy, educational attainment and occupational activities. Since it is inefficient to create a multiply imputed dataset for each outcome separately, the goal is to construct one set of imputed data that is used for all analyses.

Table 10.1: Count (percentage) of various factors for three response groups. Source: Hille et al. (2005).

	All responders		Full responders		Postal responders		Non-responders	
n	959		596		109		254	
Sex								
Boy	497	(51.8)	269	(45.1)	60	(55.0)	168	(66.1)
Girl	462	(48.2)	327	(54.9)	49	(45.0)	86	(33.9)
Origin								
Dutch	812	(84.7)	524	(87.9)	96	(88.1)	192	(75.6)
Non-Dutch	147	(15.3)	72	(12.1)	13	(11.9)	62	(24.4)
Maternal education								
Low	437	(49.9)	247	(43.0)	55	(52.9)	135	(68.2)
Medium	299	(34.1)	221	(38.5)	31	(29.8)	47	(23.7)
High	140	(16.0)	106	(18.5)	18	(17.3)	16	(8.1)
Social economic level								
Low	398	(42.2)	210	(35.5)	48	(44.4)	140	(58.8)
Medium	290	(30.9)	193	(32.6)	31	(28.7)	66	(27.7)
High	250	(26.7)	189	(31.9)	29	(26.9)	32	(13.4)
Handicap status at age 14 years								
Normal	480	(50.8)	308	(51.7)	42	(38.5)	130	(54.2)
Impairment	247	(26.1)	166	(27.9)	36	(33.0)	45	(18.8)
Mild	153	(16.2)	101	(16.9)	16	(14.7)	36	(15.0)
Severe	65	(6.9)	21	(3.5)	15	(13.8)	29	(12.1)

For each outcome, the investigator created a list of potentially relevant predictors according to the predictor selection strategy set forth in Section 6.3.2. In total, this resulted in a set of 85 unique variables. Only four of these were completely observed for all 959 children. Moreover, the information provided by the investigators was coded (in Microsoft Excel) as an 85×85 predictor matrix that is used to define the imputation model.

Figure 10.1 shows a miniature version of the predictor matrix. The dark cell indicates that the column variable is used to impute the row variable. Note the four complete variables with rows containing only zeroes. There are three blocks of variables. The first nine variables (Set 1: `geslacht–sga`) are potential confounders that should be controlled for in all analyses. The second set of variables (Set 2: `grad.t–sch910r`) are variables measured at intermediate time points that appear in specific models. The third set of variables (Set 3: `iq–occrec`) are the incomplete outcomes of primary interest collected at the age of 19 years. The imputation model is defined such that:

1. All variables in Set 1 are used as predictors to impute Set 1, to preserve relations between them;

2. All variables in Set 1 are used as predictors to impute Set 3, because all variables in Set 1 appear in the complete-data models of Set 3;

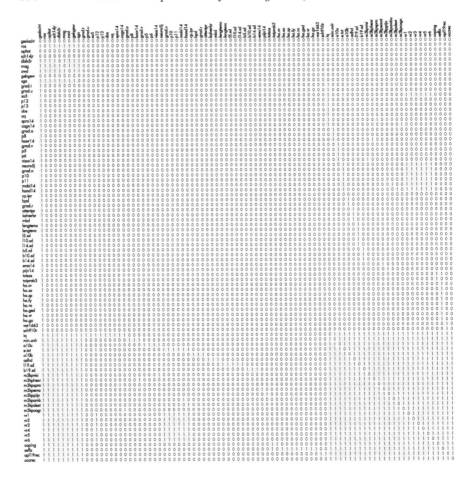

Figure 10.1: The 85 × 85 predictor matrix used in the POPS study. The gray parts signal the column variables that are used to impute the row variable.

3. All variables in Set 3 are used as predictors to impute Set 3, to preserve the relation between the variables measured at age 19;

4. Selected variables in Set 2 that appear in complete-data models are used as predictors to impute specific variables in Set 3;

5. Selected variables in Set 3 are "mirrored" to impute incomplete variables in Set 2, so as to maintain consistency between Set 2 and Set 3 variables;

6. The variable geslacht (sex) is included in all imputation models.

This setup of the predictor matrix avoids fitting unwieldy imputation models, while maintaining the relations of scientific interest.

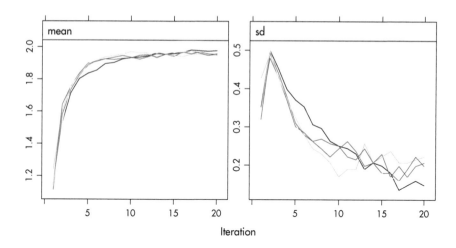

Figure 10.2: Trace lines of the MICE algorithm for the variable a10u illustrating problematic convergence.

10.1.4 A solution "that does not look good"

The actual imputations can be produced by

```
imp1 <- mice(data, pred = pred, maxit = 20,
             seed = 51121, print = FALSE)
```

The number of iterations is set to 20 because the trace lines from the MICE algorithm show strong initial trends and slow mixing.

Figure 10.2 plots the trace lines of the binary variable a10u, an indicator of visual disabilities. The behavior of these trace lines looks suspect, especially for a10u. The mean of a10u (left side) of the imputed values converges to a value near 1.9, while the standard deviation (right side) drops below that variability that is found in the data. Since the categories are coded as 1 = no problem and 2 = problem, a value of 1.9 actually implies that 90% of the nonresponders would have a problem. The observed prevalence of a10u in the full responders is only 1.5%, so 90% is clearly beyond any reasonable value.

In addition, iq and coping move into remote territories. Figure 10.3 illustrates that the imputed values for iq appear unreasonably low, whereas for coping they appear unreasonably high. What is happening here?

The phenomenon we see illustrates a weakness (or feature) of the MICE algorithm that manifests itself when the imputation model is overparametrized relative to the available information. The source of the problem lies in the imputation of the variables in Set 3, the measurements at 19 years. We specified that all variables in Set 3 should impute each other, with the idea of

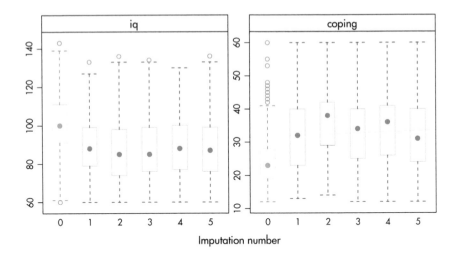

Figure 10.3: Distributions (observed and imputed) of IQ and coping score at 19 years in the POPS study for the imputation model in Figure 10.1.

preserving the multivariate relations between these variables. For 254 out of 959 children (26.5%), we do not have any information at age 19. The MICE algorithm starts out by borrowing information from the group of responders, and then quickly finds out that it can create imputations that are highly correlated. However, the imputed values do not look at all like the observed data, and are more like multivariate outliers that live in an extreme part of the data space.

There are several ways to alleviate the problem. The easiest approach is to remove the 254 nonresponders from the data. This is a sensible approach if the analysis focuses on determinants of 19-year outcomes, but it is not suitable for making inferences on the marginal outcome distribution of the entire cohort. A second approach is to simplify the model. Many of the 19-year outcomes are categorical, and we reduce the number of parameters drastically by applying predictive mean matching to these outcomes. A third approach would be to impute all outcomes as a block. This would find potential donors among the observed data, and impute all outcomes simultaneously. This removes the risk of artificially inflating the relations among outcomes, and is a promising alternative. Finally, the approach we follow here is to simplify the imputation model by removing the gray block in the lower-right part of Figure 10.1. The relation between the outcomes would then only be maintained through their relation with predictors measured at other time points. It is easy to change and rerun the model as:

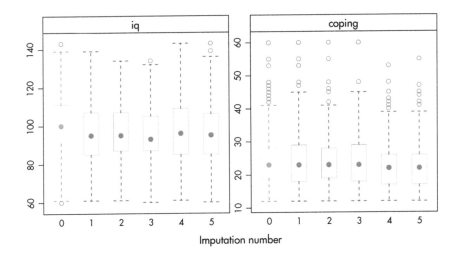

Imputation number

Figure 10.4: Distributions (observed and imputed) of IQ and coping score at 19 years in the POPS study for the simplified imputation model.

```
pred[61:86, 61:86] <- 0
imp2 <- mice(data, pred = pred, maxit = 20,
             seed = 51121, print = FALSE)
```

These statements produce imputations with marginal distributions much closer to the observed data. Also, the trace lines now show normal behavior (not shown). Convergence occurs rapidly in about 5–10 iterations.

Figure 10.4 displays the distributions of IQ and coping. The nonrespondents have slightly lower IQ scores, but not as extreme as in Figure 10.3. There are now hardly any differences in coping.

10.1.5   Results

Table 10.2 provides estimates of the percentage of three health problems, both uncorrected and corrected for selective drop-out. As expected, all estimates are adjusted upward. Note that the prevalence of visual problems tripled to 4.7% after correction. While this increase is substantial, it is well within the range of odds ratios of 2.6 and 4.4 reported by Hille et al. (2005). The adjustment shows that prevalence estimates in the whole group can be substantially higher than in the group of full responders. Hille et al. (2007) provide additional and more detailed results.

Table 10.2: Estimated percentage (95% CI) of three health problems at 19 years in the POPS study, uncorrected and corrected for selective drop-out.

	n_{obs}	Full responders		n	All children	
Severe visual handicap	690	1.4	(0.5–2.3)	959	4.7	(1.0–10.9)
Asthma, bronchitis, CARA	690	8.0	(5.9–10.0)	959	9.2	(6.9–11.2)
ADHD	666	4.7	(3.1–6.3)	959	5.7	(3.8–10.8)

10.1.6 Conclusion

Many studies are plagued by selective drop-out. Multiple imputation provides an intuitive way to adjust for drop-out, thus enabling estimation of statistics relative to the entire cohort rather than the subgroup. The method assumes MAR. The formulation of the imputation model requires some care. Section 10.1.3 outlines a simple strategy to specify the predictor matrix to fit an imputation model for multiple uses. This methodology is easily adapted to other studies.

Section 10.1.4 illustrates that multiple imputation is not without dangers. The imputations produced by the initial model were far off, which underlines the importance of diagnostic evaluation of the imputed data. A disadvantage of the approach taken to alleviate the problem is that it preserves the relations between the variables in Set 3 only insofar as they are related through their common predictors. These relations may thus be attenuated. Some alternatives were highlighted, and an especially promising one is to impute blocks of variables (cf. Section 4.7.2). Whatever is done, it is important to diagnose aberrant algorithmic behavior, and decide on an appropriate strategy to prevent it given the scientific questions at hand.

10.2 Correcting for nonresponse

This section describes how multiple imputation can be used to "make a sample representative." Weighting to known population totals is widely used to correct for nonresponse (Bethlehem, 2002; Särndal and Lundström, 2005). Imputation is an alternative to weighting. Imputation provides fine-grained control over the correction process. Provided that the imputation method is confidence proper, estimation of the correct standard errors can be done using Rubin's rules. Note however that this is not without controversy: Marker et al. (2002, p. 332) criticize multiple imputation as "difficult to apply," "to require massive amounts of computation," and question its performance for clustered datasets and unplanned analyses. Weighting and multiple imputation can also be combined, as was done in the NHANES III imputation project (Khare et al., 1993; Schafer et al., 1996).

This section demonstrates an application in the situation where the nonresponse is assumed to depend on known covariates, and where the distribution of covariates in the population is known. The sample is augmented by a set of artificial records, the outcomes in this set are multiply imputed and the whole set is analyzed. Though the application assumes random sampling, it should not be difficult to extend the basic ideas to more complex sampling designs.

10.2.1   Fifth Dutch Growth Study

The Fifth Dutch Growth Study is a cross-sectional nationwide study of height, weight and other anthropometric measurements among children 0–21 years living in the Netherlands (Schönbeck et al., 2013). The goal of the study is to provide updated growth charts that are representative for healthy children. The study is an update of similar studies performed in the Netherlands in 1955, 1965, 1980 and 1997. A strong secular trend in height has been observed over the last 150 years, making the Dutch population the tallest in the world (Fredriks et al., 2000a). The growth studies yield essential information needed to calibrate the growth charts for monitoring childhood growth and development. One of the parameters of interest is *final height*, the mean height of the population when fully grown around the age of 20 years.

The survey took place between May 2008 and October 2009. The sample was stratified into five regions: North (Groningen, Friesland, Drenthe), East (Overijssel, Gelderland, Flevoland), West (Noord-Holland, Zuid-Holland, Utrecht), South (Zeeland, Noord-Brabant, Limburg) and the four major cities (Amsterdam, Rotterdam, The Hague, Utrecht City). The way in which the children were sampled depended on age. Up to 8 years of age, measurements were performed during regular periodical health examinations. Children older than 9 years were sampled from the population register, and received a personal invitation from the local health care provider.

The total population was stratified into three ethnic subpopulations. Here we consider only the subpopulation of Dutch descent. This group consists of all children whose biological parents are born in the Netherlands. Children with growth-related diseases were excluded. The planned sample size for the Dutch subpopulation was equal to 14782.

10.2.2   Nonresponse

During data collection, it quickly became evident that the response in children older than 15 years was extremely poor, and sometimes fell even below 20%. Though substantial nonresponse was caused by lack of perceived interest by the children, we could not rule out the possibility of selective nonresponse. For example, overweight children may have been less inclined to participate. The data collection method was changed in November 2008 so that all children with a school class were measured. Once a class was selected, nonresponse of the pupils was very generally small. In addition, children were measured by

Table 10.3: Distribution of the population and the sample over five geographical regions by age. Numbers are column percentages. Source: Fifth Dutch Growth Study (Schönbeck et al., 2013).

| | 0–9 Years | | 10–13 Years | | 14–21 Years | |
Region	Population	Sample	Population	Sample	Population	Sample
North	12	7	12	11	12	4
East	24	28	24	11	24	55
South	23	27	24	31	25	21
West	21	26	20	26	20	15
City	20	12	19	22	19	4

special teams at two high schools, two universities and a youth festival. The sample was supplemented with data from two studies from Amsterdam and Zwolle.

10.2.3   Comparison to known population totals

The realized sample size was $n = 10030$ children aged 0–21 years (4829 boys, 5201 girls). The nonresponse and the changes in the design may have biased the sample. If the sample is to be representative for the Netherlands, then the distribution of measured covariates like age, sex, region or educational level should conform to known population totals. Such population totals are based on administrative sources and are available in STATLINE, the online publication system of Statistics Netherlands.

Table 10.3 compares the proportion of children within five geographical regions in the Netherlands per January 1, 2010, with the proportions in the sample. Geography is known to be related to height, with the 20-year-olds in the North being about 3 cm taller in the North (Fredriks et al., 2000a). There are three age groups. In the youngest children, the population and sample proportions are reasonably close in the East, South and West, but there are too few children from the North and the major cities. For children aged 10–13 years, there are too few children from the North and East. In the oldest children, the sample underrepresents the North and the major cities, and overrepresents the East.

10.2.4   Augmenting the sample

The idea is to augment the sample in such a way that it will be nationally representative, followed by multiple imputation of the outcomes of interest. Table 10.4 lists the number of the measured children. The table also reports the number of children needed to bring the sample close to the population distribution.

In total 1975 records are appended to the 10030 records of children who were measured. The appended data contain three complete covariates: region,

Table 10.4: Number of observed and imputed children in the sample by geographical regions and age. Source: Fifth Dutch Growth Study (Schönbeck et al., 2013).

Region	0–9 Years		10–13 Years		14–21 Years	
	n_{obs}	n_{imp}	n_{obs}	n_{imp}	n_{obs}	n_{imp}
North	389	400	200	75	143	200
East	1654	0	207	300	667	0
South	1591	0	573	0	767	0
West	1530	0	476	0	572	0
City	696	600	401	0	164	400
Total	5860	1000	1857	375	2313	600

sex and age in years. For example, for the combination (North, 0--9 years) $n_{imp} = 400$ new records are created as follows. All 400 records have the region category North. The first 200 records are boys and the last 200 records are girls. Age is drawn uniformly from the range 0–9 years. The outcomes of interest, like height and weight, are set to missing. Similar blocks of records are created for the other five categories of interest, resulting in a total of 1975 new records with complete covariates and missing outcomes.

The following R code creates a dataset of 1975 records, with four complete covariates (id, reg, sex, age) and four missing outcomes (hgt, wgt, hgt.z, wgt.z). The outcomes hgt.z and wgt.z are standard deviation scores (SDS), or Z-scores, derived from hgt and wgt, respectively, standardized for age and sex relative to the Dutch references (Fredriks et al., 2000a).

```
nimp <- c(400, 600, 75, 300, 200, 400)
regcat <- c("North", "City", "North", "East", "North", "City")
reg <- rep(regcat, nimp)
nimp2 <- floor(rep(nimp, each = 2)/2)
nimp2[5:6] <- c(38, 37)
sex <- rep(rep(c("boy", "girl"), 6), nimp2)
minage <- rep(c(0, 0, 10, 10, 14, 14), nimp)
maxage <- rep(c(10, 10, 14, 14, 21, 21), nimp)
set.seed(42444)
age <- runif(length(minage), minage, maxage)
id <- 600001:601975
data("fdgs", package = "mice")
pad <- data.frame(id, reg, age, sex,
                  hgt = NA, wgt = NA, hgt.z = NA, wgt.z = NA)
data <- rbind(fdgs, pad)
```

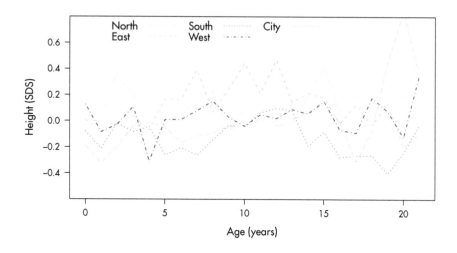

Figure 10.5: Height SDS by age and region of Dutch children. Source: Fifth Dutch Growth Study ($n = 10030$).

10.2.5　Imputation model

Regional differences in height are not constant across age, and tend to be more pronounced in older children. Figure 10.5 displays mean height standard deviation scores by age and region. Children from the North are generally the tallest, while those from the South are shortest, but the difference varies somewhat with age. Children from the major cities are short at early ages, but relatively tall in the oldest age groups. Imputation should preserve these features in the data, so we need to include at least the age by region interaction into the imputation model. In addition, we incorporate the interaction between SDS and age, so that the relation between height and weight could differ across age. The following specification uses the new `formulas` argument of `mice` to specify the interaction terms. This setup eliminates the need for passive imputation.

```
form <- list(hgt.z ~ reg + age + sex + wgt.z +
                I((age - 10) * wgt.z) + age * reg,
             wgt.z ~ reg + age + sex + hgt.z +
                I((age - 10) * hgt.z) + age * reg)
imp <- mice(data, meth = "norm", form = form, m = 10,
            maxit = 20, seed = 28107, print = FALSE)
```

Height SDS and weight SDS are is approximately normally distributed with a mean of zero and a standard deviation of 1, so we use the linear

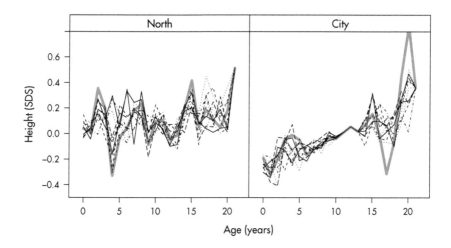

Figure 10.6: Mean height SDS by age for regions North and City, in the observed data ($n = 10030$) (blue) and 10 augmented datasets that correct for the nonresponse ($n = 12005$).

normal model method `norm` rather the `pmm`. If necessary, absolute values in centimeters (cm) and kilograms (kg) can be calculated after imputation.

Figure 10.6 displays mean height SDS per year for regions `North` and `City` in the original and augmented data. The 10 imputed datasets show patterns in mean height SDS similar to those in the observed data. Because of the lower sample size, the means for region `North` are more variable than `City`. Observe also that the rising pattern in `City` is reproduced in the imputed data. No imputations were generated for the ages 10–13 years, which explains that the means of the imputed and observed data coincide. The imputations tend to smooth out sharp peaks at higher ages due to the low number of data points.

10.2.6   Influence of nonresponse on final height

Figure 10.7 displays the mean of fitted height distribution of the original and the 10 imputed datasets. Since children from the shorter population in the South are overrepresented, the estimates of final height from the sample (183.6 cm for boys, 170.6 cm for girls) are biased downward. The estimates calculated from the imputed data vary from 183.6 to 184.1 cm (boys) and 170.6 to 171.1 cm (girls). Thus, correcting for the nonresponse leads to final height estimates that are about 2 mm higher.

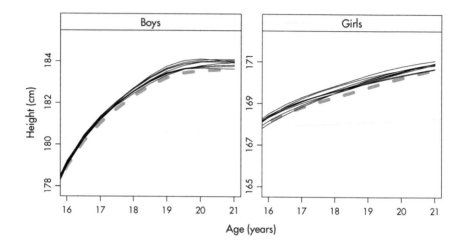

Figure 10.7: Final height estimates in Dutch boys and girls from the original sample ($n = 10030$) and 10 augmented samples ($n = 12005$) that correct for the nonresponse.

10.2.7 Discussion

The application as described here only imputes height and weight in Dutch children. It is straightforward to extend the method to impute additional outcomes, like waist or hip circumference.

The method can only correct for covariates whose distributions are known in both the sample and population. It does not work if nonresponse depends on factors for which we have no population distribution. However, if we have possession of nonresponse forms for a representative sample, we may use any covariates common to the responders and nonresponders to correct for the nonresponse using a similar methodology. The correction will be more successful if these covariates are related to the reasons for the nonresponse.

There are no accepted methods yet to calculate the number of extra records needed. Here we used 1975 new records to augment the existing 10030 records, about 16% of the total. This number of artificial records brought the covariate distribution in the augmented sample close to the population distribution without the need to discard any of the existing records. When the imbalance grows, we may need a higher percentage of augmentation. The estimates will then be based on a larger fraction of missing information, and may thus become unstable. Alternatively, we could sacrifice some of the existing records by taking a random subsample of strata that are overrepresented, but discarding data is likely to lead it less efficient estimates. It would be interesting to compare the methodology to traditional weighting approaches.

10.3 Exercises

1. *90th centile.* Repeat the analysis in Section 10.2.6 for final height. Study the effect of omitting the interaction effect from the imputation model. Are the effects on the 90th centile the same as for the mean?

2. *How many records?* Section 10.2.4 describes an application in which incomplete records are appended to create a representative sample. Develop a general strategy to determine the number of records needed to append.

Chapter 11

Longitudinal data

> *Failure of an imputation model does not damage the integrity of the entire dataset, but only the portion that is imputed.*
> Joseph L. Schafer

11.1 Long and wide format

Longitudinal data can be coded into "long" and "wide" formats. A wide dataset will have one record for each individual. The observations made at different time points are coded as different columns. In the wide format every measure that varies in time occupies a set of columns. In the long format there will be multiple records for each individual. Some variables that do not vary in time are identical in each record, whereas other variables vary across the records. The long format also needs a "time" variable that records the time in each record, and an "id" variable that groups the records from the same person.

A simple example of the wide format is

```
id age Y1 Y2
 1  14 28 22
 2  12 34 16
 3  ...
```

In the long format, this dataset looks like

```
id age  Y
 1  14 28
 1  14 22
 2  12 34
 2  12 16
 3  ...
```

Note that the concepts of long and wide are general, and also apply to cross-sectional data. For example, we have seen the long format before in

Section 5.1.3, where it referred to stacked imputed data that was produced by the `complete()` function. The basic idea is the same.

Both formats have their advantages. If the data are collected on the same time points, the wide format has no redundancy or repetition. Elementary statistical computations like calculating means, change scores, age-to-age correlations between time points, or the *t*-test are easy to do in this format. The long format is better at handling irregular and missed visits. Also, the long format has an explicit time variable available that can be used for analysis. Graphs and statistical analyses are easier in the long format.

Applied researchers often collect, store and analyze their data in the wide format. Classic ANOVA and MANOVA techniques for repeated measures and structural equation models for longitudinal data assume the wide format. Modern multilevel techniques and statistical graphs, however, work only from the long format. The distinction between the two formats is a first stumbling block for those new to longitudinal analysis.

Singer and Willett (2003) advise the data storing in both formats. The wide and the long formats can be easily converted in another by means of `gather()` and `spread()` functions on `tidyr` (Wickham and Grolemund, 2017). The wide-to-long conversion can usually be done without a problem. The long-to-wide conversion can be difficult. If individuals are seen at different times, direct conversion is impractical. The number of columns in the wide format becomes overly large, and each column contains many missing values. An ad hoc solution is to create homogeneous time groups, which then become the new columns in the wide format. Such regrouping will lead to loss of precision of the time variable. For some studies this need not be a problem, but for others it will.

Multiple imputation is somewhat more convenient in the wide format. Apart from the fact that the columns are ordered in time, there is nothing special about the imputation problem. We may thus apply techniques for single level data to longitudinal data. Section 11.2 discusses an imputation technique in the wide format in a clinical trial application with the goal of performing a statistical analysis according to the intention to treat (ITT) principle. The longitudinal character of the data helped specify the imputation model.

Multiple imputation of the longitudinal data in the long form can be done by multilevel imputation techniques. See Chapter 7 for an overview. Section 11.3 discusses multiple imputation in the long format. The application defines a common time raster for all persons. Multiple imputations are drawn for each raster point. The resulting imputed datasets can be converted to, and analyzed in, the wide format if desired. This approach is a more principled way to deal with the information loss problem discussed previously. The procedure aligns times to a common raster, hence the name *time raster imputation* (cf. Section 11.3).

11.2   SE Fireworks Disaster Study

On May 13, 2000, a catastrophic fireworks explosion occurred at SE Fireworks in Enschede, the Netherlands. The explosion killed 23 people and injured about 950. Around 500 houses were destroyed, leaving 1250 people homeless. Ten thousand residents were evacuated.

The disaster marked the starting point of a major operation to recover from the consequences of the explosion. Over the years, the neighborhood has been redesigned and rebuilt. Right after the disaster, the evacuees were relocated to improvised housing. Those in need received urgent medical care. A considerable number of residents showed signed of post-traumatic stress disorder (PTSD). This disorder is associated with flashback memories, avoidance of behaviors, places, or people that might lead to distressing memories, sleeping disorders and emotional numbing. When these symptoms persist and disrupt normal daily functioning, professional treatment is indicated.

Amidst the turmoil in the aftermath, Mediant, the disaster health aftercare center, embedded a randomized controlled trial comparing two treatments for anxiety-related disorders: eye movement desensitization and reprocessing (EMDR) (Shapiro, 2001) and cognitive behavioral therapy (CBT) (Stallard, 2006). CBT is the standard therapy. The data collection started within one year of the explosion, and lasted until the year 2004 (De Roos et al., 2011). The study included $n = 52$ children 4–18 years, as well as their parents. Children were randomized to EMDR or CBT by a flip of the coin. Each group contained 26 children.

The children received up to four individual sessions over a 4–8 week period, along with up to four parent sessions. Blind assessment took place pre-treatment (T1) and post-treatment (T2) and at 3 months follow-up (T3). The primary outcomes were the UCLA PTSD Reaction Index (PTSD-RI) (Steinberg et al., 2004), the Child Report of Post-traumatic Symptoms (CROPS) and the Parent Report of Post-traumatic Symptoms (PROPS) (Greenwald and Rubin, 1999). Treatment was stopped if children were asymptomatic according to participant and parent verbal report (both conditions), or if there was no remaining distress associated with the trauma memory, as indicated by a self-reported Subjective Units of Disturbance Scale (SUDS) of 0 (EMDR condition only).

The objective of the study was to answer the following questions:

- Is one of these treatments more effective in reducing PTSD symptoms at T2 and T3?

- Does the number of sessions needed to produce the therapeutic effect differ between the treatments?

Table 11.1: SE Fireworks Disaster Study. The UCLA PTSD Reaction Index of 52 subjects, children and parents, randomized to EMDR or CBT.

id	trt	pp	Y_1^c	Y_2^c	Y_3^c	Y_1^p	Y_2^p	Y_3^p	id	trt	pp	Y_1^c	Y_2^c	Y_3^c	Y_1^p	Y_2^p	Y_3^p
1	E	Y	–	–	–	36	35	38	32	E	N	28	17	8	40	42	33
2	C	N	45	–	–	–	–	–	33	E	N	–	–	–	38	22	25
3	E	N	–	–	–	13	19	13	34	E	N	–	–	–	17	–	–
4	C	Y	–	–	–	33	27	20	35	E	Y	50	20	–	19	1	5
5	E	Y	26	6	4	27	16	11	37	C	N	30	–	26	59	–	28
6	C	Y	8	1	2	32	15	13	38	C	Y	–	–	–	35	24	27
7	C	Y	41	26	31	–	39	39	39	E	N	–	–	–	–	–	–
8	C	N	–	–	–	24	13	35	40	E	Y	25	5	2	42	13	11
10	C	Y	35	27	14	48	23	–	41	E	Y	36	11	9	30	2	1
12	C	Y	28	15	13	45	33	36	43	E	N	17	–	–	–	–	–
13	E	Y	–	–	–	26	17	14	44	E	N	27	–	–	40	–	–
14	C	Y	33	8	9	37	7	3	45	C	Y	31	12	29	34	28	29
15	E	Y	43	–	7	25	27	1	46	C	Y	–	–	–	44	35	25
16	C	Y	50	8	35	39	21	34	47	C	Y	–	–	–	30	18	14
17	C	Y	31	21	10	32	21	19	48	E	Y	25	18	–	18	17	2
18	E	Y	30	17	16	47	28	34	49	C	N	24	23	16	44	29	34
19	E	Y	29	6	5	20	14	11	50	E	Y	31	13	9	34	18	13
20	E	Y	47	14	22	44	21	25	51	C	Y	–	–	–	52	13	13
21	C	Y	39	12	12	39	5	19	52	C	Y	30	35	28	–	44	50
23	C	Y	14	12	5	29	9	4	53	C	Y	19	33	21	36	21	21
24	E	N	27	–	–	–	–	–	54	C	N	43	–	–	48	–	–
25	E	Y	6	10	5	25	16	16	55	E	Y	64	42	35	44	31	16
28	C	Y	–	2	6	36	17	23	56	C	Y	–	–	–	37	6	9
29	E	Y	23	23	28	23	25	13	57	C	Y	31	12	–	32	26	–
30	E	Y	–	–	–	20	23	12	58	E	Y	–	–	–	49	28	25
31	C	N	15	24	26	33	36	38	59	E	Y	39	7	–	39	7	–

11.2.1　Intention to treat

Table 11.1 contains the outcome data of all subjects. The columns labeled Y_t^c contain the child data, and the columns labeled Y_t^p contain the parent data at time $t = (1, 2, 3)$. Children under the age of 6 years did not fill in the child form, so their scores are missing.

Of the 52 initial participants 14 children (8 EMDR, 6 CBT) did not follow the protocol. The majority (11) of this group did not receive the therapy, but still provided outcome measurements. The three others received therapy, but failed to provide outcome measures. The combined group is labeled as "dropout," while the other group is called the "completers" or "per-protocol" group. Figure 11.1 shows the missing data patterns for both groups.

The main reason given for dropping out was that the parents were overburdened (8). Other reasons for dropping out were: refusing to talk (1), language problems (1) and a new trauma rising to the forefront (2). One adolescent

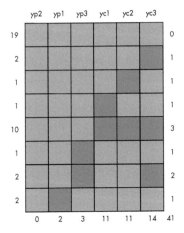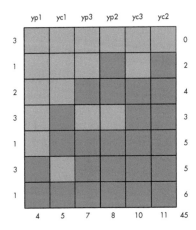

Figure 11.1: Missing data patterns for the "per-protocol" group (left) and the "drop-out" group (right).

refused treatment from a therapist not belonging to his own culture (1). One child showed spontaneous recovery before treatment started (1).

Comparison between the 14 drop-outs and the 38 completers regarding presentation at time of initial assessment yielded no significant differences in any of the demographic characteristics or number of traumatic experiences. On the symptom scales, only the mean score of the PROPS was marginally significantly higher for the drop-out group than for the treatment completers ($t = 2.09$, df $= 48$, $p = .04$).

Though these preliminary analyses are comforting, the best way to analyze the data is to the compare participants in the groups to which they were randomized, regardless of whether they received or adhered to the allocated intervention. Formal statistical testing requires random assignment to groups. The intention to treat (ITT) principle is widely recommended as the preferred approach to the analysis of clinical trials. DeMets et al. (2007) and White et al. (2011a) provide balanced discussions of pros and cons of ITT.

11.2.2 Imputation model

The major problem of the ITT principle is that some of the data that are needed are missing. Multiple imputation is a natural way to solve this problem, and thus to enable ITT analyses.

A difficulty in setting up the imputation model in the SE Fireworks Disaster Study is the large number of outcome variables relative to the number of cases. Even though the analysis of the data in Table 11.1 is already challenging, the real dataset is more complex than this. There are six additional outcome variables (e.g., the Child Behavior Checklist, or CBCL), each measured over time and similarly structured as in Table 11.1. In addition, some of the outcome measures are to be analyzed on both the subscale level and the total score level. For example, the PTSD-RI has three subscales (intrusiveness/numbing/avoidance, fear/anxiety, and disturbances in sleep and concentration and two additional summary measures (full PTSD and partial PTSD). All in all, there were 65 variables in data to be analyzed. Of these, 49 variables were incomplete. The total number of cases was 52, so in order to avoid grossly overdetermined models, the predictors of the imputation model should be selected very carefully.

A first strategy for predictor reduction was to preserve all deterministic relations columns in the incomplete data. This was done by passive imputation. For example, let $Y_{a,1}^p$, $Y_{b,1}^p$ and $Y_{c,1}^p$ represent the scores on three subscales of the PTSD parent form administered at T1. Each of these is imputed individually. The total variable Y_1^p is then imputed by mice in a deterministic way as the sum score.

A second strategy to reduce the number of predictors was to leave out other outcomes, measured at other time points. To illustrate this, a subset of the predictor matrix for imputing $Y_{a,1}^p$, $Y_{b,1}^p$ and $Y_{c,1}^p$ is:

```
vars <-c("ypa1", "ypb1", "ypc1",
         "ypa2", "ypb2", "ypc2",
         "ypa3", "ypb3", "ypc3")
fdd.pred[vars[1:3], vars]
```

	ypa1	ypb1	ypc1	ypa2	ypb2	ypc2	ypa3	ypb3	ypc3
ypa1	0	1	1	1	0	0	1	0	0
ypb1	1	0	1	0	1	0	0	1	0
ypc1	1	1	0	0	0	1	0	0	1

The conditional distribution $P(Y_{a,1}^p | Y_{b,1}^p, Y_{c,1}^p, Y_{a,2}^p, Y_{a,3}^p)$ leaves out the cross-lagged predictors $Y_{b,2}^p$, $Y_{c,2}^p$, $Y_{b,3}^p$ and $Y_{c,3}^p$. The assumption is the cross-lagged predictors are represented by through their non-cross-lagged predictors. Applying this idea consistently throughout the entire 65×65 predictor matrix brings vast reductions of the number of predictors. The largest number of predictors for any incomplete variable was 23, which still leaves degrees of freedom for residual variation.

Specifying a 65×65 predictor matrix by syntax in R is tedious and prone to error. I copied the variable names to Microsoft Excel, defined a square matrix of small cells containing zeroes, and used the menu option Conditional formatting... to define a cell color if the cell contains a "1." The option Freeze Panes was helpful for keeping variable names visible at all times. After filling in the matrix with the appropriate patterns of ones, I exported it to R to be used as argument to the mice() function. Excel is convenient for setting up large, patterned imputation models.

The imputations were generated as

```
meth <- make.method(fdd)
meth["yc1"] <- "~I(yca1 + ycb1 + ycc1)"
meth["yc2"] <- "~I(yca2 + ycb2 + ycc2)"
meth["yc3"] <- "~I(yca3 + ycb3 + ycc3)"
meth["yp1"] <- "~I(ypa1 + ypb1 + ypc1)"
meth["yp2"] <- "~I(ypa2 + ypb2 + ypc2)"
meth["yp3"] <- "~I(ypa3 + ypb3 + ypc3)"
imp <- mice(fdd, pred = fdd.pred, meth = meth, maxit = 20,
            seed = 54434, print = FALSE)
```

11.2.3   Inspecting imputations

For plotting purposes we need to convert the imputed data into long form. In R this can be done as follows:

```
lowi <- mice::complete(imp, "long", inc=TRUE)
lowi <- data.frame(lowi,cbcl2=NA, cbin2=NA,cbex2=NA)
lolo <- reshape(lowi, idvar = 'id',
                varying = 11:ncol(lowi),
                direction = "long",
                new.row.names = 1:(nrow(lowi)*3),
                sep="")
lolo <- lolo[order(lolo$.imp, lolo$id, lolo$time),]
row.names(lolo) <- 1:nrow(lolo)
```

This code executes two wide-to-long transformations in succession. The data are imputed in wide format. The call to `complete()` writes the $m + 1$ imputed stacked datasets to `lowi`, which stands for "long-wide." The `data.frame()` statement appends three columns to the data with missing CBCL scores, since the CBCL was not administered at time point 2. The `reshape()` statement interprets everything from column 11 onward as time-varying variables. As long as the variables are labeled consistently, `reshape()` will be smart enough to identify groups of columns that belong together, and stack them in the double-long format `lolo`. Finally, the result is sorted such that the original data with `lolo$.imp==0` are stored as the first block.

Figure 11.2 plots the profiles from 13 subjects with a missing score on Y_1^p, Y_2^p or Y_3^p in Table 11.1. Some profiles are partially imputed. Examples are subjects 7 (missing T1) and 37 (missing T2). Other profiles are missing entirely, and are thus completely imputed. Examples are subjects 2 and 43. Similar plots can be made for other outcomes. In general, the imputed profiles look similar to the completely observed profiles (not shown).

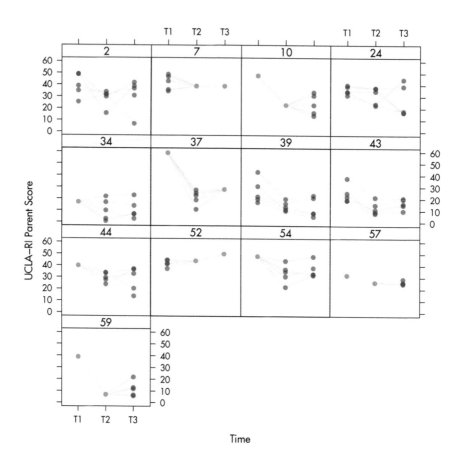

Figure 11.2: Plot of the multiply imputed data of the 13 subjects with one
or more missing values on PTSD-RI parent form.

11.2.4   Complete-data model

In the absence of missing data, we would have liked to perform a classical
repeated measures MANOVA as in Potthoff and Roy (1964). This method
construct derived variables that represent time as polynomial contrasts that
can be tested. An appealing feature of the method is that the covariances
among the repeated measures can take any form.

Let y_{ikt} denote the measurement of individual i ($i = 1, \ldots, n_k$) in group k
(CBT or EMDR) at time point t ($t = 1, \ldots, n_t$). In the SE Fireworks Disaster
Study data, we have $n_k = 26$ and $n_t = 3$. All subjects have been measures at
the same time points. The model represents the time trend in each group by

a linear and quadratic trend as

$$y_{ikt} = \beta_{k0} + t\beta_{k1} + t^2\beta_{k2} + e_{ikt} \qquad (11.1)$$

where the subject residual e_i has an arbitrary covariance 3×3 matrix Σ that is common to both groups. This model has six β parameters, three for each treatment group. To answer the first research question, we would be interested in testing the null hypotheses $\beta_{11} = \beta_{21}$ and $\beta_{12} = \beta_{22}$, i.e., whether the linear and quadratic trends are different between the treatment groups.

Potthoff and Roy (1964) showed how this model can be transformed into the usual MANOVA model and be fitted by standard software. Suppose that the repeated measures are collected in variables Y1, Y2 and Y3. In SPSS we can use the GLM command to test for the hypothesis of linear and quadratic time trends, and for the hypothesis that these trends are different between CBT and EMDR groups. Though application of the method is straightforward for complete data, it cannot be used directly for the SE Fireworks Disaster Study, because of the missing data.

The mids object created by mice() can be exported as a multiply imputed dataset to SPSS by means of the mids2spss() function. If the data came originally from SPSS it is also possible to merge the imputed data with the original data by means of the UPDATE command. SPSS will recognize an imported multiply imputed dataset, and execute the analysis m times in parallel. It can also provide the pooled statistics. Note that pooling requires a license to the Missing Values module.

Unfortunately, in SPSS 18.0 pooling is not implemented for GLM. As a solution, I stored the results by means of the OMS command in SPSS and shipped the output back to R for further analysis. I then applied a yet unpublished procedure for pooling F-tests to the datasets stored by the OMS command. In this way, pooling procedures that are not built into SPSS can be done with mice.

Of course, I could have saved myself the trouble of exporting the imputed data to SPSS and performed all analyses in R. That would, however, lock out the investigator from her own data. With the new pooling facilities investigators can now do their own data analysis on multiply imputed data. Some re-exporting is therefore worthwhile.

An alternative could have been to create the multiply imputed datasets within SPSS. This option was not possible for these data because the MULTIPLE IMPUTATION command in SPSS does not support predictor selection and passive imputation. With a bit of conversion between software packages, it is possible to have best of both worlds.

11.2.5 Results from the complete-data model

Figures 11.3 and 11.4 show the development of the mean level of PTSD complaint according to the PTSD-RI. All curves display a strong downward trend between start of treatment (T1) and end of treatment (T2), which is

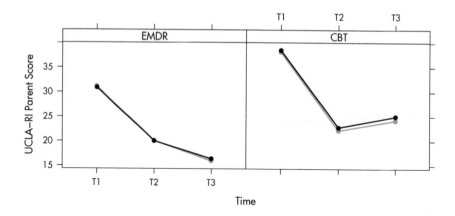

Figure 11.3: Mean levels of PTSD-RI parent form for the completely observed profiles (blue) and all profiles (black) in the EMDR and CBT groups.

presumably caused by the EMDR and CBT therapies. The shape between end of treatment (T2) and follow-up (T3) differs somewhat for the group, suggesting that EMDR has better long-term effects, but this difference was not statistically significant. Also note that the complete-case analysis and the analysis based on ITT are in close agreement with each other here.

We will not go into details here to answer the second research question as stated on page 313. It is of interest to note that EMDR needed fewer sessions to achieve its effect. The original publication (De Roos et al., 2011) contains the details.

11.3   Time raster imputation

Longitudinal analysis has become virtually synonymous with mixed effects modeling. Following the influential work of Laird and Ware (1982) and Jennrich and Schluchter (1986), this approach characterizes individual growth trajectories by a small number of random parameters. The differences between individuals are expressed in terms of these parameters.

In some applications, it is natural to consider *change scores*. Change scores are however rather awkward within the context of mixed effects models. This section introduces *time raster imputation*, a new method to generate imputations on a regular time raster from irregularly spaced longitudinal data. The imputed data can then be used to calculate change scores or age-to-age correlations, or apply quantitative techniques designed for repeated measures.

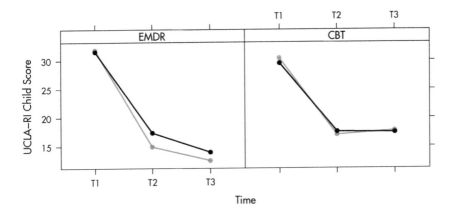

Figure 11.4: Mean levels of PTSD-RI Child Form for the completely observed profiles (blue) and all profiles (black) in the EMDR and CBT groups.

11.3.1   Change score

Let Y_1 and Y_2 represent repeated measurements of the same object at times T_1 and T_2 where $T_1 < T_2$. The difference $\Delta = Y_2 - Y_1$ is the most direct measure of change over time. Willett (1989, p. 588) characterized the change score as an "intuitive, unbiased, and computationally-simple measure of individual growth."

One would expect that modern books on longitudinal data would take the change score as their starting point. That is not the case. The change score is fully absent from most current books on longitudinal analysis. For example, there is no entry "change score" in the index of Verbeke and Molenberghs (2000), Diggle et al. (2002), Walls and Schafer (2006) or Fitzmaurice et al. (2009). Singer and Willett (2003, p. 10) do discuss the change score, but they quickly dismiss it on the basis that a study with only two time points cannot reveal the shape of a person's growth trajectory.

The change score, once the centerpiece of longitudinal analysis, has disappeared from the methodological literature. I find this is somewhat unfortunate as the parameters in the mixed effects model are more difficult to interpret than the change score. Moreover, classic statistical techniques, like the paired t-test or split-plot ANOVA, are built on the change score. There is a gap between modern mixed effects models and classical linear techniques for change scores and repeated measures data.

Calculating a mean change score is only sensible if different persons are measured at the same time points. When the data are observed at irregular times, there is no simple way to calculate change scores. Calculating change scores from the person parameters of the mixed effects model is technically trivial, but such scores are difficult to interpret. The person parameters are fitted values that have been smoothed. Deriving a change score as the difference between the fitted curve of the person at T_1 and T_2 results in values that are closer to zero than those derived from data that have been observed.

This section describes a technique that inserts pseudo time points to the observed data of each person. The outcome data at these supplementary time points are multiply imputed. The idea is that the imputed data can be analyzed subsequently by techniques for change scores and repeated measures.

The imputation procedure is akin to the process needed to print a photo in a newspaper. The photo is coded as points on a predefined raster. At the microlevel there could be information loss, but the scenery is essentially unaffected. Hence the name *time raster imputation*. My hope is that this method will help bridge the gap between modern and classic approaches to longitudinal data.

11.3.2   Scientific question: Critical periods

The research was motivated by the question: *At what ages do children become overweight?* Knowing the answer to this question may provide handles for preventive interventions to counter obesity.

Dietz (1994) suggested the existence of three *critical periods* for obesity at adult age: the prenatal period, the period of adiposity rebound (roughly around the age of 5–6 years), and adolescence. Obesity that begins at these periods is expected to increase the risk of persistent obesity and its complications. Overviews of studies on critical periods are given by Cameron and Demerath (2002) and Lloyd et al. (2010).

In the sequel, we use the body mass index (BMI) as a measure of overweight. BMI will be analyzed in standard deviation scores (SDS) using the relevant Dutch references (Fredriks et al., 2000a,b). Our criterion for being overweight in adulthood is defined as BMI SDS ≥ 1.3.

As an example, imagine an 18-year-old person with a BMI SDS equal to +1.5 SD. How did this person end up at 1.5 SD? If we have the data, we can plot the measurements against age, and study the individual track. The BMI SDS trajectory may provide key insights into development of overweight and obesity.

Figure 11.5 provides an overview of five theoretical BMI SDS trajectories that the person might have followed. These are:

1. *Long critical period.* A small but persistent centile crossing across the entire age range. In this case, everything (or nothing) is a critical period.

2. *No critical period.* The person is born with a BMI SDS of 1.5 SD and this remains unaltered throughout age.

3. *Short early.* There is a large increase between ages 2y and 5y. We would surely interpret the period 2y–5y is as a critical period for this person.

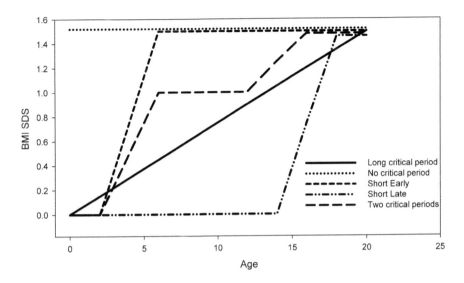

Figure 11.5: Five theoretical BMI SDS trajectories for a person age 18 years
with a BMI SDS = 1.5 SD.

4. *Short late.* This is essentially the same as before, but shifted forward in time.

5. *Two critical periods.* Here the total increase of 1.5 SD is spread over two periods. The first occurs at 2y–5y with an increase of 1.0 SD. The second at 12y–15y with an increase of 0.5 SD.

In practice, mixing between these and other forms will occur.

The objective is to identify any periods during childhood that contribute to an increase in overweight at adult age. A period is "critical" if

1. change differs between those who are and are not later overweight; and

2. change is associated with the outcome after correction for the measure at the end of the period.

Both need to hold. In order to solve the problem of irregular age spacing, De Kroon et al. (2010) use the *broken stick model*, a piecewise linear growth curve fitted, as a means to describe individual growth curves at fixed times.

This section extends this methodology by generating imputations according to the broken stick model. The multiply imputed values are then used to estimate difference scores and regression models that throw light on the question of scientific interest.

11.3.3   Broken stick model♠

In a sample of n persons $i = 1, \ldots, n$, we assume that there are n_i measurement occasions for person i. Let y_i represent the $n_i \times 1$ vector containing the SDS values obtained for person i. Let t_i represent the $n_i \times 1$ vector with the ages at which the measurements were made.

The broken stick model requires the user to specify an ordered set of k break ages, collected in the vector $\kappa = (\kappa_1, \ldots, \kappa_k)$. The set should cover the entire range of the measured ages, so $\kappa_1 \leq \min(t_i)$ and $\kappa_k \geq \max(t_i)$ for all i. It is convenient to set κ_1 and κ_k to rounded values just below and above the minimum and maximum ages in the data, respectively. De Kroon et al. (2010) specified nine break ages: birth (0d), 8 days (8d), 4 months (4m), 1 year (1y), 2 years (2y), 6 years (6y), 10 years (10y), 18 years (18y) and 29 years (29y).

Without loss of information, the time points t_i of person i are represented by a B-spline of degree 1, with knots specified by κ. More specifically, the vector t_i is recoded as the $n_i \times k$ design matrix $X_i = (x_{1i}, \ldots, x_{ki})$. We refer to Ruppert et al. (2003, p. 59) for further details. For the set of break ages we calculate the B-splines matrix in R by the bs() function from the splines package as follows:

```
library(splines)
data <- tbc

### specify break ages
brk <- c(0, 8/365, 1/3, 1, 2, 6, 10, 18, 29)
k <- length(brk)

### calculate B-spline
X <- bs(data$age,
        knots = brk,
        B = c(brk[1],brk[k]+0.0001),
        degree = 1)
X <- X[,-(k+1)]
dimnames(X)[[2]] <- paste("x",1:ncol(X),sep="")
data <- cbind(data,X)
round(head(X,3),2)
```

```
      x1    x2    x3 x4 x5 x6 x7 x8 x9
[1,] 1.00  0.00  0.00  0  0  0  0  0  0
[2,] 0.27  0.73  0.00  0  0  0  0  0  0
[3,] 0.00  0.83  0.17  0  0  0  0  0  0
```

Matrix X has only two nonzero elements in each row. Each row sums to 1. If an observed age coincides with a break age, the corresponding entry is equal to 1, and all remaining elements are zero. In the data example, this occurs in the first record, at birth. A small constant of 0.0001 was added to the last break age. This was done to accommodate for a pseudo time point with an exact age of 29 years, which will be inserted later in Section 11.3.6.

The measurements y_i for person i are modeled by the linear mixed effects model

$$
\begin{aligned}
y_i &= X_i(\beta + \beta_i) + \epsilon_i \qquad\qquad (11.2) \\
&= X_i \gamma_i + \epsilon_i
\end{aligned}
$$

where $\gamma_i = \beta + \beta_i$. The $k \times 1$ column vector β contains k fixed-effect coefficients common to all persons. The vector β_i contains k subject-specific random effect coefficients for person i. The vector ϵ_i contains n_i subject-specific residuals.

We make the usual assumption that $\gamma_i \sim N(\beta, \Omega)$, i.e., the random coefficients of the subjects have a multivariate normal distribution with global mean β and an unstructured covariance Ω. We also assume that the residuals are independently and normally distributed as $\epsilon_i \sim N(0, \sigma^2 I(n_i))$ where σ^2 is a common variance parameter. The covariances between β_i and e_i are assumed to be zero.

Since the rows of the B-spline basis all sum to 1, the intercept is implicit. In fact, one could interpret the model as a special form of the random intercept model, where the intercept is represented by a B-spline rather than by the usual column of ones.

The model prescribes that growth follows a straight line between the break ages. In this application, we are not so much interested in what happens *within* the age interval of each period. Rogosa and Willett (1985) contrasted the analysis of individual differences based on change scores with the analysis of individual differences based on multilevel parameters. They concluded that in general the analysis of change scores is inferior to the parameter approach. The exception is when growth is assumed to follow a straight line within the interval of interest. In that case, the change score approach and the mixed effects model are interchangeable (Rogosa and Willett, 1985, p. 225). The straight line assumption is often reasonable in epidemiological studies if the time interval is short (Hui and Berger, 1983). For extra detail, we could add an extra break age within the interval.

The function `lmer()` from the `lme4` package fits the model. Change scores can be calculated from the fixed and random effects as follows:

```
library(lme4)
fit <- lmer(wgt.z ~ 0 + x1 + x2 + x3 + x4 + x5 +
            x6 + x7 + x8 + x9 + (0 + x1 + x2 +
            x3 + x4 + x5 + x6 + x7 + x8 + x9 | id),
         data = data)

### calculate size and increment per person
tsiz <- t(ranef(fit)$id) + fixef(fit)
```

```
tinc <- diff(tsiz)
```

```
round(head(t(tsiz)), 2)
```

The $\hat{\gamma}_i$ estimates are found in the variable `tsiz`. Let $\hat{\delta}_{ik} = \hat{\gamma}_{i,j+1} - \hat{\gamma}_{i,j}$ with $j = 1, \ldots, k-1$ denote the successive differences (or increments) of the elements in $\hat{\gamma}_i$. These are found in the variable `tinc`. We may interpret $\hat{\delta}_i$ as the expected change scores for person i.

The first criterion for a critical period is that change differs between those who are and are not later overweight. A simple analysis for this criterion is the Student's t-test applied to $\hat{\delta}_{ik}$ for every period k. The correlations between $\hat{\delta}_{ik}$ at successive k were generally higher than 0.5, so we analyzed unconditional change scores (Jones and Spiegelhalter, 2009). The second criterion for a critical period involves fitting two regression models, both of which have final BMI SDS at adulthood, denoted by γ_i^{adult}, as their outcome. The two models are:

$$\gamma_i^{\text{adult}} = \hat{\gamma}_{i,j+1}\zeta_{j+1} + \epsilon_i \tag{11.3}$$
$$\gamma_i^{\text{adult}} = \hat{\gamma}_{i,j+1}\eta_{j+1} + \hat{\gamma}_j\eta_j + \varepsilon_i \tag{11.4}$$

which are fitted for $j = 1, \ldots, k-2$. The parameter of scientific interest is the added value of including η_j.

11.3.4 Terneuzen Birth Cohort

The Terneuzen Birth Cohort consists of all ($n = 2604$) newborns in Terneuzen, the Netherlands, between 1977 and 1986. The most recent measurements were made in the year 2005, so the data spans an age range of 0–29 years. Height and weight were measured throughout this age range. More details on the measurement procedures and the data can be found in De Kroon et al. (2008, 2010).

Suppose the model is fitted to weight SDS. The parameters γ_i can be interpreted as attained weight SDS relative to the reference population. This allows us to represent the observed trajectory of each child in a condensed way by k numbers. The values in $\hat{\gamma}_i$ are the set of most likely weight SDS values at each break age, given all true measurements we have of child i. This implies that if the child has very few measurements, the estimates will be close to the global mean. When taken together, the values $\hat{\gamma}_i$ form the broken stick.

Figure 11.6 displays Weight SDS against age for six selected individuals. Child 1259 has a fairly common pattern. This child starts off near the average, but then steadily declines, apart from a blip around 10 months. Child 2447 is fairly constant, but had a major valley near the age of 4 months, perhaps because of a temporary illness. Child 7019 is also typical. The pattern hovers around the mean. Observe that no data beyond 10 years are available for this child. Child 7460 experienced a substantial change in the height/weight

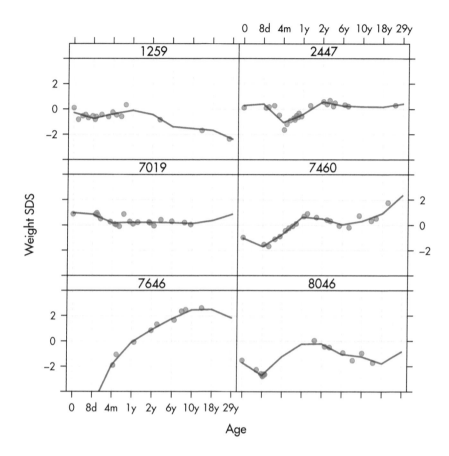

Figure 11.6: Broken stick trajectories for Weight SDS from six selected individuals from the Terneuzen cohort.

proportions during the first year. Child 7646 was born prematurely with a gestational age of 32 weeks. This individual has an unusually large increase in weight between birth and puberty. Child 8046 is aberrant with an unusually large number of weight measurements around the age of 8 days, but was subsequently not measured for about 1.5 years.

Figure 11.6 also displays the individual broken stick estimates for each outcome as a line. Observe that the model follows the individual data points very well. De Kroon et al. (2010) analyzed these estimates by the methods described at the end of Section 11.3.2, and found that the periods 2y–6y and 10y–18y were most relevant for developing later overweight.

11.3.5   Shrinkage and the change score♠

Thus far we have looked at the problem from a prediction perspective. This is a useful first step, but it does not address all aspects. The $\hat{\beta}_i$ estimate in the mixed effects model combines the person-specific ordinary least squares (OLS) estimate of β_i with the grand mean $\hat{\beta}$. The amount of shrinkage toward the grand mean depends on three factors: the number of data points n_i, the residual variance estimate $\hat{\sigma}^2$ around the fitted broken stick, and the variance estimate $\hat{\omega}_j^2$ for the jth random effect. If $n_i = \sigma^2/\hat{\omega}_j^2$ then $\hat{\beta}_i$ is halfway between $\hat{\beta}$ and the OLS estimate of β_i. If $n_i < \hat{\sigma}^2/\omega_j^2$ then $\hat{\beta}_i$ is closer to the global mean, while $n_i > \hat{\sigma}^2/\hat{\omega}_j^2$ implies that $\hat{\beta}_i$ is closer to the OLS-estimate. We refer to Gelman and Hill (2007, p. 394) for more details.

Shrinkage will stabilize the estimates of persons with few data points. Shrinkage also implies that the same $\hat{\gamma}_i = \hat{\beta} + \hat{\beta}_i$ can correspond to quite different data trajectories. Suppose profile A is an essentially flat and densely measured trajectory just above the mean. Profile B, on the other hand, is a sparse and highly variable trajectory far above the mean. Due to differential shrinkage, profiles A and B can have the same $\hat{\gamma}_i$ estimates. As a consequence, shrinkage will affect the change scores $\hat{\delta}_i$. For both profiles A and B the estimated change scores $\hat{\delta}_i$ are approximately zero at every period. For profile A this is reasonable since the profile itself is flat. In profile B we would expect to see substantial variation in $\hat{\delta}_i$ if the data had been truly measured. Yet, shrinkage has dampened $\hat{\gamma}_i$, and thus made $\hat{\delta}_i$ closer to zero than if calculated from observed data.

It is not quite known whether this effect is a problem in this application. It is likely that dampening of $\hat{\delta}_i$ will bias the result in the conservative direction, and hence primarily affects statistical power. The next section explores an alternative based on multiple imputation. The idea is to insert the break ages into the data, and impute the corresponding outcome data.

11.3.6   Imputation

The measured outcomes are denoted by Y_{obs}, e.g., weight SDS. For the moment, we assume that the Y_{obs} are coded in long format and complete, though neither is an essential requirement. For each person i we append k records, each of which corresponds to a break age. In R we first set up a time warping model that connect real age to warped age, and then integrate the new ages into the data.

```
  obs   sup  pred
32845 15705     0
```

The function `appendbreak()` is a custom function of about 20 lines in mice specific to the Terneuzen Birth Cohort data. It copies the first available record of the ith person k times, updates administrative and age variables, sets the outcome variables to `NA`, appends the result to the original data and sorts the result with respect to `id` and `age`. The real data are thus mingled with the supplementary records with missing outcomes. The first few records of `data2` look like this:

```
head(data2)
```

```
     id occ nocc first  typ   age sex hgt.z  wgt.z  bmi.z ao   x1
4     4   0      19  TRUE obs 0.000   1  0.33  0.195  0.666  0 1.00
42    4  NA      19 FALSE sup 0.000   1    NA     NA     NA  0 1.00
5     4   1      19 FALSE obs 0.016   1    NA -0.666     NA  0 0.27
4.1   4  NA      19 FALSE sup 0.022   1    NA     NA     NA  0 0.00
6     4   2      19 FALSE obs 0.076   1  0.71  0.020 -0.381  0 0.00
7     4   3      19 FALSE obs 0.104   1  0.18  0.073  0.075  0 0.00
       x2   x3 x4 x5 x6 x7 x8 x9 age2
4    0.00 0.00  0  0  0  0  0  0  0.0
42   0.00 0.00  0  0  0  0  0  0  0.0
5    0.73 0.00  0  0  0  0  0  0  2.6
4.1  1.00 0.00  0  0  0  0  0  0  3.6
6    0.83 0.17  0  0  0  0  0  0  4.3
7    0.74 0.26  0  0  0  0  0  0  4.6
```

Multiple imputation must take into account that the data are clustered within persons. The setup for `mice()` requires some care, so we discuss each step in detail.

```
Y <- c("hgt.z", "wgt.z", "bmi.z")
meth <- make.method(data2)
meth[1:length(meth)] <- ""
meth[Y] <- "2l.pan"
```

These statements specify that only `hgt.z`, `wgt.z` and `bmi.z` need to be imputed. For these three outcomes we request the elementary imputation function `mice.impute.2l.pan()`, which is designed to impute data with two levels. See Section 7.6 for more detail.

```
pred <- make.predictorMatrix(data2)
pred[1:nrow(pred), 1:ncol(pred)] <- 0
pred[Y, "id"] <- (-2)
pred[Y, "sex"] <- 1
pred[Y, paste("x", 2:9, sep = "")] <- 1
pred[Y[1], Y[2]] <- 1
pred[Y[2], Y[1]] <- 1
pred[Y[3], Y[1:2]] <- 1
```

The setup of the predictor matrix needs some care. We first empty all entries from the variable `pred`. The statement `pred[Y, "id"] <- (-2)` defines variable `id` as the class variable. The statement `pred[Y, "sex"] <- 1` specifies `sex` as a fixed effect, as usual, while `pred[Y, paste("x", 1:9, sep = "")] <- 2` sets the B-spline basis as a random effect, as in Equation 11.2. The remaining three statement specify the Y_2 is a random effects predictor of Y_1 (and vice versa), and both Y_1 and Y_2 are random effects predictors of Y_3. Note that Y_3 (BMI SDS) is not a predictor of Y_1 or Y_2 in order to prevent the type of convergence problems explained in Section 6.5.2. Note also that age is not included in order to evade duplication with its B-spline coding. In summary, there are 12 random effects (9 for age and 3 for the outcomes), one class variable, and one fixed effect.

The actual imputations are produced by

```
imp.1745 <- mice(data2, meth = meth, pred = pred, m = 10,
                 maxit = 10, seed = 52711, print = FALSE)
```

There are over 48000 records. This call takes about 30 minutes to complete, which is much longer than the other applications discussed in this book. In the year 2012 the same problem still took over 10 hours, so there is certainly progress.

Figure 11.7 displays ten multiply imputed trajectories for the six persons displayed in Figure 11.6. The general impression is that the imputed trajectory follows the data quite well. At ages where the are many data points (e.g., in period 0d–1y in person 1259 or in period 8d–1y in person 7460) the curves are quite close, indicating a relatively large certainty. On the other hand, at locations where data are sparse (e.g., the period 10y–29y in person 7019, or the period 8d–2y in person 8046) the curves diverge, indicating a large amount of uncertainty about the imputation. This effect is especially strong at the edges of the age range. Incidentally, we noted that the end effects are less pronounced for larger sample sizes.

It is also interesting to study whether imputation preserves the relation between height, weight and BMI. Figure 11.8 is a scattergram of height SDS and weight SDS split according to age that superposes the imputations on the observed data in the period after the break point. In general the relation in the observed data is preserved in the imputed data. Note that the imputations become more variable for regions with fewer data. This is especially visible at the panel in the upper-right corner at age 29y, where there were no data at all. Similar plots can be made in combination with BMI SDS. In general, the data in these plots all behave as one would expect.

11.3.7 Complete-data model

Table 11.2 provides a comparison of the mean changes observed under the broken stick model and under time raster imputation. The estimates are very

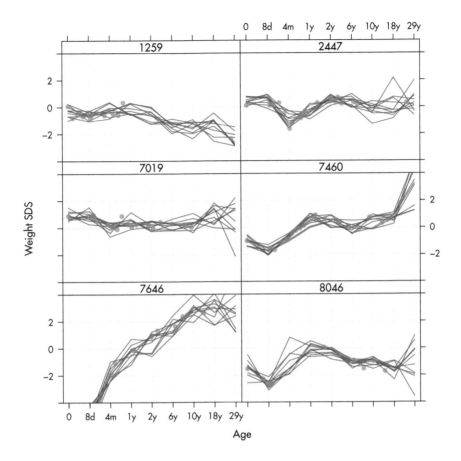

Figure 11.7: Ten multiply imputed trajectories of weight SDS for the same persons as in Figure 11.6 (in red). Also shown are the data points (in blue).

similar, so the mean change estimated under both methods is similar. The p-values in the broken stick method are generally more optimistic relative to multiple imputation, which is due to the fact that the broken stick model ignores the uncertainty about the estimates.

There is also an effect on the correlations. In general, the age-to-age correlations of the broken stick method are higher than the raster imputations.

Table 11.3 provides the age-to-age correlation matrix of BMI SDS estimated from 1745 cases from the Terneuzen Birth Cohort. Apart from the peculiar values for the age of 8 days, the correlations decrease as the period between time points increases. The values for the broken stick method are higher because these do not incorporate the uncertainty of the estimates.

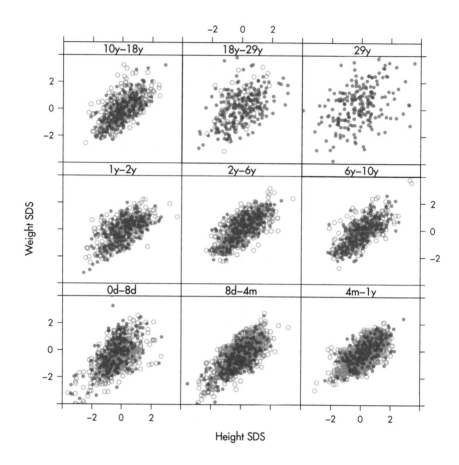

Figure 11.8: The relation between height SDS and weight SDS in the observed (blue) and imputed (red) longitudinal trajectories. The imputed data occur exactly at the break ages. The observed data come from the period immediately after the break age. No data beyond 29 years were observed, so the upper-right panel contains no observed data.

11.4   Conclusion

This chapter described techniques for imputing longitudinal data in both the wide and long formats. Some things are easier in the wide format, e.g., change scores or imputing data, while other procedures are easier in the long format, e.g., graphics and advanced statistical modeling. It is therefore useful to have both formats available.

Table 11.2: Mean change per period, split according to adult overweight (AO) ($n = 124$) and no adult overweight (NAO) ($n = 486$) for the broken stick method and for multiple imputation of the time raster.

		Broken stick				
Time raster imputation						
Period	*NAO*	*AO*	*p-value*	*NAO*	*AO*	*p-value*
0d–8d	−0.88	−0.80	0.214	−0.93	−0.82	0.335
8d–4m	−0.32	−0.34	0.811	−0.07	−0.11	0.745
4m–1y	0.42	0.62	0.006	0.35	0.58	0.074
1y–2y	0.22	0.28	0.242	0.24	0.26	0.884
2y–6y	−0.36	−0.10	<0.001	−0.35	−0.06	0.026
6y–10y	0.05	0.34	<0.001	−0.01	0.31	0.029
10y–18y	0.09	0.52	<0.001	0.17	0.68	0.009

Table 11.3: Age-to-age correlations of BMI SDS the broken stick estimates (lower triangle) and raster imputations (upper triangle) for the Terneuzen Birth Cohort ($n = 1745$).

Age	0d	8d	4m	1y	2y	6y	10y	18y
0d	–	0.64	0.20	0.21	0.18	0.17	0.16	0.11
8d	0.75	–	0.30	0.17	0.20	0.20	0.15	0.13
4m	0.28	0.44	–	0.39	0.30	0.29	0.20	0.16
1y	0.28	0.23	0.65	–	0.55	0.40	0.31	0.23
2y	0.31	0.33	0.46	0.76	–	0.56	0.36	0.23
6y	0.31	0.36	0.46	0.59	0.79	–	0.62	0.42
10y	0.26	0.26	0.35	0.47	0.55	0.89	–	0.53
18y	0.23	0.26	0.29	0.37	0.40	0.72	0.89	–

The methodology for imputing data in the wide format is not really different from that of cross-sectional data. When possible, always try to convert the data into the wide format before imputation. If the data have been observed at irregular time points, as in the Terneuzen Birth Cohort, conversion of the data into the wide format is not possible, however, and imputation can be done in the long format by multilevel imputation.

This chapter introduced time raster imputation, a technique for converting data with an irregular age spacing into the wide format by means of imputation. Time rastering seems to work well in the sense that the generated trajectories follow the individual trajectories. The technique is still experimental and may need further refinement before it can be used routinely.

The current method inserts missing data at the full time grid, and thus imputes data even at time points where there are real observations. One obvious improvement would be to strip such points from the grid so that they are not imputed. For example, in the Terneuzen Birth Cohort this means that we would always take observed birth weight when it is measured.

Another potential improvement is to use the OLS estimates within each cluster as the center of the posterior predictive distribution rather than their shrunken versions. This would decrease within cluster variability in the imputations, and increase between cluster variability. It is not yet clear how to deal with clusters with only a few time points, but this modification is likely to produce age-to-age correlations that are most faithful to the data.

Finally, the selection of the data could be much stricter. The analysis of the Terneuzen Birth Cohort data used a very liberal inclusion criterion that requires a minimum of only three data points across the entire age range. Sparse trajectories will have large imputation variances, and may thus bias the age-to-age correlations toward zero. As a preliminary rule of thumb, there should be at least one, and preferably two or more, measurements per period.

11.5   Exercises

1. *Potthoff–Roy, wide format imputation.* Potthoff and Roy (1964) published classic data on a study in 16 boys and 11 girls, who at ages 8, 10, 12, and 14 had the distance (mm) from the center of the pituitary gland to the pteryomaxillary fissure measured. Changes in pituitary-pteryomaxillary distances during growth is important in orthodontic therapy. The goals of the study were to describe the distance in boys and girls as simple functions of age, and then to compare the functions for boys and girls. The data have been reanalyzed by many authors including Jennrich and Schluchter (1986), Little and Rubin (1987), Pinheiro and Bates (2000), Verbeke and Molenberghs (2000) and Molenberghs and Kenward (2007).

   - Take the version from Little and Rubin (1987) in which nine entries have been made missing. The missing data have been created such that children with a low value at age 8 are more likely to have a missing value at age 10. Use `mice()` to impute the missing entries under the normal model using $m = 100$.

   - For each missing entry, summarize the distribution of the 100 imputations. Determine the interquartile range of each distribution. If the imputations fit the data, how many of the original values you expect to fall within this range? How many actually do?

   - Produce a `lattice` graph of the nine imputed trajectories that clearly shows the range of the imputed values.

2. *Potthoff–Roy comparison.* Use the multiply imputed data from the previous exercise, and apply a linear mixed effects model with an unstructured

mean and an unstructured covariance. See Molenberghs and Kenward (2007, ch. 5) for a discussion of the setup. Discuss advantages and disadvantages of the analysis of the multiply imputed data compared to direct likelihood.

3. *Potthoff–Roy, long format imputation.* Do this exercise with the complete Potthoff–Roy data. Warning: This exercise requires good data handling skills and some patience.

- Calculate the broken stick estimates for each child using 8, 10, 12 and 14 as the break ages. Make a graph like Figure 11.6. Each data point has exactly one parameter, so the fit could be perfect in principle. Why doesn't that happen? Which two children show the largest discrepancies between the data and the model?

- Compare the age-to-age correlation matrix of the broken stick estimates to the original data. Why are these correlation matrices different?

- How would you adapt the analysis such that the age-to-age correlation matrix of the broken stick estimates would reproduce the age-to-age correlation matrix of the original data. Hint: Think of a simpler form of multilevel analysis.

- Multiply impute the data according to the method used in Section 11.3.6, and produce a display like Figure 11.7 for children 1, 7, 20, 21, 22 and 24.

- Compare the age-to-age correlation matrix from the imputed data to that of the original data. Are these different? How? Calculate the correlation matrix after deleting the data from the two children who showed the largest discrepancy in the broken stick model. Did this help?

- How would you adapt the imputation method for the longitudinal data so that its correlation matrix is close to that of the original?

Part IV

Extensions

Chapter 12

Conclusion

The missing-data problem is different.
Thomas Permutt

This closing chapter starts with a description of the limitations and pitfalls of multiple imputation. Section 12.2 provides reporting guidelines for applications. Section 12.3 gives an overview of applications that were omitted from the book. Section 12.4 contains some speculations about possible future developments.

12.1 Some dangers, some do's and some don'ts

Any statistical technique has limitations and pitfalls, and multiple imputation is no exception. This books emphasizes the virtues of being flexible, but this comes at a price. The next sections outline some dangers, do's and don'ts.

12.1.1 Some dangers

The major danger of the technique is that it may provide nonsensical or even misleading results if applied without appropriate care or insight. Multiple imputation is not a simple technical fix for the missing data. Scientific and statistical judgment comes into play at various stages: during diagnosis of the missing data problem, in the setup of a good imputation model, during validation of the quality of the generated synthetic data and in combining the repeated analyses. While software producers attempt to set defaults that will work in a large variety of cases, we cannot simply hand over our scientific decisions to the software. We need to open the black box, and adjust the process when appropriate.

The MICE algorithm is univariate optimal, but not necessarily multivariate optimal. There is no clear theoretical rationale for convergence of the multivariate algorithm. The main justification of the MICE algorithm rests on simulation studies. The research on this topic is intensifying. Even though the results obtained thus far are reassuring, at this moment it is not possible to outline in advance the precise conditions that would guarantee convergence for some set of conditionally specified models.

Another danger occurs if the imputation model is uncongenial (Meng, 1994; Schafer, 2003). Uncongeniality can occur if the imputation model is specified as more restrictive than the complete-data model, or if it fails to account for important factors in the missing data mechanism. Both types of omissions introduce biased and possibly inefficient estimates. The other side of the coin is that multiple imputation can be more efficient if the imputer uses information that is not accessible to the analyst. The statistical infererences may become more precise than those in maximum likelihood, a property known as *superefficiency* (Rubin, 1996).

There are many data-analytic situations for which we do not yet know the appropriate way to generate imputations. For example, it is not yet clear how design factors of a complex sampling design, e.g., a stratified cluster sample, should be incorporated into the imputation model. Also, relatively little is known about how to impute nested and hierarchical data, or autocorrelated data that form time series. These problems are not inherent limitations of multiple imputation, but of course they may impede the practical application of the imputation techniques for certain types of data.

12.1.2 Some do's

Constructing good imputation models requires analytic skills. The following list of do's summarizes some of the advice given in this book.

- Find out the reasons for the missing data;
- Include the outcome variable(s) in the imputation model;
- Include factors that govern the missingness in the imputation model;
- Impute categorical data by techniques for categorical data;
- Remove response indicators from the imputation model;
- Aim for a scope broad enough for all analyses;
- Set the random seed to enhance reproducible results;
- Break any direct feedback loops that arise in passive imputation;
- Inspect the trace lines for slow convergence;
- Inspect the imputed data;
- Evaluate whether the imputed data could have been real data if they had not been missing;
- Take $m = 5$ for model building, and increase afterward if needed;

- Specify simple MNAR models for sensitivity analysis;

- Impute by proper imputation methods;

- Impute by robust hot deck-like models like predictive mean matching;

- Reduce a large imputation model into smaller components;

- Transform statistics toward approximate normality before pooling;

- Assess critical assumptions about the missing data mechanism;

- Eliminate badly connected, uninteresting variables (low influx, low outflux) from the imputation model;

- Take obvious features like non-negativity, functional relations and skewed distributions in the data into account in the imputation model;

- Use more flexible (e.g., nonlinear or nonparametric) imputation models;

- Perform and pool the repeated analyses per dataset;

- Describe potential departures from MAR;

- Report accurately and concisely.

12.1.3   Some don'ts

Do not:

- Use multiple imputation if simpler methods are valid;

- Take predictions as imputations;

- Impute blindly;

- Put too much faith in the defaults;

- Average the multiply imputed data;

- Create imputations using a model that is more restrictive than needed;

- Uncritically accept imputations that are very different from the observed data.

12.2 Reporting

Section 1.1.2 noted that the attitude toward missing data is changing. Many aspects related to missing data could potentially affect the conclusions drawn for the statistical analysis, but not all aspects are equally important. This leads to the question: *What should be reported from an analysis with missing data?*

Guidelines to report the results of a missing data analysis have been given by Sterne et al. (2009), Enders (2010), National Research Council (2010) and Mackinnon (2010). These sources vary in scope and comprehensiveness, but they also exhibit a great deal of overlap and consensus. Section 12.2.1 combines some of the material found in the three sources.

Reviewers or editors may be unfamiliar with, or suspicious of, newer approaches to handling missing data. Substantive researchers are therefore often wary about using advanced statistical methods in their reports. Though this concern is understandable,

> ... resorting to flawed procedures in order to avoid criticism from an uninformed reviewer or editor is a poor reason for avoiding sophisticated missing data methodology (Enders, 2010, p. 340)

Until reviewers and referees become more familiar with the newer methods, a better approach is to add well-chosen and concise explanatory notes. On the other hand, editors and reviewers are increasingly expecting applied researchers to do multiple imputation, even when the authors had good reasons for not doing it (e.g., less than 5% incomplete cases) (Ian White, personal communication).

The natural place to report about the missing data in a manuscript is the paragraph on the statistical methodology. As scientific articles are often subject to severe space constraints, part of the report may need to go into supplementary online materials instead of the main text. Since the addition of explanatory notes increases the number of words, there needs to be some balance between the material that goes into the main text and the supplementary material. In applications that requires novel methods, a separate paper may need to be written by the team's statistician. For example, Van Buuren et al. (1999) explained the imputation methodology used in the substantive paper by Boshuizen et al. (1998). In general, the severity of the missing data problem and the method used to deal with the problem needs to be part of the main paper, whereas the precise modeling details could be relegated to the appendix or to a separate methodological paper.

The following list contains questions that need to be answered when using multiple imputation. Evaluate each question carefully, and report the answers.

1. *Amount of missing data:* What is the number of missing values for each variable of interest? What is the number of cases with complete data for the analyses of interest? If people drop out at various time points, break down the number of participants per occasion.

2. *Reasons for missingness:* What is known about the reasons for missing data? Are the missing data intentional? Are the reasons possibly related to the outcome measurements? Are the reasons related to other variables in the study?

3. *Consequences:* Are there important differences between individuals with complete and incomplete data? Do these groups differ in mean or spread on the key variables? What are the consequences if complete-case analysis is used?

4. *Method:* What method is used to account for missing data (e.g., complete-case analysis, multiple imputation)? Which assumptions were made (e.g., missing at random)? How were multivariate missing data handled?

5. *Software:* What multiple imputation software is used? Which settings differ from the default?

6. *Number of imputed datasets:* How many imputed datasets were created and analyzed?

7. *Imputation model:* Which variables were included in the imputation model? Was any form of automatic variable predictor used? How were non-normally distributed and categorical variables imputed? How were design features (e.g., hierarchical data, complex samples, sampling weights) taken into account?

8. *Derived variables:* How were derived variables (transformations, recodes, indices, interaction terms, and so on) taken into account?

9. *Diagnostics:* How has convergence been monitored? How do the observed and imputed data compare? Are imputations plausible in the sense that they could have been plausibly measured if they had not been missing?

10. *Pooling:* How have the repeated estimates been combined (pooled) into the final estimates? Have any statistics been transformed for pooling?

11. *Complete-case analysis:* Do multiple imputation and complete-case analysis lead to similar similar conclusions? If not, what might explain the difference?

12. *Sensitivity analysis:* Do the variables included in the imputation model make the missing at random assumption plausible? Are the conclusions affected if imputations are generated under a plausible nonignorable model?

If space is limited, the main text can be restricted to a short summary of points 1, 2, 4, 5, 6 and 11, whereas the remaining points are addressed in a appendix or online supplement. Section 12.2.2 contains an example template.

For clinical trials, reporting in the main text should be extended by point 12, conform to recommendation 15 of National Research Council (2010). Moreover, the study protocol should specify the statistical methods for handling missing data in advance, and their associated assumptions should be stated in a way that can be understood by clinicians (National Research Council, 2010, recommendation 9).

12.2.2   Template

Enders (2010, pp. 340–343) provides four useful templates for reporting the results of a missing data analysis. These templates include explanatory notes for uninformed editors and reviewers. It is straightforward to adapt the template text to other settings. Below I provide a template loosely styled after Enders that I believe captures the essentials needed to report multiple imputation in the statistical paragraph of the main text.

> The percentage of missing values across the nine variables varied between 0 and 34%. In total 1601 out of 3801 records (42%) were incomplete. Many girls had no score because the nurse felt that the measurement was "unnecessary," or because the girl did not give permission. Older girls had many more missing data. We used multiple imputation to create and analyze 40 multiply imputed datasets. Methodologists currently regard multiple imputation as a state-of-the-art technique because it improves accuracy and statistical power relative to other missing data techniques. Incomplete variables were imputed under fully conditional specification, using the default settings of the mice 3.0.0 package (Van Buuren and Groothuis-Oudshoorn, 2011). The parameters of substantive interest were estimated in each imputed dataset separately, and combined using Rubin's rules. For comparison, we also performed the analysis on the subset of complete cases.

This text is about 135 words. If this is too long, then the sentences that begin with "Methodologists" and "For comparison" can be deleted. In the paragraphs that describe the results we can add the following sentence:

> Table 1 gives the missing data rates of each variable.

In addition, if complete-case analysis is included, then we need to summarize it. For example:

> We obtained similar results when the analysis was restricted to the complete cases only. Multiple imputation was generally more efficient as can be seen from the shorter confidence intervals and lower p-values in Table X.

It is also possible that the two analyses lead to diametrically opposed conclusions. Since a well-executed multiple imputation is theoretically superior to complete-case analysis, we should give multiple imputation more weight. It would be comforting though to have an explanation of the discrepancy.

The template texts can be adapted as needed. In addition obtain inspiration from good articles in your own field that apply multiple imputation.

12.3 Other applications

Chapters 9–11 illustrated several applications of multiple imputation. This section briefly reviews some other applications. These underscore the general nature and broad applicability of multiple imputation.

12.3.1 Synthetic datasets for data protection

Many governmental agencies make microdata available to the public. One of the major practical issues is that the identity of anonymous respondents can be disclosed through the data they provide. Rubin (1993) suggested publishing fully synthetic microdata instead of the real data, with the obvious advantage of zero disclosure risk. The released synthetic data should reproduce the essential features of confidential microdata.

Raghunathan et al. (2003) and Reiter (2005a) demonstrated the practical application of the idea. Real and synthetic records can be mixed, resulting in partially synthetic data. Recent advances can be found in Reiter (2009), Drechsler and Reiter (2010), Reiter et al. (2014) and Loong and Rubin (2017). Yu et al. (2017) present an application to protect confidentiality in the Californian Cancer Registry.

12.3.2 Analysis of coarsened data

Many datasets contain data that are partially missing. Some values are known accurately, but others are only known to lie within a certain range. Heitjan and Rubin (1991) proposed a general theory for data coarsening processes that includes rounding, heaping, censoring and missing data as special

cases. See also Gill et al. (1997) for a slightly more extended model. Heitjan and Rubin (1990) provided an application where age is misreported, and
the amount of misreporting increases with age itself. Such problems with the
data can be handled by multiple imputation of true age, given reported age
and other personal factors. Heitjan (1993) discussed various other biomedical
examples and an application to data from the Stanford Heart Transplant Program. Related work on measurement error is available from several sources
(Brownstone and Valletta, 1996; Ghosh-Dastidar and Schafer, 2003; Yucel
and Zaslavsky, 2005; Cole et al., 2006; Glickman et al., 2008). Goldstein and
Carpenter (2015) formulated joint models for three types of coarsened data.

12.3.3   File matching of multiple datasets

Statistical file matching, or data fusion, attempts to integrate two or
more datasets with different units observed on common variables. Rubin and
Schenker (1986a) considered file matching as a missing data problem, and
suggested multiple imputation as a solution. Moriarity and Scheuren (2003)
developed modifications that were found to improve the procedure. Further
relevant work can be found in the books by Rässler (2002), D'Orazio et al.
(2006) and Herzog et al. (2007).

The imputation techniques proposed to date were developed from the multivariate normal model. Application of the MICE algorithm under conditional
independence is straightforward. Rässler (2002) compared MICE to several
alternatives, and found MICE to work well under normality and conditional
independence. If the assumption of conditional independence does not hold,
we may bring prior information into MICE by appending a third data file that
contains records with data that embody the prior information. Sections 6.5.2
and 9.4.5 put this idea into practice in a different context. This techique
can perform file matching for mixed continuous-discrete data under any data
coded prior.

12.3.4   Planned missing data for efficient designs

Lengthy questionnaires increase the missing data rate and can make a
study expensive. An alternative is to cut up a long questionnaire into separate forms, each of which is considerably shorter than the full version. The
split questionnaire design (Raghunathan and Grizzle, 1995) poses certain restrictions on the selection of the forms, thus enabling analysis by multiple
imputation. Gelman et al. (1998) provide additional techniques for the related problem of analysis of multiple surveys. The loss of efficiency depends
on the strengths of the relations between form and can be compensated for
by a larger initial sample size. Graham et al. (2006) and Graham (2012) are
excellent resources for methodology based on planned missing data. Little and
Rhemtulla (2013) and Rhemtulla and Hancock (2016) discuss applications in
child development and educational research.

12.3.5 Adjusting for verification bias

Partial verification bias in diagnostic accuracy studies may occur if not all patients are assessed by the reference test (golden standard). Bias occurs if the group of patients is selective, e.g., when only those that score on a previous test are measured. Multiple imputation has been suggested as a way to correct for this bias (Harel and Zhou, 2006; De Groot et al., 2008). The classic Begg-Greenes method may be used only if the missing data mechanism is known and simple. For more complex situations De Groot et al. (2011) and Naaktgeboren et al. (2016) recommend multiple imputation.

12.4 Future developments

Multiple imputation is not a finished product or algorithm. New applications call for innovative ways to implement the key ideas. This section identifies some areas where further research could be useful.

12.4.1 Derived variables

Section 6.4 describes techniques to generate imputations for interactions, sum scores, quadratic terms and other derived variables. Many datasets contain derived variables of some form. The relations between the variables need to be maintained if imputations are to be plausible. There are no fail-safe methods for drawing imputations that preserve the proper interactions in the substantive models. One promising area of development is the rise of model-based forms of imputation that essentially take the substantive model as leading (cf. Section 4.5.5). It would be interesting to study how well model-based techniques can cope with derived variables of various sorts.

12.4.2 Algorithms for blocks and batches

In some applications it is useful to generalize the variable-by-variable scheme of the MICE algorithm to blocks. A block can contain just one variable, but also groups of variables. An imputation model is specified for each block, and the algorithm iterates over the blocks. Starting with `mice 3.0`, the user can specify blocks of variables that are structurally related, such as dummy variables, semi-continuous variables, bracketed responses, compositions, item subsets, and so on. It would be useful to obtain experience with the practical application of this facility, which could stimulate specification of the imputation model on a higher level of abstraction, and allow for mixes of joint and conditional models.

Likewise, it may be useful to define batches, groups of records that form logical entities. For example, batches could consist of different populations, time points, classes, and so on. Imputation models can be defined per batch, and iteration takes place over the batches. Javaras and Van Dyk (2003) proposed algorithms for blocks using joint modeling. Zhu (2016) discusses alternatives within an FCS context. The incorporation of blocks and batches will allow for tremendous flexibility in the specification of imputation models. Such techniques require a keen database administration strategy to ensure that the predictors needed at any point are completed.

12.4.3   Nested imputation

In some applications it can be useful to generate different numbers of imputations for different variables. Rubin (2003) described an application that used fewer imputations for variables that were expensive to impute. Alternatively, we may want to impute a dataset that has already been multiply imputed, for example, to impute some additional variables while preserving the original imputations. The technique of using different numbers of imputations is known as nested multiple imputation (Shen, 2000). Nested multiple imputation also has potential applications for modeling different types of missing data (Harel, 2009). Nested multiple imputation has theoretical advantages, but it would be good to develop a good understanding of its added value in typical use cases.

12.4.4   Better trials with dynamic treatment regimes

Adaptive treatment designs follow patients over time, and can re-randomize them to alternate treatments conditional on previous outcomes. The design poses several challenges, and it is possible to address these in a coherent way by multiple imputation (Shortreed et al., 2014). As adaptive designs become more widely used, better methology with dynamic treatment regimes could result in huge savings, and at the same time, make designs more ethical.

12.4.5   Distribution-free pooling rules

Rubin's theory is based on a convenient summary of the sampling distribution by the mean and the variance. There seems to be no intrinsic limitation in multiple imputation that would prevent it from working for more elaborate summaries. Suppose that we summarize the work for more elaborate summaries. Suppose that we summarize the distribution of the parameter estimates for each completed dataset by a dense set of quantiles. As before, there will be within- and between-variability as a result of the sampling and missing data mechanisms, respectively. The problem of how to combine these two types of distribution into the appropriate total distribution has not yet been solved. If we would be able to construct the total distribution, this would permit precise distribution-free statistical inference from incomplete data.

12.4.6 Improved diagnostic techniques

The key problem in multiple imputation is how to generate good imputations. The ultimate criterion of a good imputation is that it is confidence proper with respect to the scientific parameters of interest. Diagnostic methods are intermediate tools to evaluate the plausibility of a set of imputations. Section 6.6 discussed several techniques, but these may be laborious for datasets involving many variables. It would be useful to have informative summary measures that can signal whether "something is wrong" with the imputed data. Multiple measures are likely to be needed, each of which is sensitive to a particular aspect of the data.

12.4.7 Building block in modular statistics

Multiple imputation requires a well-defined function of the population data, an adequate missing data mechanism, and an idea of the parameters that will be estimated from the imputed data. The technique is an attempt to separate the missing data problem from the complete-data problem, so that both can be addressed independently. This helps in simplifying statistical analyses that are otherwise difficult to optimize or interpret.

The modular nature of multiple imputation helps our understanding. Aided by the vast computational possibilities, statistical models are becoming more and more complex nowadays, up to the point that the models outgrow the capabilities of our minds. The modular approach to statistics starts from a series of smaller models, each dedicated to a particular task. The main intellectual task is to arrange these models in a sensible way, and to link up the steps to provide an overall solution. Compared to the one-big-model-for-everything approach, the modular strategy may sacrifice some optimality. On the other hand, the analytic results are easier to track, as we can inspect what happened after each step, and thus easier to understand. And that is what matters in the end.

12.5 Exercises

1. *Do's:* Take the list of do's in Section 12.1.2. For each item on the list, answer the following questions:

   (a) Why is it on the list?
   (b) Which is the most relevant section in the book?
   (c) Can you order the list of elements from most important to least important?
   (d) What were your reasons for picking the top three?

(e) And why did you pick the bottom three?

(f) Could you make suggestions for new items that should be on the list?

2. *Don'ts:* Repeat the previous exercise for the list of don'ts in Section 12.1.3.

3. *Template:* Adapt the template for the `mammalsleep` data in `mice`. Be sure to include your major assumptions and decisions.

4. *Nesting:* Develop an extension of the `mids` object in `mice` that allows for nested multiple imputation. Try to build upon the existing `mids` object.

References

Aarts, E., Van Buuren, S., and Frank, L. E. (2010). *A novel method to obtain the treatment effect assessed for a completely randomized design: Multiple imputation of unobserved potential outcomes.* Master thesis. University of Utrecht, Utrecht.

Abayomi, K., Gelman, A., and Levy, M. (2008). Diagnostics for multivariate imputations. *Journal of the Royal Statistical Society C*, 57(3):273–291.

Agresti, A. (1990). *Categorical Data Analysis.* John Wiley & Sons, New York.

Aitkin, M., Francis, B., Hinde, J., and Darnell, R. (2009). *Statistical Modelling in R.* Oxford University Press, Oxford.

Akande, O., Li, F., and Reiter, J. P. (2017). An empirical comparison of multiple imputation methods for categorical data. *The American Statistician*, 71(2):162–170.

Ake, C. F. (2005). Rounding after multiple imputation with non-binary categorical covariates. In *Proceedings of the SAS Users Group International (SUGI)*, volume 30, pages 112–30.

Akl, E. A., Shawwa, K., Kahale, L. A., Agoritsas, T., Brignardello-Petersen, R., Busse, J. W., Carrasco-Labra, A., Ebrahim, S., Johnston, B. C., Neumann, I., Sola, I., Sun, X., Vandvik, P., Zhang, Y., Alonso-Coello, P., and Guyatt, G. H. (2015). Reporting missing participant data in randomised trials: Systematic survey of the methodological literature and a proposed guide. *BMJ Open*, 5(12):e008431.

Albert, A. and Anderson, J. A. (1984). On the existence of maximum likelihood estimates in logistic regression models. *Biometrika*, 71(1):1–10.

Allan, F. E. and Wishart, J. (1930). A method of estimating the yield of a missing plot in field experiment work. *Journal of Agricultural Science*, 20(3):399–406.

Allison, P. D. (2005). Imputation of categorical variables with PROC MI. In *Proceedings of the SAS Users Group International (SUGI)*, volume 30, pages 113–30.

Allison, P. D. (2010). *Survival Analysis Using SAS: A Practical Guide.* SAS Press, Cary, NC, 2nd edition.

Allison, T. and Cicchetti, D. (1976). Sleep in mammals: Ecological and constitutional correlates. *Science*, 194(4266):732–734.

Andridge, R. R. (2011). Quantifying the impact of fixed effects modeling of clusters in multiple imputation for cluster randomized trials. *Biometrical Journal*, 53(1):57–74.

Andridge, R. R. and Little, R. J. A. (2010). A review of hot deck imputation for survey non-response. *International Statistical Review*, 78(1):40–64.

Angrist, J. D. (2004). Treatment effect heterogeneity in theory and practice. *The Economic Journal*, 114(494):C52–C83.

Arnold, B. C., Castillo, E., and Sarabia, J. M. (1999). *Conditional Specification of Statistical Models*. Springer, New York.

Arnold, B. C., Castillo, E., and Sarabia, J. M. (2002). Exact and near compatibility of discrete conditional distributions. *Computational Statistics & Data Analysis*, 40(2):231–252.

Arnold, B. C. and Press, S. J. (1989). Compatible conditional distributions. *Journal of the American Statistical Association*, 84(405):152–156.

Asparouhov, T. and Muthén, B. O. (2010). Multiple imputation with m*plus*. *M*plus *Web Notes*.

Audigier, V., Husson, F., and Josse, J. (2016). Multiple imputation for continuous variables using a Bayesian principal component analysis. *Journal of Statistical Computation and Simulation*, 86(11):2140–2156.

Audigier, V., Husson, F., and Josse, J. (2017). MIMCA: Multiple imputation for categorical variables with multiple correspondence analysis. *Statistics and Computing*, 27(2):501–518.

Audigier, V. and Resche-Rigon, M. (2018). *micemd: Multiple Imputation by Chained Equations with Multilevel Data*. R package version 1.2.0.

Audigier, V., White, I. R., Jolani, S., Debray, T. P. A., Quartagno, M., Carpenter, J. R., Van Buuren, S., and Resche-Rigon, M. (2018). Multiple imputation for multilevel data with continuous and binary variables. *Statistical Science*, 33(2):160–183.

Austin, P. C. (2008). Bootstrap model selection had similar performance for selecting authentic and noise variables compared to backward variable elimination: a simulation study. *Journal of Clinical Epidemiology*, 61(10):1009–1017.

Aylward, D. S., Anderson, R. A., and Nelson, T. D. (2010). Approaches to handling missing data within developmental and behavioral pediatric research. *Journal of Developmental & Behavioral Pediatrics*, 31(1):54–60.

Bang, K. and Robins, J. M. (2005). Doubly robust estimation in missing data and causal inference models. *Biometrics*, 61(4):962–972.

Bárcena, M. J. and Tusell, F. (2000). Tree-based algorithms for missing data imputation. In Bethlehem, J. G. and Van der Heijden, P. G. M., editors, *COMPSTAT 2000: Proceedings in Computational Statistics*, pages 193–198, Heidelberg, Germany. Physica-Verlag.

Barnard, J. and Rubin, D. B. (1999). Small-sample degrees of freedom with multiple imputation. *Biometrika*, 86(4):948–955.

Bartlett, J. W. and Keogh, R. (2018). *smcfcs: Multiple Imputation of Co-variates by Substantive Model Compatible Fully Conditional Specification*. http://www.missingdata.org.uk, http://thestatsgeek.com.

Bartlett, J. W., Seaman, S. R., White, I. R., and Carpenter, J. R. (2015). Multiple imputation of covariates by fully conditional specification: Accom-modating the substantive model. *Statistical Methods in Medical Research*, 24(4):462–487.

Bartlett, M. S. (1978). *An Introduction to Stochastic Processes*. Press Syndi-cate of the University of Cambridge, 3rd edition.

Bates, D. M., Mächler, M., Bolker, B., and Walker, S. (2015). Fitting linear mixed-effects models using lme4. *Journal of Statistical Software*, 67(1):1–48.

Beaton, A. E. (1964). The use of special matrix operations in statistical cal-culus. Research Bulletin RB-64-51, Educational Testing Service, Princeton, NJ.

Bebchuk, J. D. and Betensky, R. A. (2000). Multiple imputation for simple estimation of the hazard function based on interval censored data. *Statistics in Medicine*, 19(3):405–419.

Beddo, V. (2002). *Applications of Parallel Programming in Statistics*. PhD thesis, University of California, Los Angeles.

Belin, T. R., Hu, M. Y., Young, A. S., and Grusky, O. (1999). Performance of a general location model with an ignorable missing-data assumption in a mul-tivariate mental health services study. *Statistics in Medicine*, 18(22):3123–3135.

Bernaards, C. A., Belin, T. R., and Schafer, J. L. (2007). Robustness of a multivariate normal approximation for imputation of incomplete binary data. *Statistics in Medicine*, 26(6):1368–1382.

Besag, J. (1974). Spatial interaction and the statistical analysis of lattice systems. *Journal of the Royal Statistical Society B*, 36(2):192–236.

Bethlehem, J. G. (2002). Weighting adjustments for ignorable nonresponse. In Groves, R. M., Dillman, D. A., Eltinge, J. L., and Little, R. J. A., editors, *Survey Nonresponse*, chapter 18, pages 275–287. John Wiley & Sons, New York.

Bodner, T. E. (2008). What improves with increased missing data imputations? *Structural Equation Modeling*, 15(4):651–675.

Bondarenko, I. and Raghunathan, T. E. (2010). Multiple imputation for causal inference. In *Section on Survey Research Methods - JSM 2010*, pages 3934–3944. American Statistical Association, Alexandria VA.

Bondarenko, I. and Raghunathan, T. E. (2016). Graphical and numerical diagnostic tools to assess suitability of multiple imputations and imputation models. *Statistics in medicine*, 35(17):3007–3020.

Boshuizen, H. C., Izaks, G. J., Van Buuren, S., and Ligthart, G. J. (1998). Blood pressure and mortality in elderly people aged 85 and older: Community based study. *British Medical Journal*, 316(7147):1780–1784.

Box, G. E. P. and Tiao, G. C. (1973). *Bayesian Inference in Statistical Analysis*. John Wiley & Sons, New York.

Brand, J. E. and Xie, Y. (2010). Who benefits most from college? Evidence for negative selection in heterogeneous economic returns to higher education. *American Sociological Review*, 75(2):273–302.

Brand, J. P. L. (1999). *Development, Implementation and Evaluation of Multiple Imputation Strategies for the Statistical Analysis of Incomplete Data Sets*. PhD thesis, Erasmus University, Rotterdam.

Brand, J. P. L., Van Buuren, S., Groothuis-Oudshoorn, C. G. M., and Gelsema, E. S. (2003). A toolkit in SAS for the evaluation of multiple imputation methods. *Statistica Neerlandica*, 57(1):36–45.

Brandsma, H. P. and Knuver, J. W. M. (1989). Effects of school and classroom characteristics on pupil progress in language and arithmetic. *International Journal of Educational Research*, 13(7):777–788.

Breiman, L., Friedman, J., Olshen, R., and Stone, C. (1984). *Classification and Regression Trees*. Wadsworth Publishing, New York.

Brick, J. M. and Kalton, G. (1996). Handling missing data in survey research. *Statistical Methods in Medical Research*, 5(3):215–238.

Brooks, S. P. and Gelman, A. (1998). General methods for monitoring convergence of iterative simulations. *Journal of Computational and Graphical Statistics*, 7(4):434–455.

Brownstone, D. and Valletta, R. G. (1996). Modeling earnings measurement error: A multiple imputation approach. *Review of Economics and Statistics*, 78(4):705–717.

Bryk, A. S. and Raudenbush, S. W. (1992). *Hierarchical Linear Models*. Sage, Newbury Park, CA.

Burgette, L. F. and Reiter, J. P. (2010). Multiple imputation for missing data via sequential regression trees. *American Journal of Epidemiology*, 172(9):1070–1076.

Burton, A. and Altman, D. G. (2004). Missing covariate data within cancer prognostic studies: A review of current reporting and proposed guidelines. *British Journal of Cancer*, 91(1):4–8.

Cameron, N. L. and Demerath, E. W. (2002). Critical periods in human growth and their relationship to diseases of aging. *American Journal of Physical Anthropology*, Suppl 35:159–184.

Carpenter, J. R., Goldstein, H., and Kenward, M. G. (2011). REALCOM-IMPUTE software for multilevel multiple imputation with mixed response types. *Journal of Statistical Software*, 45(5):1–14.

Carpenter, J. R. and Kenward, M. G. (2013). *Multiple Imputation and its Applications*. John Wiley & Sons, Chichester, UK.

Casella, G. and George, E. I. (1992). Explaining the Gibbs sampler. *The American Statistician*, 46(3):167–174.

Chen, H. Y. (2011). Compatibility of conditionally specified models. *Statistics and Probability Letters*, 80(7-8):670–677.

Chen, H. Y., Xie, H., and Qian, Y. (2011). Multiple imputation for missing values through conditional semiparametric odds ratio models. *Biometrics*, 67(3):799–809.

Chen, L. and Sun, J. (2010). A multiple imputation approach to the analysis of interval-censored failure time data with the additive hazards model. *Computational Statistics & Data Analysis*, 54(4):1109–1116.

Chen, Q. and Wang, S. (2013). Variable selection for multiply-imputed data with application to dioxin exposure study. *Statistics in Medicine*, 32(21):3646–3659.

Cheung, M. W. L. (2007). Comparison of methods of handling missing time-invariant covariates in latent growth models under the assumption of missing completely at random. *Organizational Research Methods*, 10(4):609–634.

Chevalier, A. and Fielding, A. (2011). An introduction to anchoring vignettes. *Journal of the Royal Statistical Society A*, 174(3):569–574.

Chung, Y., Rabe-Hesketh, S., Dorie, V., Gelman, A., and Liu, J. (2013). A nondegenerate penalized likelihood estimator for variance parameters in multilevel models. *Psychometrika*, 78(4):685–709.

Clogg, C. C., Rubin, D. B., Schenker, N., Schultz, B., and Weidman, L. (1991). Multiple imputation of industry and occupation codes in census public-use samples using Bayesian logistic regression. *Journal of the American Statistical Association*, 86(413):68–78.

Cochran, W. G. (1977). *Sampling Techniques*. John Wiley & Sons, New York, 3rd edition.

Cole, S. R., Chu, H., and Greenland, S. (2006). Multiple imputation for measurement error correction. *International Journal of Epidemiology*, 35:1074–1081.

Cole, T. J. and Green, P. J. (1992). Smoothing reference centile curves: The LMS method and penalized likelihood. *Statistics in Medicine*, 11(10):1305–1319.

Collins, L. M., Schafer, J. L., and Kam, C. M. (2001). A comparison of inclusive and restrictive strategies in modern missing data procedures. *Psychological Methods*, 6(3):330–351.

Conversano, C. and Cappelli, C. (2003). Missing data incremental imputation through tree based methods. In Härdle, W. and Rönz, B., editors, *COMPSTAT 2002: Proceedings in Computational Statistics*, pages 455–460, Heidelberg, Germany. Physica-Verlag.

Conversano, C. and Siciliano, R. (2009). Incremental tree-based missing data imputation with lexicographic ordering. *Journal of Classification*, 26(3):361–379.

Cowles, M. K. and Carlin, B. P. (1996). Markov chain Monte Carlo convergence diagnostics: A comparative review. *Journal of the American Statistical Association*, 91(434):883–904.

Creel, D. V. and Krotki, K. (2006). Creating imputation classes using classification tree methodology. In *Proceeding of the Joint Statistical Meeting 2006, ASA Section on Survey Research Methods*, pages 2884–2887, Alexandria, VA. American Statistical Association.

Daniels, M. J. and Hogan, J. W. (2008). *Missing Data in Longitudinal Studies. Strategies for Bayesian Modeling and Sensitivity Analysis*. Chapman & Hall/CRC, Boca Raton, FL.

Dauphinot, V., Wolff, H., Naudin, F., Guéguen, R., Sermet, C., Gaspoz, J., and Kossovsky, M. (2008). New obesity body mass index threshold for self-reported data. *Journal of Epidemiology and Community Health*, 63(2):128–132.

De Groot, J. A. H., Janssen, K. J. M., Zwinderman, A. H., Bossuyt, P. M. M., Reitsma, J. B., and Moons, K. G. M. (2011). Adjusting for partial verification bias in diagnostic accuracy studies: A comparison of methods. *Annals of Epidemiology*, 21(2):139–148.

De Groot, J. A. H., Janssen, K. J. M., Zwinderman, A. H., Moons, K. G. M., and Reitsma, J. B. (2008). Multiple imputation to correct for partial verification bias: A revision of the literature. *Statistics in Medicine*, 27(28):5880–5889.

De Jong, R. (2012). *Robust Multiple Imputation*. PhD thesis, University of Hamburg, Hamburg, Germany.

De Jong, R., Van Buuren, S., and Spiess, M. (2016). Multiple imputation of predictor variables using generalized additive models. *Communications in Statistics - Simulation and Computation*, 45(3):968–985.

De Kroon, M. L. A., Renders, C. M., Kuipers, E. C., Van Wouwe, J. P., Van Buuren, S., De Jonge, G. A., and Hirasing, R. A. (2008). Identifying metabolic syndrome without blood tests in young adults - the Terneuzen birth cohort. *European Journal of Public Health*, 18(6):656–660.

De Kroon, M. L. A., Renders, C. M., Van Wouwe, J. P., Van Buuren, S., and Hirasing, R. A. (2010). The Terneuzen birth cohort: BMI changes between 2 and 6 years correlate strongest with adult overweight. *PloS ONE*, 5(2):e9155.

De Leeuw, E. D., Hox, J. J., and Dillman, D. A. (2008). *International Handbook of Survey Methodology*. Lawrence Erlbaum Associates, New York.

De Leeuw, J. and Meijer, E. (2008). *Handbook of Multilevel Analysis*. Springer, New York.

De Roos, C., Greenwald, R., Den Hollander-Gijsman, M., Noorthoorn, E., Van Buuren, S., and De Jong, A. (2011). A randomized comparison of cognitive behavioral therapy (CBT) and eye movement desensitization and reprocessing (EMDR) in disaster-exposed children. *European Journal of Psychotraumatology*, 2:5694.

De Waal, T., Pannekoek, J., and Scholtus, S. (2011). *Handbook of Statistical Data Editing and Imputation*. John Wiley & Sons, Hoboken, NJ.

Delord, M. and Génin, E. (2016). Multiple imputation for competing risks regression with interval-censored data. *Journal of Statistical Computation and Simulation*, 86(11):2217–2228.

DeMets, D. L., Cook, T. D., and Roecker, E. (2007). Selected issues in the analysis. In Cook, T. and DeMets, D., editors, *Introduction to Statistical Methods for Clinical Trials*, chapter 11, pages 339–376. Chapman & Hall /CRC, Boca Raton, FL.

Demirtas, H. (2009). Rounding strategies for multiply imputed binary data. *Biometrical Journal*, 51(4):677–688.

Demirtas, H. (2010). A distance-based rounding strategy for post-imputation ordinal data. *Journal of Applied Statistics*, 37(3):489–500.

Demirtas, H., Freels, S. A., and Yucel, R. M. (2008). Plausibility of multivariate normality assumption when multiply imputing non-Gaussian continuous outcomes: A simulation assessment. *Journal of Statistical Computation and Simulation*, 78(1):69–84.

Demirtas, H. and Hedeker, D. (2008a). Imputing continuous data under some non-Gaussian distributions. *Statistica Neerlandica*, 62(2):193–205.

Demirtas, H. and Hedeker, D. (2008b). Multiple imputation under power polynomials. *Communications in Statistics - Simulation and Computation*, 37(8):1682–1695.

Dempster, A. P., Laird, N. M., and Rubin, D. B. (1977). Maximum likelihood estimation from incomplete data via the EM algorithm (with discussion). *Journal of the Royal Statistical Society B*, 39(1):1–38.

Dempster, A. P. and Rubin, D. B. (1983). Introduction. In *Incomplete Data in Sample Surveys*, volume 2, pages 3–10, New York. Academic Press.

Díaz-Ordaz, K., Kenward, M. G., Cohen, A., Coleman, C. L., and Eldridge, S. (2014). Are missing data adequately handled in cluster randomised trials? A systematic review and guidelines. *Clinical Trials*, 11(5):590–600.

Dietz, W. H. (1994). Critical periods in childhood for the development of obesity. *American Journal of Clinical Nutrition*, 59(5):955–959.

Diggle, P. J., Heagerty, P., Liang, K. Y., and Zeger, S. L. (2002). *Analysis of Longitudinal Data*. Clarendon Press, Oxford, 2nd edition.

Dillman, D. A., Smyth, J. D., and Melani Christian, L. (2008). *Internet, Mail, and Mixed-Mode Surveys: The Tailored Design Method*. John Wiley & Sons, New York, 3rd edition.

Doove, L. L., Van Buuren, S., and Dusseldorp, E. (2014). Recursive partitioning for missing data imputation in the presence of interaction effects. *Computational Statistics & Data Analysis*, 72:92–104.

Dorans, N. J. (2007). Linking scores from multiple health outcome instruments. *Quality of Life Research*, 16(Suppl 1):85–94.

D'Orazio, M., Di Zio, M., and Scanu, M. (2006). *Statistical Matching: Theory and Practice*. John Wiley & Sons, Chichester, UK.

Dorey, F. J., Little, R. J. A., and Schenker, N. (1993). Multiple imputation for threshold-crossing data with interval censoring. *Statistics in Medicine*, 12(17):1589–1603.

Drechsler, J. (2015). Multiple imputation of multilevel missing data: Rigor versus simplicity. *Journal of Educational and Behavioral Statistics*, 40(1):69–95.

Drechsler, J. and Reiter, J. P. (2010). Sampling with synthesis: A new approach for releasing public use census microdata. *Journal of the American Statistical Association*, 105(492):1347–1357.

Eekhout, I., De Vet, H. C. W., De Boer, M. R., Twisk, J. W. R., and Heymans, M. W. (2018). Passive imputation and parcel summaries are both valid to handle missing items in studies with many multi-item scales. *Statistical Methods in Medical Research*, 27(4):1128–1140.

Eekhout, I., Wiel, M. A., and Heymans, M. W. (2017). Methods for significance testing of categorical covariates in logistic regression models after multiple imputation: Power and applicability analysis. *BMC Medical Research Methodology*, 17(1):129.

Efron, B. and Tibshirani, R. J. (1993). *An Introduction to the Bootstrap*. Chapman & Hall, London.

El Adlouni, S., Favre, A.-C., and Bobée, B. (2006). Comparison of methodologies to assess the convergence of Markov chain Monte Carlo methods. *Computational Statistics & Data Analysis*, 50(10):2685–2701.

Enders, C. K. (2010). *Applied Missing Data Analysis*. Guilford Press, New York.

Enders, C. K., Keller, B. T., and Levy, R. (2018). A fully conditional specification approach to multilevel imputation of categorical and continuous variables. *Psychological Methods*, dx.doi.org/10.1037/met0000148.

Enders, C. K. and Mansolf, M. (2018). Assessing the fit of structural equation models with multiply imputed data. *Psychological Methods*, 23(1):76–93.

Enders, C. K., Mistler, S. A., and Keller, B. T. (2016). Multilevel multiple imputation: A review and evaluation of joint modeling and chained equations imputation. *Psychological Methods*, 21(2):222–240.

Enders, C. K. and Tofighi, D. (2007). Centering predictor variables in cross-sectional multilevel models: A new look at an old issue. *Psychological Methods*, 12(2):121.

Erler, N. S., Rizopoulos, D., Jaddoe, V. W. V., Franco, O. H., and Lesaffre, E. M. (2018). Bayesian imputation of time-varying covariates in linear mixed models. *Statistical Methods in Medical Research*, to appear.

Erler, N. S., Rizopoulos, D., Van Rosmalen, J., Jaddoe, V. W. V., Franco, O. H., and Lesaffre, E. M. (2016). Dealing with missing covariates in epidemiologic studies: A comparison between multiple imputation and a full Bayesian approach. *Statistics in Medicine*, 35(17):2955–2974.

Fay, R. E. (1992). When are inferences from multiple imputation valid? In *ASA 1992 Proceedings of the Survey Research Methods Section*, pages 227–232, Alexandria, VA.

Fay, R. E. (1996). Alternative paradigms for the analysis of imputed survey data. *Journal of the American Statistical Association*, 91(434):490–498.

Firth, D. (1993). Bias reduction of maximum likelihood estimates. *Biometrika*, 80(1):27–38.

Fisher, R. A. (1925). *Statistical methods for research workers*. Oliver & Boyd, Edinburgh, London.

Fitzmaurice, G. M., Davidian, M., Verbeke, G., and Molenberghs, G. (2009). *Longitudinal Data Analysis*. Chapman & Hall/CRC, Boca Raton, FL.

Fitzmaurice, G. M., Laird, N. M., and Ware, J. H. (2011). *Applied Longitudinal Analysis, Second Edition*. John Wiley & Sons, New York.

Ford, B. L. (1983). An overview of hot-deck procedures. In Madow, W., Olkin, I., and Rubin, D. B., editors, *Incomplete Data in Sample Surveys*, volume 2, chapter 14, pages 185–207. Academic Press.

Fredriks, A. M., Van Buuren, S., Burgmeijer, R. J. F., Meulmeester, J. F., Beuker, R. J., Brugman, E., Roede, M. J., Verloove-Vanhorick, S. P., and Wit, J. M. (2000a). Continuing positive secular growth change in The Netherlands 1955–1997. *Pediatric Research*, 47(3):316–323.

Fredriks, A. M., Van Buuren, S., Wit, J. M., and Verloove-Vanhorick, S. P. (2000b). Body index measurements in 1996–7 compared with 1980. *Archives of Disease in Childhood*, 82(2):107–112.

Gaffert, P., Koller-Meinfelder, F., and Bosch, V. (2016). *Towards an mi-proper predictive mean matching*, volume Working Paper. University of Bamberg, Bamberg, Germany.

Galimard, J. E., Chevret, S., Protopopescu, C., and Resche-Rigon, M. (2016). A multiple imputation approach for MNAR mechanisms compatible with Heckman's model. *Statistics in Medicine*, 35(17):2907–2920.

Gelfand, A. E. and Smith, A. F. M. (1990). Sampling-based approaches to calculating marginal densities. *Journal of the American Statistical Association*, 85(410):398–409.

Gelman, A. (2004). Parameterization and Bayesian modeling. *Journal of the American Statistical Association*, 99(466):537–545.

Gelman, A., Carlin, J. B., Stern, H. S., and Rubin, D. B. (2004). *Bayesian Data Analysis*. Chapman & Hall/CRC, London, 2nd edition.

Gelman, A. and Hill, J. (2007). *Data Analysis Using Regression and Multilevel/Hierarchical Models*. Cambridge University Press, Cambridge.

Gelman, A., Jakulin, A., Grazia Pittau, M., and Su, Y. S. (2008). A weakly informative default prior distribution for logistic and other regression models. *Annals of Applied Statistics*, 2(4):1360–1383.

Gelman, A., King, G., and Liu, C. (1998). Not asked and not answered: Multiple imputation for multiple surveys. *Journal of the American Statistical Association*, 93(443):846–857.

Gelman, A. and Meng, X.-L., editors (2004). *Applied Bayesian Modeling and Causal Inference from Incomplete-Data Perspectives*. John Wiley & Sons, Chichester, UK.

Gelman, A. and Speed, T. P. (1993). Characterizing a joint probability distribution by conditionals. *Journal of the Royal Statistical Society B*, 55(1):185–188.

Geskus, R. B. (2001). Methods for estimating the AIDS incubation time distribution when date of seroconversion is censored. *Statistics in Medicine*, 20(5):795–812.

Ghosh-Dastidar, B. and Schafer, J. L. (2003). Multiple edit/multiple imputation for multivariate continuous data. *Journal of the American Statistical Association*, 98(464):807–817.

Gibson, N. M. and Olejnik, S. (2003). Treatment of missing data at the second level of hierarchical linear models. *Educational and Psychological Measurement*, 63(2):204–238.

Gilks, W. R. (1996). Full conditional distributions. In Gilks, W. R., Richardson, S., and Spiegelhalter, D. J., editors, *Markov Chain Monte Carlo in Practice*, pages 75–88. Chapman & Hall, London.

Gill, R. D., Van der Laan, M. L., and Robins, J. M. (1997). Coarsening at random: Characterizations, conjectures and counter-examples. In Lin, D. Y. and Fleming, T. R., editors, *Proceedings of the First Seattle Conference on Biostatistics*, pages 255–294, Berlin. Springer-Verlag.

Gleason, T. C. and Staelin, R. (1975). A proposal for handling missing data. *Psychometrika*, 40(2):229–252.

Glickman, M. E., He, Y., Yucel, R. M., and Zaslavsky, A. M. (2008). Misreporting, missing data, and multiple imputation: Improving accuracy of cancer registry databases. *Chance*, 21(3):55–58.

Glynn, R. J., Laird, N. M., and Rubin, D. B. (1986). Selection modeling versus mixture modeling with nonignorable nonresponse. In Wainer, H., editor, *Drawing Inferences from Self-Selected Samples*, pages 115–142. Springer-Verlag.

Glynn, R. J. and Rosner, B. (2004). Multiple imputation to estimate the association between eyes in disease progression with interval-censored data. *Statistics in Medicine*, 23(21):3307–3318.

Goldstein, H. (2011a). Bootstrapping in multilevel models. In Hox, J. and Roberts, J., editors, *The Handbook of Advanced Multilevel Analysis*, chapter 9, pages 163–171. Routledge, Milton Park, UK.

Goldstein, H. (2011b). *Multilevel Statistical Models*. John Wiley & Sons, Chichester, UK, 4th edition.

Goldstein, H. and Carpenter, J. R. (2015). Multilevel multiple imputation. In Molenberghs, G., Fitzmaurice, G. M., Kenward, M. G., Tsiatis, A. A., and Verbeke, G., editors, *Handbook of Missing Data Methodology*, pages 295–316. Chapman & Hall/CRC Press, Boca Raton, FL.

Goldstein, H., Carpenter, J. R., and Browne, W. J. (2014). Fitting multilevel multivariate models with missing data in responses and covariates that may include interactions and non-linear terms. *Journal of the Royal Statistical Society: Series A*, 177(2):553–564.

Goldstein, H., Carpenter, J. R., Kenward, M. G., and Levin, K. A. (2009). Multilevel models with multivariate mixed response types. *Statistical Modelling*, 9(3):173–179.

Gomes, M., Gutacker, N., Bojke, C., and Street, A. (2016). Addressing missing data in patient-reported outcome measures (PROMS): Implications for the use of PROMS for comparing provider performance. *Health Economics*, 25(5):515–528.

Gonzalez, J. M. and Eltinge, J. L. (2007). Multiple matrix sampling: A review. In *ASA 2007 Proceedings of the Section on Survey Research Methods*, pages 3069–3075, Alexandria, VA.

Goodman, L. A. (1970). The multivariate analysis of qualitative data: Interactions among multiple classifications. *Journal of the American Statistical Association*, 65(329):226–256.

Gorber, S. C., Tremblay, M., Moher, D., and Gorber, B. (2007). A comparison of direct vs. self-report measures for assessing height, weight and body mass index: A systematic review. *Obesity Reviews*, 8(4):307–326.

Gordon, M. (2014). Parallel computation of multiple imputation by using `mice` R package. *https://stackoverflow.com/questions/24040280/parallel-computation-of-multiple-imputation-by-using-mice-r-package*.

Graham, J. W. (2012). *Missing Data: Analysis and Design*. Springer, New York.

Graham, J. W., Olchowski, A. E., and Gilreath, T. D. (2007). How many imputations are really needed? Some practical clarifications of multiple imputation theory. *Preventive Science*, 8(3):206–213.

Graham, J. W., Taylor, B. J., Olchowski, A. E., and Cumsille, P. E. (2006). Planned missing data designs in psychological research. *Psychological Methods*, 11(4):323–343.

Greenland, S. and Finkle, W. D. (1995). A critical look at methods for handling missing covariates in epidemiologic regression analyses. *American Journal of Epidemiology*, 142(12):1255–1264.

Greenwald, R. and Rubin, A. (1999). Brief assessment of children's post-traumatic symptoms: Development and preliminary validation of parent and child scales. *Research on Social Work Practice*, 9(1):61–75.

Groenwold, R. H. H., White, I. R., Donders, A. R. T., Carpenter, J. R., Altman, D. G., and Moons, K. G. M. (2012). Missing covariate data in clinical research: When and when not to use the missing-indicator method for analysis. *Canadian Medical Association Journal*, 184(11):1265–1269.

Groothuis-Oudshoorn, C. G. M., Van Buuren, S., and Van Rijckevorsel, J. L. A. (1999). Flexible multiple imputation by chained equations of the AVO-95 survey. Technical Report PG/VGZ/00.045, TNO Prevention and Health, Leiden.

Groves, R. M., Fowler Jr., F. J., Couper, M. P., Lepkowski, J. M., Singer, E., and Tourangeau, R. (2009). *Survey Methodology*. John Wiley & Sons, New York, 2nd edition.

Grund, S., Lüdtke, O., and Robitzsch, A. (2016a). Multiple imputation of missing covariate values in multilevel models with random slopes: A cautionary note. *Behavior Research Methods*, 48(2):640–649.

Grund, S., Lüdtke, O., and Robitzsch, A. (2016b). Pooling ANOVA results from multiply imputed datasets. *Methodology*, 12(3):75–88.

Grund, S., Lüdtke, O., and Robitzsch, A. (2018a). Multiple imputation of missing data at level 2: A comparison of fully conditional and joint modeling in multilevel designs. *Journal of Educational and Behavioral Statistics*, doi.org/10.3102/1076998617738087.

Grund, S., Lüdtke, O., and Robitzsch, A. (2018b). Multiple imputation of missing data for multilevel models: Simulations and recommendations. *Organizational Research Methods*, 21(1):111–149.

Grund, S., Robitzsch, A., and Lüdtke, O. (2018c). *mitml: Tools for Multiple Imputation in Multilevel Modeling*. R package version 0.3-5.7.

Gutman, R. and Rubin, D. B. (2015). Estimation of causal effects of binary treatments in unconfounded studies. *Statistics in Medicine*, 34(26):3381–3398.

Hand, D. J., Daly, F., Lunn, A. D., McConway, K. J., and Ostrowski, E. (1994). *A Handbook of Small Data Sets*. Chapman & Hall, London.

Hardt, J., Herke, M., Brian, T., and Laubach, W. (2013). Multiple imputation of missing data: A simulation study on a binary response. *Open Journal of Statistics*, 3(5).

Hare, R. D., Clark, D., Grann, M., and Thornton, D. (2000). Psychopathy and the predictive validity of the PCL-R: An international perspective. *Behavioral Sciences & the Law*, 18(5):623–645.

Harel, O. (2009). The estimation of r^2 and adjusted r^2 in incomplete data sets using multiple imputation. *Journal of Applied Statistics*, 36(10):1109–1118.

Harel, O. and Zhou, X. H. (2006). Multiple imputation for correcting verification bias. *Statistics in Medicine*, 25(22):3769–3786.

Harkness, J. A., Van de Vijver, F. J. R., and Mohler, P. P., editors (2002). *Cross-Cultural Survey Methods*. John Wiley & Sons, New York.

Harrell, F. E. (2001). *Regression Modeling Strategies*. Springer-Verlag, New York.

Harvey, A. C. (1981). *The Econometric Analysis of Time Series*. Philip Allen, Oxford.

He, Y. (2006). *Missing Data Imputation for Tree-Based Models*. PhD thesis, University of California, Los Angeles, CA.

He, Y. and Raghunathan, T. E. (2006). Tukey's *gh* distribution for multiple imputation. *The American Statistician*, 60(3):251–256.

Heckerman, D., Chickering, D. M., Meek, C., Rounthwaite, R., and Kadie, C. (2001). Dependency networks for inference, collaborative filtering, and data visualisation. *Journal of Machine Learning Research*, 1(1):49–75.

Heckman, J. J. (1976). The common structure of statistical models of truncation, sample selection and limited dependent variables and a simple estimator for such models. *Annals of Economic and Social Measurement*, 5(4):475–492.

Heinze, G. and Schemper, M. (2002). A solution to the problem of separation in logistic regression. *Statistics in Medicine*, 21(16):2409–2419.

Heitjan, D. F. (1993). Ignorability and coarse data: Some biomedical examples. *Biometrics*, 49(4):1099–1109.

Heitjan, D. F. and Little, R. J. A. (1991). Multiple imputation for the fatal accident reporting system. *Journal of the Royal Statistical Society C*, 40(1):13–29.

Heitjan, D. F. and Rubin, D. B. (1990). Inference from coarse data via multiple imputation with application to age heaping. *Journal of the American Statistical Association*, 85(410):304–314.

Heitjan, D. F. and Rubin, D. B. (1991). Ignorability and coarse data. *Annals of Statistics*, 19(4):2244–2253.

Herrington, D. M., Howard, T. D., Hawkins, G. A., Reboussin, D. M., Xu, J., Zheng, S. L., Brosnihan, K. B., Meyers, D. A., and Bleecker, E. R. (2002). Estrogen-receptor polymorphisms and effects of estrogen replacement on high-density lipoprotein cholesterol in women with coronary disease. *New England Journal of Medicine*, 346(13):967–974.

Herzog, T. N. and Rubin, D. B. (1983). Using multiple imputations to handle nonresponse in sample surveys. In Madow, W., Olkin, I., and Rubin, D. B., editors, *Incomplete Data in Sample Surveys*, volume 2, chapter 15, pages 209–245. Academic Press.

Herzog, T. N., Scheuren, F. J., and Winkler, W. E. (2007). *Data Quality and Record Linking Techniques*. Springer, New York.

Heymans, M. W., Van Buuren, S., Knol, D. L., Van Mechelen, W., and De Vet, H. C. W. (2007). Variable selection under multiple imputation using the bootstrap in a prognostic study. *BMC Medical Research Methodology*, 7:33.

Hill, P. W. and Goldstein, H. (1998). Multilevel modeling of educational data with cross-classification and missing identification for units. *Journal of Educational and Behavioral Statistics*, 23(2):117–128.

Hille, E. T. M., Elbertse, L., Bennebroek Gravenhorst, J., Brand, R., and Verloove-Vanhorick, S. P. (2005). Nonresponse bias in a follow-up study of 19-year-old adolescents born as preterm infants. *Pediatrics*, 116(5):662–666.

Hille, E. T. M., Weisglas-Kuperus, N., Van Goudoever, J. B., Jacobusse, G. W., Ens-Dokkum, M. H., De Groot, L., Wit, J. M., Geven, W. B., Kok, J. H., De Kleine, M. J. K., Kollée, L. A. A., Mulder, A. L. M., Van Straaten, H. L. M., De Vries, L. S., Van Weissenbruch, M. M., and Verloove-Vanhorick, S. P. (2007). Functional outcomes and participation in young adulthood for very preterm and very low birth weight infants: The

Dutch project on preterm and small for gestational age infants at 19 years of age. *Pediatrics*, 120(3):587–595.

Holland, P. W. (1986). Statistics and causal inference. *Journal of the American Statistical Association*, 81(396):945–960.

Holland, P. W. (2007). A framework and history for score linking. In Dorans, N. J., Pommerich, M., and Holland, P. W., editors, *Linking and Aligning Scores and Scales*, chapter 2, pages 5–30. Springer, New York.

Holland, P. W. and Wainer, H., editors (1993). *Differential Item Functioning.* Lawrence Erlbaum Associates, Hillsdale, NJ.

Hopke, P. K., Liu, C., and Rubin, D. B. (2001). Multiple imputation for multivariate data with missing and below-threshold measurements: Time-series concentrations of pollutants in the Arctic. *Biometrics*, 57(1):22–33.

Hopman-Rock, M., Dusseldorp, E., Chorus, A. M. J., Jacobusse, G. W., Rütten, A., and Van Buuren, S. (2012). Response conversion for improving comparability of international physical activity data. *Journal of Physical Activity & Health*.

Horton, N. J. and Kleinman, K. P. (2007). Much ado about nothing: A comparison of missing data methods and software to fit incomplete data regression models. *The American Statistician*, 61(1):79–90.

Horton, N. J., Lipsitz, S. R., and Parzen, M. (2003). A potential for bias when rounding in multiple imputation. *The American Statistician*, 57(4):229–232.

Hosmer, D. W. and Lemeshow, S. (2000). *Applied Logistic Regression.* John Wiley & Sons, New York, 2nd edition.

Hosmer, D. W., Lemeshow, S., and May, S. (2008). *Applied Survival Analysis: Regression Modeling of Time to Event Data.* John Wiley & Sons, Hoboken, NJ, 2nd edition.

Hox, J. J., Moerbeek, M., and Van de Schoot, R. (2018). *Multilevel Analysis: Techniques and Applications. Third Edition.* Routledge, New York.

Hron, K., Templ, M., and Filzmoser, P. (2010). Imputation of missing values for compositional data using classical and robust methods. *Computational Statistics & Data Analysis*, 54(12):3095–3107.

Hsu, C. H. (2007). Multiple imputation for interval censored data with auxiliary variables. *Statistics in Medicine*, 26(4):769–781.

Hsu, C. H., Taylor, J. M. G., and Hu, C. (2015). Analysis of accelerated failure time data with dependent censoring using auxiliary variables via nonparametric multiple imputation. *Statistics in Medicine*, 34(19):2768–2780.

Hsu, C. H., Taylor, J. M. G., Murray, S., and Commenges, D. (2006). Survival analysis using auxiliary variables via non-parametric multiple imputation. *Statistics in Medicine*, 25(20):3503–3517.

Hughes, R. A., White, I. R., Seaman, S. R., Carpenter, J. R., Tilling, K., and Sterne, J. A. C. (2014). Joint modelling rationale for chained equations. *BMC Medical Research Methodology*, 14(1):28.

Hui, S. L. and Berger, J. O. (1983). Empirical Bayes estimation of rates in longitudinal studies. *Journal of the American Statistical Association*, 78(384):753–760.

Imai, K. and Ratkovic, M. (2013). Estimating treatment effect heterogeneity in randomized program evaluation. *Annals of Applied Statistics*, 7(1):443–470.

Imbens, G. W. and Rubin, D. B. (2015). *Causal Inference in Statistics, Social, and Biomedical Sciences*. Cambridge University Press, Cambridge, U.K.

Ip, E. H. and Wang, Y. J. (2009). Canonical representation of conditionally specified multivariate discrete distributions. *Journal of Multivariate Analysis*, 100(6):1282–1290.

Ishwaran, H., Kogalur, U. B., Blackstone, E. H., and Lauer, M. S. (2008). Random survival forests. *Annals of Applied Statistics*, 2(3):841–860.

Izaks, G. J., Van Houwelingen, H. C., Schreuder, G. M., and Ligthart, G. J. (1997). The association between human leucocyte antigens (HLA) and mortality in community residents aged 85 and older. *Journal of the American Geriatrics Society*, 45(1):56–60.

Jackson, D., White, I. R., Seaman, S. R., Evans, H., Baisley, K., and Carpenter, J. R. (2014). Relaxing the independent censoring assumption in the cox proportional hazards model using multiple imputation. *Statistics in Medicine*, 33(27):4681–4694.

James, I. R. and Tanner, M. A. (1995). A note on the analysis of censored regression data by multiple imputation. *Biometrics*, 51(1):358–362.

Javaras, K. N. and Van Dyk, D. A. (2003). Multiple imputation for incomplete data with semicontinuous variables. *Journal of the American Statistical Association*, 98(463):703–715.

Jeličić, H., Phelps, E., and Lerner, R. M. (2009). Use of missing data methods in longitudinal studies: The persistence of bad practices in developmental psychology. *Developmental Psychology*, 45(4):1195–1199.

Jennrich, R. I. and Schluchter, M. D. (1986). Unbalanced repeated-measures models with structured covariance matrices. *Biometrics*, 42(4):805–820.

Jolani, S. (2012). *Dual Imputation Strategies for Analyzing Incomplete Data.* PhD thesis, University of Utrecht, Utrecht.

Jolani, S. (2018). Hierarchical imputation of systematically and sporadically missing data: An approximate Bayesian approach using chained equations. *Biometrical Journal*, 60(2):333–351.

Jolani, S., Debray, T. P. A., Koffijberg, H., Van Buuren, S., and Moons, K. G. M. (2015). Imputation of systematically missing predictors in an individual participant data meta-analysis: A generalized approach using MICE. *Statistics in Medicine*, 34(11):1841–1863.

Jones, H. E. and Spiegelhalter, D. J. (2009). Accounting for regression-to-the-mean in tests for recent changes in institutional performance: Analysis and power. *Statistics in Medicine*, 30(12):1645–1667.

Kaciroti, N. A. and Raghunathan, T. E. (2014). Bayesian sensitivity analysis of incomplete data: Bridging pattern-mixture and selection models. *Statistics in Medicine*, 33(27):4841–4857.

Kang, J. D. Y. and Schafer, J. L. (2007). Demystifying double robustness: A comparison of alternative strategies for estimating a population mean from incomplete data. *Statistical Science*, 22(4):523–539.

Karahalios, A., Baglietto, L., Carlin, J. B., English, D. R., and Simpson, J. A. (2012). A review of the reporting and handling of missing data in cohort studies with repeated assessment of exposure measures. *BMC Medical Research Methodology*, 12:96.

Kasim, R. M. and Raudenbush, S. W. (1998). Application of Gibbs sampling to nested variance components models with heterogeneous within-group variance. *Journal of Educational and Behavioral Statistics*, 23(2):93–116.

Katsikatsou, M., Moustaki, I., Yang-Wallentin, F., and Jöreskog, K. (2012). Pairwise likelihood estimation for factor analysis models with ordinal data. *Computational Statistics & Data Analysis*, 56(12):4243–4258.

Keller, B. T. and Enders, C. K. (2017). *Blimp Users Guide 1.0.* Los Angeles, CA.

Kennickell, A. B. (1991). Imputation of the 1989 survey of consumer finances: Stochastic relaxation and multiple imputation. *ASA 1991 Proceedings of the Section on Survey Research Methods*, pages 1–10.

Kenward, M. G. and Molenberghs, G. (2009). Last observation carried forward: A crystal ball? *Journal of Biopharmaceutical Statistics*, 19(5):872–888.

Kenward, M. G. and Molenberghs, G. (2015). A perspective and historical overview on selection, pattern-mixture and shared parameter models. In Molenberghs, G., Fitzmaurice, G. M., Kenward, M. G., Tsiatis, A. A., and Verbeke, G., editors, *Handbook of Missing Data Methodology*, pages 53–89. Chapman & Hall/CRC Press, Boca Raton, FL.

Khare, M., Little, R. J. A., Rubin, D. B., and Schafer, J. L. (1993). Multiple imputation of NHANES III. In *ASA 1993 Proceedings of the Survey Research Methods Section*, volume 1, pages 297–302, Alexandria, VA.

Kim, J. K., Brick, J. M., Fuller, W. A., and Kalton, G. (2006). On the bias of the multiple-imputation variance estimator in survey sampling. *Journal of the Royal Statistical Society B*, 68(3):509–521.

Kim, J. K. and Shao, J. (2013). *Statistical Methods for Handling Incomplete Data*. Chapman & Hall/CRC Press, Boca Raton, FL.

Kim, S., Sugar, C., and Belin, T. R. (2015). Evaluating model-based imputation methods for missing covariates in regression models with interactions. *Statistics in Medicine*, 34(11):1876–1888.

King, G., Honaker, J., Joseph, A., and Scheve, K. (2001). Analyzing incomplete political science data: An alternative algorithm for multiple imputation. *American Political Science Review*, 95(1):49–69.

King, G., Murray, C. J. L., Salomon, J. A., and Tandon, A. (2004). Enhancing the validity and cross-cultural comparability of measurement in survey research. *American Political Science Review*, 98(1):191–207.

Klebanoff, M. A. and Cole, S. R. (2008). Use of multiple imputation in the epidemiologic literature. *American Journal of Epidemiology*, 168(4):355–357.

Kleinbaum, D. G. and Klein, M. B. (2005). *Survival Analysis: A Self-Learning Text*. Springer-Verlag, New York, 2nd edition.

Kleinke, K. (2017). Multiple imputation under violated distributional assumptions: A systematic evaluation of the assumed robustness of predictive mean matching. *Journal of Educational and Behavioral Statistics*, 42(4):371–404.

Kleinke, K. and Reinecke, J. (2013). Multiple imputation of incomplete zero-inflated count data. *Statistica Neerlandica*, 67(3):311–336.

Kleinke, K. and Reinecke, J. (2015). Multiple imputation of multilevel count data. In Engel, U., Jann, B., Lynn, P., Scherpenzeel, A., and Sturgis, P., editors, *Improving survey methods: Lessons from recent research*, pages 381–396. Routledge, New York.

Knol, M. J., Janssen, K. J. M., Donders, A. R. T., Egberts, A. C. G., Heerdink, E. R., Grobbee, D. E., Moons, K. G. M., and Geerlings, M. I. (2010). Unpredictable bias when using the missing indicator method or complete case analysis for missing confounder values: An empirical example. *Journal of Clinical Epidemiology*, 63:728–736.

Kolen, M. J. and Brennan, R. L. (1995). *Test Equating: Methods and Practices*. Springer, New York.

Koller-Meinfelder, F. (2009). *Analysis of Incomplete Survey Data – Multiple Imputation via Bayesian Bootstrap Predictive Mean Matching*. PhD thesis, University of Bamberg, Bamberg, Germany.

Kreft, I. G., De Leeuw, J., and Aiken, L. S. (1995). The effect of different forms of centering in hierarchical linear models. *Multivariate Behavioral Research*, 30(1):1–21.

Kropko, J., Goodrich, B., Gelman, A., and Hill, J. (2014). Multiple imputation for continuous and categorical data: Comparing joint multivariate normal and conditional approaches. *Political Analysis*, 22(4):497–519.

Krul, A., Daanen, H. A. M., and Choi, H. (2010). Self-reported and measured weight, height and body mass index (BMI) in Italy, The Netherlands and North America. *European Journal of Public Health*, 21(4):414–419.

Kunkel, D. and Kaizar, E. E. (2017). A comparison of existing methods for multiple imputation in individual participant data meta-analysis. *Statistics in Medicine*, 36(22):3507–3532.

Kuo, K.-L., Song, C.-C., and Jiang, T. J. (2017). Exactly and almost compatible joint distributions for high-dimensional discrete conditional distributions. *Journal of Multivariate Analysis*, 157:115–123.

Lagaay, A. M., Van der Meij, J. C., and Hijmans, W. (1992). Validation of medical history taking as part of a population based survey in subjects aged 85 and over. *British Medical Journal*, 304(6834):1091–1092.

Laird, N. M. and Ware, J. H. (1982). Random-effects models for longitudinal data. *Biometrics*, 38(4):963–974.

Lam, K. F., Tang, O. Y., and Fong, D. Y. T. (2005). Estimating the proportion of cured patients in a censored sample. *Statistics in Medicine*, 24(12):1865–1879.

Lam, P. K. (2013). *Estimating Individual Causal Effects*. PhD thesis, Harvard University, Cambridge MA.

Lange, K. L., Little, R. J. A., and Taylor, J. M. G. (1989). Robust statistical modeling using the t distribution. *Journal of the American Statistical Association*, 84(408):881–896.

Lee, H., Rancourt, E., and Särndal, C. E. (1994). Experiments with variance estimation from survey data with imputed values. *Journal of Official Statistics*, 10(3):231–243.

Lee, K. J. and Carlin, J. B. (2010). Multiple imputation for missing data: Fully conditional specification versus multivariate normal imputation. *American Journal of Epidemiology*, 171(5):624–632.

Lee, K. J., Galati, J. C., Simpson, J. A., and Carlin, J. B. (2012). Comparison of methods for imputing ordinal data using multivariate normal imputation: a case study of non-linear effects in a large cohort study. *Statistics in Medicine*, 31(30):4164–4174.

Lee, M., Rahbar, M. H., Brown, M., Gensler, L., Weisman, M., Diekman, L., and Reveille, J. D. (2018). A multiple imputation method based on weighted quantile regression models for longitudinal censored biomarker data with missing values at early visits. *BMC Medical Research Methodology*, 18(1):8.

Lesaffre, E. M. and Albert, A. (1989). Partial separation in logistic discrimination. *Journal of the Royal Statistical Society B*, 51(1):109–116.

Li, F., Baccini, M., Mealli, F., Zell, E. R., Frangakis, C. E., and Rubin, D. B. (2014). Multiple imputation by ordered monotone blocks with application to the anthrax vaccine research program. *Journal of Computational and Graphical Statistics*, 23(3):877–892.

Li, F., Yu, Y., and Rubin, D. B. (2012). Imputing missing data by fully conditional models: Some cautionary examples and guideline. *Duke University Department of Statistical Science Discussion Paper*, 11-24.

Li, K.-H. (1988). Imputation using Markov chains. *Journal of Statistical Computation and Simulation*, 30(1):57–79.

Li, K.-H., Meng, X.-L., Raghunathan, T. E., and Rubin, D. B. (1991a). Significance levels from repeated p-values with multiply-imputed data. *Statistica Sinica*, 1(1):65–92.

Li, K.-H., Raghunathan, T. E., and Rubin, D. B. (1991b). Large-sample significance levels from multiply imputed data using moment-based statistics and an F reference distribution. *Journal of the American Statistical Association*, 86(416):1065–1073.

Liaw, A. and Wiener, M. (2002). Classification and regression by randomForest. *R News*, 2(3):18–22.

Licht, C. (2010). *New methods for generating significance levels from multiply-imputed data*. PhD thesis, University of Bamberg, Bamberg, Germany.

Lipsitz, S. R., Parzen, M., and Zhao, L. P. (2002). A degrees-of-freedom approximation in multiple imputation. *Journal of Statistical Computation and Simulation*, 72(4):309–318.

Little, R. J. A. (1988). Missing-data adjustments in large surveys (with discussion). *Journal of Business Economics and Statistics*, 6(3):287–301.

Little, R. J. A. (1992). Regression with missing X's: A review. *Journal of the American Statistical Association*, 87(420):1227–1237.

Little, R. J. A. (1993). Pattern-mixture models for multivariate incomplete data. *Journal of the American Statistical Association*, 88(421):125–134.

Little, R. J. A. (1995). Modeling the drop-out mechanism in repeated-measures studies. *Journal of the American Statistical Association*, 90(431):1112–1121.

Little, R. J. A. (2009). Selection and pattern-mixture models. In Fitzmaurice, G. M., Davidian, M., Verbeke, G., and Molenberghs, G., editors, *Longitudinal data analysis*, chapter 18, pages 409–431. CRC Press, Boca Raton, FL.

Little, R. J. A. (2013). In praise of simplicity not mathematistry! Ten simple powerful ideas for the statistical scientist. *Journal of the American Statistical Association*, 108(502):359–369.

Little, R. J. A. and Rubin, D. B. (1987). *Statistical Analysis with Missing Data*. John Wiley & Sons, New York.

Little, R. J. A. and Rubin, D. B. (2002). *Statistical Analysis with Missing Data*. John Wiley & Sons, New York, 2nd edition.

Little, R. J. A., Rubin, D. B., and Zangeneh, S. Z. (2017). Conditions for ignoring the missing-data mechanism in likelihood inferences for parameter subsets. *Journal of the American Statistical Association*, 112(517):314–320.

Little, T. D. and Rhemtulla, M. (2013). Planned missing data designs for developmental researchers. *Child Development Perspectives*, 7(4):199–204.

Liu, C. (1993). Barlett's decomposition of the posterior distribution of the covariance for normal monotone ignorable missing data. *Journal of Multivariate Analysis*, 46(2):198–206.

Liu, C. (1995). Missing data imputation using the multivariate t distribution. *Journal of Multivariate Analysis*, 53(1):139–158.

Liu, C. and Rubin, D. B. (1998). Ellipsoidally symmetric extensions of the general location model for mixed categorical and continuous data. *Biometrika*, 85(3):673–688.

Liu, J., Gelman, A., Hill, J., Su, Y. S., and Kropko, J. (2013). On the stationary distribution of iterative imputations. *Biometrika*, 101(1):155–173.

Liu, L. X., Murray, S., and Tsodikov, A. (2011). Multiple imputation based on restricted mean model for censored data. *Statistics in Medicine*, 30(12):1339–1350.

Liu, Y. and Enders, C. K. (2017). Evaluation of multi-parameter test statistics for multiple imputation. *Multivariate Behavioral Research*, 53(3):371–390.

Liu, Y., Wang, Y., Feng, Y., and Wall, M. M. (2016). Variable selection and prediction with incomplete high-dimensional data. *The Annals of Applied Statistics*, 10(1):418.

Liublinska, V. and Rubin, D. B. (2014). Sensitivity analysis for a partially missing binary outcome in a two-arm randomized clinical trial. *Statistics in Medicine*, 33(24):4170–4185.

Lloyd, L. J., Langley-Evans, S. C., and McMullen, S. (2010). Childhood obesity and adult cardiovascular disease risk: A systematic review. *International Journal of Obesity*, 34(1):18–28.

Long, Q. and Johnson, B. A. (2015). Variable selection in the presence of missing data: Resampling and imputation. *Biostatistics*, 16(3):596–610.

Loong, B. and Rubin, D. B. (2017). Multiply-imputed synthetic data: Advice to the imputer. *Journal of Official Statistics*, 33(4):1005–1019.

Lüdtke, O., Marsh, H. W., Robitzsch, A., Trautwein, U., Asparouhov, T., and Muthén, B. O. (2008). The multilevel latent covariate model: A new, more reliable approach to group-level effects in contextual studies. *Psychological Methods*, 13(3):203.

Lüdtke, O., Robitzsch, A., and Grund, S. (2017). Multiple imputation of missing data in multilevel designs: A comparison of different strategies. *Psychological Methods*, 22(1):141–165.

Lyles, R. H., Fan, D., and Chuachoowong, R. (2001). Correlation coefficient estimation involving a left censored laboratory assay variable. *Statistics in Medicine*, 20(19):2921–2933.

Lynn, H. S. (2001). Maximum likelihood inference for left-censored HIV RNA data. *Statistics in Medicine*, 20(1):33–45.

MacKay, D. J. C. (2003). *Information Theory, Inference, and Learning Algorithms*. Cambridge University Press, Cambridge.

Mackinnon, A. (2010). The use and reporting of multiple imputation in medical research – A review. *Journal of Internal Medicine*, 268(6):586–593.

Madow, W. G., Olkin, I., and Rubin, D. B., editors (1983). *Incomplete Data in Sample Surveys*, volume 2. Academic Press, New York.

Mallinckroth, C. H. (2013). *Preventing and Treating Missing Data in Longitudinal Clinical Trials: A Practical Guide*. Cambridge University Press, Cambridge, UK.

Marino, M., Buxton, O. M., and Li, Y. (2017). Covariate selection for multilevel models with missing data. *Stat*, 6(1):31–46.

Marker, D. A., Judkins, D. R., and Winglee, M. (2002). Large-scale imputation for complex surveys. In Groves, R. M., Dillman, D. A., Eltinge, J. L., and Little, R. J. A., editors, *Survey Nonresponse*, chapter 22, pages 329–341. John Wiley & Sons, New York.

Marsh, H. W. (1998). Pairwise deletion for missing data in structural equation models: Nonpositive definite matrices, parameter estimates, goodness of fit, and adjusted sample sizes. *Structural Equation Modeling*, 5(1):22–36.

Marshall, A., Altman, D. G., and Holder, R. L. (2010a). Comparison of imputation methods for handling missing covariate data when fitting a Cox proportional hazards model: A resampling study. *BMC Medical Research Methodology*, 10:112.

Marshall, A., Altman, D. G., Royston, P., and Holder, R. L. (2010b). Comparison of techniques for handling missing covariate data within prognostic modelling studies: A simulation study. *BMC Medical Research Methodology*, 10:7.

Marshall, A., Billingham, L. J., and Bryan, S. (2009). Can we afford to ignore missing data in cost-effectiveness analyses? *European Journal of Health Economics*, 10(1):1–3.

Matsumoto, D. and Van de Vijver, F. J. R., editors (2010). *Cross-Cultural Research Methods in Psychology*. Cambridge University Press, Cambridge.

McCullagh, P. and Nelder, J. A. (1989). *Generalized Linear Models*. Chapman & Hall, New York, 2nd edition.

McCulloch, C. E. and Searle, S. R. (2001). *Generalized, Linear, and Mixed Models*. John Wiley & Sons, New York.

McKnight, P. E., McKnight, K. M., Sidani, S., and Figueredo, A. J. (2007). *Missing Data: A Gentle Introduction*. Guilford Press, New York.

Meng, X.-L. (1994). Multiple imputation with uncongenial sources of input (with discusson). *Statistical Science*, 9(4):538–573.

Meng, X.-L. and Rubin, D. B. (1992). Performing likelihood ratio tests with multiply-imputed data sets. *Biometrika*, 79(1):103–111.

Miettinen, O. S. (1985). *Theoretical Epidemiology: Principles of Occurence Research in Medicine*. John Wiley & Sons, New York.

Mislevy, R. J. (1991). Randomization-based inferences about latent variables from complex samples. *Psychometrika*, 1991(2):177–196.

Mistler, S. A. and Enders, C. K. (2017). A comparison of joint model and fully conditional specification imputation for multilevel missing data. *Journal of Educational and Behavioral Statistics*, 42(4):432–466.

Molenberghs, G., Fitzmaurice, G. M., Kenward, M. G., Tsiatis, A. A., and Verbeke, G. (2015). *Handbook of Missing Data Methodology*. Chapman & Hall/CRC Press, Baco Raton, FL.

Molenberghs, G. and Kenward, M. G. (2007). *Missing Data in Clinical Studies*. John Wiley & Sons, Chichester.

Molenberghs, G. and Verbeke, G. (2005). *Models for Discrete Longitudinal Data*. Springer, New York.

Moons, K. G. M., Donders, A. R. T., Stijnen, T., and Harrell, F. E. (2006). Using the outcome for imputation of missing predictor values was preferred. *Journal of Clinical Epidemiology*, 59(10):1092–1101.

Morgan, S. L. and Harding, D. J. (2006). Matching estimators of causal effects prospects and pitfalls in theory and practice. *Sociological Methods & Research*, 35(1):3–60.

Moriarity, C. and Scheuren, F. J. (2003). Note on Rubin's statistical matching using file concatenation with adjusted weights and multiple imputations. *Journal of Business Economics and Statistics*, 21(1):65–73.

Morris, T. P., White, I. R., and Royston, P. (2014). Tuning multiple imputation by predictive mean matching and local residual draws. *BMC Medical Research Methodology*, 14:75.

Musoro, J. Z., Zwinderman, A. H., Puhan, M. A., Ter Riet, G., and Geskus, R. B. (2014). Validation of prediction models based on lasso regression with multiply imputed data. *BMC Medical Research Methodology*, 14(1):116.

Muthén, B. O., Muthén, L. K., and Asparouhov, T. (2016). *Regression and Mediation Analysis Using M*plus. Muthén & Muthén, Los Angeles, CA.

Naaktgeboren, C. A., De Groot, J. A. H., Rutjes, A. W. S., Bossuyt, P. M. M., Reitsma, J. B., and Moons, K. G. M. (2016). Anticipating missing reference standard data when planning diagnostic accuracy studies. *British Medical Journal*, 352:i402.

National Research Council (2010). *The Prevention and Treatment of Missing Data in Clinical Trials*. The National Academies Press, Washington, D.C.

Netten, A. P., Dekker, F. W., Rieffe, C., Soede, W., Briaire, J. J., and Frijns, J. H. (2017). Missing data in the field of otorhinolaryngology and head & neck surgery: Need for improvement. *Ear and Hearing*, 38(1):1–6.

Neyman, J. (1923). On the application of probability theory to agricultural experiments. essay on principles. section 9. *Annals of Agricultural Sciences*, pages 1–51.

Neyman, J. (1935). Statistical problems in agricultural experimentation (with discussion). *Journal of the Royal Statistical Society, Series B*, Suppl.(2):107–180.

Nguyen, C. D., Carlin, J. B., and Lee, K. J. (2017). Model checking in multiple imputation: an overview and case study. *Emerging Themes in Epidemiology*, 14(1):8.

Nielsen, S. F. (2003). Proper and improper multiple imputation. *International Statistical Review*, 71(3):593–627.

O'Kelly, M. and Ratitch, B. (2014). *Clinical Trials with Missing Data: A Guide for Practitioners*. John Wiley & Sons, Chichester, UK.

Olkin, I. and Tate, R. F. (1961). Multivariate correlation models with discrete and continuous variables. *Annals of Mathematical Statistics*, 32(2):448–465.

Olsen, M. K. and Schafer, J. L. (2001). A two-part random effects model for semicontinuous longitudinal data. *Journal of the American Statistical Association*, 96(454):730–745.

Orchard, T. and Woodbury, M. A. (1972). A missing information principle: Theory and applications. In *Proceedings of the Sixth Berkeley Symposium on Mathematical Statistics and Probability*, volume 1, pages 697–715.

Palmer, M. J., Mercieca-Bebber, R., King, M., Calvert, M., Richardson, H., and Brundage, M. (2018). A systematic review and development of a classification framework for factors associated with missing patient-reported outcome data. *Clinical Trials*, 15(1):95–106.

Pan, W. (2000). A multiple imputation approach to Cox regression with interval-censored data. *Biometrics*, 56(1):199–203.

Pan, W. (2001). A multiple imputation approach to regression analysis for doubly censored data with application to AIDS studies. *Biometrics*, 57(4):1245–1250.

Parker, R. (2010). *Missing Data Problems in Machine Learning*. VDM Verlag Dr. Müller, Saarbrücken, Germany.

Peng, Y., Little, R. J. A., and Raghunathan, T. E. (2004). An extended general location model for causal inferences from data subject to noncompliance and missing values. *Biometrics*, 60(3):598–607.

Permutt, T. (2016). Sensitivity analysis for missing data in regulatory submissions. *Statistics in Medicine*, 35(17):2876–2879.

Peugh, J. L. and Enders, C. K. (2004). Missing data in educational research: A review of reporting practices and suggestions for improvement. *Review of Educational Research*, 74(4):525–556.

Piesse, A., Alvarez-Rojas, L., Judkins, D. R., and Shadish, W. R. (2010). Causal inference using semi-parametric imputation. In *Section on Survey Research Methods - JSM 2010*, pages 1085–1096. American Statistical Association, Alexandria, VA.

Pinheiro, J. C. and Bates, D. M. (2000). *Mixed-Effects Models in S and S-PLUS*. Spinger, New York.

Plumpton, C. O., Morris, T. P., Hughes, D. A., and White, I. R. (2016). Multiple imputation of multiple multi-item scales when a full imputation model is infeasible. *BMC Research Notes*, 9(1):45.

Potthoff, R. F. and Roy, S. N. (1964). A generalized multivariate analysis of variance model usefully especially for growth curve problems. *Biometrika*, 51(3):313–326.

Powney, M., Williamson, P., Kirkham, J., and Kolamunnage-Dona, R. (2014). A review of the handling of missing longitudinal outcome data in clinical trials. *Trials*, 15:237.

Quartagno, M. and Carpenter, J. R. (2016). Multiple imputation for ipd meta-analysis: Allowing for heterogeneity and studies with missing covariates. *Statistics in Medicine*, 35(17):2938–2954.

Quartagno, M. and Carpenter, J. R. (2017). *jomo: A package for Multilevel Joint Modelling Multiple Imputation*.

Rabe-Hesketh, S., Skrondal, A., and Pickles, A. (2002). Reliable estimation of generalized linear mixed models using adaptive quadrature. *The Stata Journal*, 2(1):1–21.

Raghunathan, T. E. (2015). *Missing Data Analysis in Practice*. Chapman & Hall/CRC, Boca Raton, FL.

Raghunathan, T. E. and Grizzle, J. E. (1995). A split questionnaire survey design. *Journal of the American Statistical Association*, 90(429):54–63.

Raghunathan, T. E., Lepkowski, J. M., Van Hoewyk, J., and Solenberger, P. W. (2001). A multivariate technique for multiply imputing missing values using a sequence of regression models. *Survey Methodology*, 27(1):85–95.

Raghunathan, T. E., Reiter, J. P., and Rubin, D. B. (2003). Multiple imputation for statistical disclosure limitation. *Journal of Official Statistics*, 19(1):1–16.

Rao, J. N. K. (1996). On variance estimation with imputed survey data. *Journal of the American Statistical Association*, 91(434):499–505.

Rässler, S. (2002). *Statistical Matching. A Frequentist Theory, Practical Applications, and Alternative Bayesian Approaches*. Springer, New York.

Raudenbush, S. W. and Bryk, A. S. (2002). *Hierarchical linear models: Applications and data analysis methods. Second edition.*, volume 1. Sage, Thousand Oaks, CA.

Reiter, J. P. (2005a). Releasing multiply imputed, synthetic public use microdata: An illustration and empirical study. *Journal of the Royal Statistical Society A*, 168(1):185–205.

Reiter, J. P. (2005b). Using CART to generate partially synthetic public use microdata. *Journal of Official Statistics*, 21(3):7–30.

Reiter, J. P. (2007). Small-sample degrees of freedom for multi-component significance tests with multiple imputation for missing data. *Biometrika*, 94(2):502–508.

Reiter, J. P. (2009). Using multiple imputation to integrate and disseminate confidential microdata. *International Statistical Review*, 77(2):179–195.

Reiter, J. P., Wang, Q., and Zhang, B. (2014). Bayesian estimation of disclosure risks for multiply imputed, synthetic data. *Journal of Privacy and Confidentiality*, 6(1):2.

Renn, S. D. (2005). *Expository Dictionary of Bible Words: Word Studies for Key English Bilble Words Based on the Hebrew and Greek Texts*. Hendrickson Publishers, Peabody, MA.

Resche-Rigon, M. and White, I. R. (2018). Multiple imputation by chained equations for systematically and sporadically missing multilevel data. *Statistical Methods in Medical Research*, doi.org/10.1177/0962280216666564.

Rezvan, H. P., Lee, K. J., and Simpson, J. A. (2015). The rise of multiple imputation: a review of the reporting and implementation of the method in medical research. *BMC Medical Research Methodology*, 15:30.

Rhemtulla, M. and Hancock, G. R. (2016). Planned missing data designs in educational psychology research. *Educational Psychologist*, 51(3-4):305–316.

Rigby, R. A. and Stasinopoulos, D. M. (2005). Generalized additive models for location, scale and shape,(with discussion). *Applied Statistics*, 54:507–554.

Rigby, R. A. and Stasinopoulos, D. M. (2006). Using the Box–Cox t distribution in GAMLSS to model skewness and kurtosis. *Statistical Modelling*, 6(3):209–229.

Roberts, G. O. (1996). Markov chain concepts related to sampling algorithms. In Gilks, W. R., Richardson, S., and Spiegelhalter, D. J., editors, *Markov Chain Monte Carlo in Practice*, pages 45–57. Chapman & Hall, London.

Robins, J. M. and Wang, N. (2000). Inference for imputation estimators. *Biometrika*, 87(1):113–124.

Robinson, D. (2017). `broom 0.4.2`: Convert statistical analysis objects into tidy data frames. *R package*.

Robitzsch, A., Grund, S., and Henke, T. (2017). `miceadds`: *Some additional multiple imputation functions, especially for* `mice`. R package version 2.7-19.

Robitzsch, A. and Lüdtke, O. (2018). `mdmb`: *Model Based Treatment of Missing Data*. R package version 0.6-11.

Rogosa, D. R. and Willett, J. B. (1985). Understanding correlates of change by modeling individual differences in growth. *Psychometrika*, 50(2):203–228.

Rosseel, Y. (2012). `lavaan`: An R package for structural equation modeling. *Journal of Statistical Software*, 48(2):1–36.

Rothwell, P. M. (2005). Subgroup analysis in randomised controlled trials: importance, indications, and interpretation. *The Lancet*, 365(9454):176–186.

Royston, P. (2004). Multiple imputation of missing values. *Stata Journal*, 4(3):227–241.

Royston, P. (2007). Multiple imputation of missing values: further update of ice, with an emphasis on interval censoring. *Stata Journal*, 7(4):445–464.

Royston, P. (2009). Multiple imputation of missing values: Further update of ice, with an emphasis on categorical variables. *Stata Journal*, 9(3):466–477.

Rubin, D. B. (1974). Estimating causal effects of treatments in randomized and nonrandomized studies. *Journal of Educational Psychology*, 66(5):688–701.

Rubin, D. B. (1976). Inference and missing data. *Biometrika*, 63(3):581–590.

Rubin, D. B. (1986). Statistical matching using file concatenation with adjusted weights and multiple imputations. *Journal of Business Economics and Statistics*, 4(1):87–94.

Rubin, D. B. (1987a). *Multiple Imputation for Nonresponse in Surveys*. John Wiley & Sons, New York.

Rubin, D. B. (1987b). A noniterative sample/importance resampling alternative to the data augmentation algorithm for creating a few imputations when the fractions of missing information are modest. *Journal of the American Statistical Association*, 82(398):543–546.

Rubin, D. B. (1993). Discussion: Statistical disclosure limitation. *Journal of Official Statistics*, 9(2):461–468.

Rubin, D. B. (1994). Comments on "Missing data, imputation, and the bootstrap" by Bradley Efron. *Journal of the American Statistical Association*, 89(426):485–488.

Rubin, D. B. (1996). Multiple imputation after 18+ years. *Journal of the American Statistical Association*, 91(434):473–489.

Rubin, D. B. (2000). Causal inference without counterfactuals: Comment. *Journal of the American Statistical Association*, 95(450):435–438.

Rubin, D. B. (2003). Nested multiple imputation of NMES via partially incompatible MCMC. *Statistica Neerlandica*, 57(1):3–18.

Rubin, D. B. (2004a). The design of a general and flexible system for handling nonresponse in sample surveys. *The American Statistician*, 58(4):298–302.

Rubin, D. B. (2004b). Direct and indirect causal effects via potential outcomes. *Scandinavian Journal of Statistics*, 31(2):161–170.

Rubin, D. B. (2005). Causal inference using potential outcomes: Design, modeling, decisions. *Journal of the American Statistical Association*, 100(469):322–331.

Rubin, D. B. and Schafer, J. L. (1990). Efficiently creating multiple imputations for incomplete multivariate normal data. In *ASA 1990 Proceedings of the Statistical Computing Section*, pages 83–88, Alexandria, VA.

Rubin, D. B. and Schenker, N. (1986a). Efficiently simulating the coverage properties of interval estimates. *Journal of the Royal Statistical Society C*, 35(2):159–167.

Rubin, D. B. and Schenker, N. (1986b). Multiple imputation for interval estimation from simple random samples with ignorable nonresponse. *Journal of the American Statistical Association*, 81(394):366–374.

Ruppert, D., Wand, M. P., and Carroll, R. J. (2003). *Semiparametric Regression*. Cambridge University Press, Cambridge.

Saar-Tsechansky, M. and Provost, F. (2007). Handling missing values when applying classification models. *Journal of Machine Learning Research*, 8:1625–1657.

Sabbaghi, A. and Rubin, D. B. (2014). Comments on the Neyman-Fisher controversy and its consequences. *Statistical Science*, 29(2):267–284.

Salfran, D. and Spiess, M. (2017). *ImputeRobust: Robust Multiple Imputation with Generalized Additive Models for Location Scale and Shape*. R package version 1.2.

Salomon, J. A., Tandon, A., and Murray, C. J. L. (2004). Comparability of self rated health: Cross sectional multi-country survey using anchoring vignettes. *British Medical Journal*, 328(7434):258.

Särndal, C. E. and Lundström, S. (2005). *Estimation in Surveys with Nonresponse*. John Wiley & Sons, New York.

Särndal, C. E., Swensson, B., and Wretman, J. (1992). *Model Assisted Survey Sampling*. Springer-Verlag, New York.

Sauerbrei, W. and Schumacher, M. (1992). A bootstrap resampling procedure for model building: Application to the Cox regression model. *Statistics in Medicine*, 11(16):2093–2109.

Schafer, J. L. (1997). *Analysis of Incomplete Multivariate Data*. Chapman & Hall, London.

Schafer, J. L. (2003). Multiple imputation in multivariate problems when the imputation and analysis models differ. *Statistica Neerlandica*, 57(1):19–35.

Schafer, J. L., Ezzati-Rice, T. M., Johnson, W., Khare, M., Little, R. J. A., and Rubin, D. B. (1996). The NHANES III multiple imputation project. In *ASA 1996 Proceedings of the Survey Research Methods Section*, pages 28–37, Alexandria, VA.

Schafer, J. L. and Graham, J. W. (2002). Missing data: Our view of the state of the art. *Psychological Methods*, 7(2):147–177.

Schafer, J. L. and Olsen, M. K. (1998). Multiple imputation for multivariate missing-data problems: A data analyst's perspective. *Multivariate Behavioral Research*, 33(4):545–571.

Schafer, J. L. and Schenker, N. (2000). Inference with imputed conditional means. *Journal of the American Statistical Association*, 95(449):144–154.

Schafer, J. L. and Yucel, R. M. (2002). Computational strategies for multivariate linear mixed-effects models with missing values. *Journal of Computational and Graphical Statistics*, 11(2):437–457.

Scharfstein, D. O., Rotnitzky, A., and Robins, J. M. (1999). Adjusting for nonignorable drop-out using semiparametric nonresponse models (with discussion). *Journal of the American Statistical Association*, 94(448):1096–1120.

Schenker, N., Raghunathan, T. E., and Bondarenko, I. (2010). Improving on analyses of self-reported data in a large-scale health survey by using information from an examination-based survey. *Statistics in Medicine*, 29(5):533–545.

Schenker, N. and Taylor, J. M. G. (1996). Partially parametric techniques for multiple imputation. *Computational Statistics & Data Analysis*, 22(4):425–446.

Scheuren, F. J. (2004). Introduction to history corner. *The American Statistician*, 58(4):290–291.

Scheuren, F. J. (2005). Multiple imputation: How it began and continues. *The American Statistician*, 59(4):315–319.

Schönbeck, Y., Talma, H., Van Dommelen, P., Bakker, B., Buitendijk, S. E., HiraSing, R. A., and Van Buuren, S. (2013). The world's tallest nation has stopped growing taller: The height of Dutch children from 1955 to 2009. *Pediatric Research*, 73(3):371–377.

Schouten, R. M., Lugtig, P. L., and Vink, G. (2018). Generating missing values for simulation purposes: A multivariate amputation procedure. Working paper, University of Utrecht.

Schouten, R. M. and Vink, G. (2017). Wrapper function `parlMICE`. page https://gerkovink.github.io/parlMICE/Vignette_parlMICE.html.

Schulz, K. F., Altman, D. G., and Moher, D. (2010). CONSORT 2010 statement: Updated guidelines for reporting parallel group randomised trials. *British Medical Journal*, 340:c332.

Scott, M. A., Shrout, P. E., and Weinberg, S. L. (2013). Multilevel model notation - establishing the commonalities. In Scott, M. A., Simonoff, J. A., and Marx, B. D., editors, *The SAGE Handbook of Multilevel Modeling*, pages 21–38. SAGE, Los Angeles, CA.

Scott Long, J. (1997). *Regression Models for Categorical and Limited Dependent Variables*. Sage, Thousand Oaks, CA.

Seaman, S. R., Bartlett, J. W., and White, I. R. (2012). Multiple imputation of missing covariates with non-linear effects and interactions: An evaluation of statistical methods. *BMC Medical Research Methodology*, 12(1):46.

Seaman, S. R. and Hughes, R. A. (2018). Relative efficiency of joint-model and full-conditional-specification multiple imputation when conditional models are compatible: The general location model. *Statistical Methods in Medical Research*, doi.org/10.1177/0962280216665872.

Shadish, W. R., Cook, T. D., and Campbell, D. T. (2001). *Experimental and Quasi-Experimental Designs for Generalized Causal Inference*. Wadsworth Publishing, Florence, KY, 2nd edition.

Shah, A. D., Bartlett, J. W., Carpenter, J. R., Nicholas, O., and Hemingway, H. (2014). Comparison of random forest and parametric imputation models for imputing missing data using MICE: A CALIBER study. *American Journal of Epidemiology*, 179(6):764–774.

Shapiro, F. (2001). *EMDR: Eye Movement Desensitization of Reprocessing: Basic Principles, Protocols and Procedures*. Guilford Press, New York, 2nd edition.

Shen, Z. (2000). *Nested Multiple Imputation*. PhD thesis, Department of Statistics, Harvard University, Cambridge, MA.

Shortreed, S. M., Laber, E., Scott Stroup, T., Pineau, J., and Murphy, S. A. (2014). A multiple imputation strategy for sequential multiple assignment randomized trials. *Statistics in Medicine*, 33(24):4202–4214.

Si, Y. and Reiter, J. P. (2013). Nonparametric Bayesian multiple imputation for incomplete categorical variables in large-scale assessment surveys. *Journal of Educational and Behavioral Statistics*, 38(5):499–521.

Siciliano, R., Aria, M., and D'Ambrosio, A. (2006). Boosted incremental tree-based imputation of missing data. In Zani, S., Cerioli, A., Riani, M., and Vichi, M., editors, *Data Analysis, Classification and the Forward Search*, pages 271–278. Springer, Berlin.

Siddique, J. and Belin, T. R. (2008). Multiple imputation using an iterative hot-deck with distance-based donor selection. *Statistics in Medicine*, 27(1):83–102.

Singer, J. D. and Willett, J. B. (2003). *Applied Longitudinal Data Analysis: Modeling Change and Event Occurrence*. Oxford University Press, Oxford.

Smink, W. (2016). *Towards estimation of individual causal effects: The use of a prior for the correlation between potential outcomes*. Master thesis. University of Utrecht, Utrecht.

Snijders, T. A. B. and Bosker, R. J. (2012). *Multilevel Analysis. An Introduction to Basic and Advanced Multilevel Modeling. Second Edition*. Sage Publications Ltd., London.

Song, J. and Belin, T. R. (2004). Imputation for incomplete high-dimensional multivariate normal data using a common factor model. *Statistics in Medicine*, 23(18):2827–2843.

Sovilj, D., Eirola, E., Miche, Y., Björk, K.-M., Nian, R., Akusok, A., and Lendasse, A. (2016). Extreme learning machine for missing data using multiple imputations. *Neurocomputing*, 174:220–231.

Speidel, M., Drechsler, J., and Sakshaug, J. W. (2017). Biases in multilevel analyses caused by cluster-specific fixed-effects imputation. *Behavior Research Methods*, pages 1–17.

Stallard, P. (2006). Psychological interventions for post-traumatic reactions in children and young people: a review of randomised controlled trials. *Clinical Psycholological Review*, 26(7):895–911.

Stasinopoulos, D. M. and Rigby, R. A. (2007). Generalized additive models for location scale and shape (GAMLSS) in R. *Journal of Statistical Software*, 23(7):1–46.

Stasinopoulos, D. M., Rigby, R. A., Heller, G. Z., Voudouris, V., and De Bastiani, F. (2017). *Flexible Regression and Smoothing*. CRC Press, Boca Raton, FL.

Steel, R. J., Wang, N., and Raftery, A. E. (2010). Inference from multiple imputation for missing data using mixtures of normals. *Statistical Methodology*, 7(10):351–365.

Steinberg, A. M., Brymer, M. J., Decker, K. B., and Pynoos, R. S. (2004). The University of California at Los Angeles post-traumatic stress disorder reaction index. *Current Psychiatry Reports*, 6(2):96–100.

Stekhoven, D. J. and Bühlmann, P. (2011). `MissForest`: non-parametric missing value imputation for mixed-type data. *Bioinformatics*, 28(1):112–118.

Sterne, J. A. C., White, I. R., Carlin, J. B., Spratt, M., Royston, P., Kenward, M. G., Wood, A. M., and Carpenter, J. R. (2009). Multiple imputation for missing data in epidemiological and clinical research: potential and pitfalls. *British Medical Journal*, 338:b2393.

Steyerberg, E. W. (2009). *Clinical Prediction Models*. Springer, New York.

Su, Y. S., Gelman, A., Hill, J., and Yajimi, M. (2011). Multiple imputation with diagnostics (mi) in R: Opening windows into the black box. *Journal of Statistical Software*, 45(2).

Subramanian, S. (2009). The multiple imputations based Kaplan–Meier estimator. *Statistics and Probability Letters*, 79(18):1906–1914.

Subramanian, S. (2011). Multiple imputations and the missing censoring indicator model. *Journal of Multivariate Analysis*, 102(1):105–117.

Sullivan, T. R., White, I. R., Salter, A. B., Ryan, P., and Lee, K. J. (2018). Should multiple imputation be the method of choice for handling missing data in randomized trials? *Statistical Methods in Medical Research*, doi.org/10.1177/0962280216683570.

Taljaard, M., Donner, A., and Klar, N. (2008). Imputation strategies for missing continuous outcomes in cluster randomized trials. *Biometrical Journal*, 50(3):329–345.

Tang, L., Unüntzer, J., Song, J., and Belin, T. R. (2005). A comparison of imputation methods in a longitudinal randomized clinical trial. *Statistics in Medicine*, 24(14):2111–2128.

Tanner, M. A. and Wong, W. H. (1987). The calculation of posterior distributions by data augmentation (with discussion). *Journal of the American Statistical Association*, 82(398):528–550.

Taylor, J. M. G., Cooper, K. L., Wei, J. T., Sarma, A. V., Raghunathan, T. E., and Heeringa, S. G. (2002). Use of multiple imputation to correct for nonresponse bias in a survey of urologic symptoms among African-American men. *American Journal of Epidemiology*, 156(8):774–782.

Tempelman, D. C. G. (2007). *Imputation of Restricted Data*. PhD thesis, University of Groningen, Groningen.

Templ, M., Hron, K., and Filzmoser, P. (2011a). *robCompositions: An R-package for Robust Statistical Analysis of Compositional Data*, chapter 25, pages 341–355. Wiley-Blackwell.

Templ, M., Kowarik, A., and Filzmoser, P. (2011b). Iterative stepwise regression imputation using standard and robust methods. *Computational Statistics & Data Analysis*, 55(10):2793–2806.

Therneau, T. M., Atkinson, B., and Ripley, B. D. (2017). *rpart: Recursive Partitioning and Regression Trees*. R package version 4.1-11.

Tian, G.-L., Tan, M. T., Ng, K. W., and Tang, M.-L. (2009). A unified method for checking compatibility and uniqueness for finite discrete conditional distributions. *Communications in Statistics - Theory and Methods*, 38(1):115–129.

Tibshirani, R. J. (1996). Regression shrinkage and selection via the lasso. *Journal of the Royal Statistical Society. Series B (Methodological)*, pages 267–288.

Tierney, L. (1996). Introduction to general state-space Markov chain theory. In Gilks, W. R., Richardson, S., and Spiegelhalter, D. J., editors, *Markov Chain Monte Carlo in Practice*, chapter 4, pages 59–74. Chapman & Hall, London.

Tutz, G. and Ramzan, S. (2015). Improved methods for the imputation of missing data by nearest neighbor methods. *Computational Statistics & Data Analysis*, 90:84–99.

US Bureau of the Census (1957). *United States Census of Manufactures, 1954, Vol II, Industry Statistics, Part 1, General Summary and Major Groups 20 to 28*. US Bureau of the Census, Washington, D.C.

Vach, W. (1994). *Logistic Regression with Missing Values in the Covariates*. Springer-Verlag, Berlin.

Vach, W. and Blettner, M. (1991). Biased estimation of the odds ratio in case-control studies due to the use of ad hoc methods of correcting for missing values for confounding variables. *American Journal of Epidemiology*, 134(8):895–907.

Van Belle, G. (2002). *Statistical Rules of Thumb*. John Wiley & Sons, New York.

Van Bemmel, T., Gussekloo, J., Westendorp, R. G. J., and Blauw, G. J. (2006). In a population-based prospective study, no association between high blood pressure and mortality after age 85 years. *Journal of Hypertension*, 24(2):287–292.

Van Buuren, S. (2007a). Multiple imputation of discrete and continuous data by fully conditional specification. *Statistical Methods in Medical Research*, 16(3):219–242.

Van Buuren, S. (2007b). Worm plot to diagnose fit in quantile regression. *Statistical Modelling*, 7(4):363–376.

Van Buuren, S. (2010). Item imputation without specifying scale structure. *Methodology*, 6(1):31–36.

Van Buuren, S. (2011). Multiple imputation of multilevel data. In Hox, J. and Roberts, J., editors, *The Handbook of Advanced Multilevel Analysis*, chapter 10, pages 173–196. Routledge, Milton Park, UK.

Van Buuren, S. (2012). *Flexible Imputation of Missing Data*. Chapman & Hall/CRC, Boca Raton, FL.

Van Buuren, S., Boshuizen, H. C., and Knook, D. L. (1999). Multiple imputation of missing blood pressure covariates in survival analysis. *Statistics in Medicine*, 18(6):681–694.

Van Buuren, S., Brand, J. P. L., Groothuis-Oudshoorn, C. G. M., and Rubin, D. B. (2006). Fully conditional specification in multivariate imputation. *Journal of Statistical Computation and Simulation*, 76(12):1049–1064.

Van Buuren, S., Eyres, S., Tennant, A., and Hopman-Rock, M. (2003). Assessing comparability of dressing disability in different countries by response conversion. *European Journal of Public Health*, 13(3 SUPPL.):15–19.

Van Buuren, S., Eyres, S., Tennant, A., and Hopman-Rock, M. (2005). Improving comparability of existing data by response conversion. *Journal of Official Statistics*, 21(1):53–72.

Van Buuren, S. and Fredriks, A. M. (2001). Worm plot: A simple diagnostic device for modelling growth reference curves. *Statistics in Medicine*, 20(8):1259–1277.

Van Buuren, S. and Groothuis-Oudshoorn, C. G. M. (1999). Flexible multivariate imputation by MICE. Technical Report PG/VGZ/99.054, TNO Prevention and Health, Leiden.

Van Buuren, S. and Groothuis-Oudshoorn, C. G. M. (2000). Multivariate imputation by chained equations: MICE V1.0 user's manual. Technical Report PG/VGZ/00.038, TNO Prevention and Health, Leiden.

Van Buuren, S. and Groothuis-Oudshoorn, C. G. M. (2011). `mice`: Multivariate imputation by chained equations in R. *Journal of Statistical Software*, 45(3):1–67.

Van Buuren, S. and Ooms, J. C. L. (2009). Stage line diagram: An age-conditional reference diagram for tracking development. *Statistics in Medicine*, 28(11):1569–1579.

Van Buuren, S. and Tennant, A., editors (2004). *Response Conversion for the Health Monitoring Program*, volume 04 145. TNO Quality of Life, Leiden.

Van Buuren, S. and Van Rijckevorsel, J. L. A. (1992). Imputation of missing categorical data by maximizing internal consistency. *Psychometrika*, 57(4):567–580.

Van Buuren, S., Van Rijckevorsel, J. L. A., and Rubin, D. B. (1993). Multiple imputation by splines. In *Bulletin of the International Statistical Institute*, volume II (CP), pages 503–504.

Van der Palm, D. W., Van der Ark, L. A., and Vermunt, J. K. (2016a). A comparison of incomplete-data methods for categorical data. *Statistical Methods in Medical Research*, 25(2):754–774.

Van der Palm, D. W., Van der Ark, L. A., and Vermunt, J. K. (2016b). Divisive latent class modeling as a density estimation method for categorical data. *Journal of Classification*, 33:52–72.

Van Deth, J. W., editor (1998). *Comparative Politics. The Problem of Equivalence*. Routledge, London.

Van Ginkel, J. R. and Kroonenberg, P. M. (2014). Analysis of variance of multiply imputed data. *Multivariate Behavioral Research*, 49(1):78–91.

Van Ginkel, J. R., Van der Ark, L. A., and Sijtsma, K. (2007). Multiple imputation for item scores when test data are factorially complex. *British Journal of Mathematical and Statistical Psychology*, 60(2):315–337.

Van Praag, B. M. S., Dijkstra, T. K., and Van Velzen, J. (1985). Least-squares theory based on general distributional assumptions with an application to the incomplete observations problem. *Psychometrika*, 50(1):25–36.

Van Wouwe, J. P., Lanting, C. I., Van Dommelen, P., Treffers, P. E., and Van Buuren, S. (2009). Breastfeeding duration related to practised contraception in The Netherlands. *Acta Paediatrica*, 98(1):86–90.

Vandenbroucke, J. P., Von Elm, E., Altman, D. G., Gotzsche, P. C., Mulrow, C. D., Pocock, S. J., Poole, C., Schlesselman, J. J., and Egger, M. (2007). Strengthening the reporting of observational studies in epidemiology (STROBE): Explanation and elaboration. *Annals of Internal Medicine*, 147(8):W163–94.

Vateekul, P. and Sarinnapakorn, K. (2009). Tree-based approach to missing data imputation. In *2009 IEEE International Conference on Data Mining Workshops*, pages 70–75. IEEE Computer Society.

Venables, W. N. and Ripley, B. D. (2002). *Modern Applied Statistics with S*. Springer-Verlag, New York, 4th edition.

Verbeke, G. and Molenberghs, G. (2000). *Linear Mixed Models for Longitudinal Data*. Springer, New York.

Vergouwe, Y., Royston, P., Moons, K. G. M., and Altman, D. G. (2010). Development and validation of a prediction model with missing predictor data: A practical approach. *Journal of Clinical Epidemiology*, 63(2):205–214.

Verloove-Vanhorick, S. P., Verwey, R. A., Brand, R., Bennebroek Gravenhorst, J., Keirse, M. J. N. C., and Ruys, J. H. (1986). Neonatal mortality risk in relation to gestational age and birthweight: Results of a national survey of preterm and very-low-birthweight infants in The Netherlands. *Lancet*, 1(8472):55–57.

Vermunt, J. K., Van Ginkel, J. R., Van der Ark, L. A., and Sijtsma, K. (2008). Multiple imputation of incomplete categorical data using latent class analysis. *Sociological Methodology*, 38(1):369–397.

Viallefont, V., Raftery, A. E., and Richardson, S. (2001). Variable selection and Bayesian model averaging in case-control studies. *Statistics in Medicine*, 20(21):3215–3230.

Vidotto, D. (2018). *Bayesian latent class models for the multiple imputation of cross-sectional, multilevel and longitudinal categorical data*. PhD thesis, Tilburg University, Tilburg, The Netherlands.

Vidotto, D., Vermunt, J. K., and Kaptein, M. C. (2015). Multiple imputation of missing categorical data using latent class models: State of art. *Psychological Test and Assessment Modelling*, 57(4):542–576.

Vink, G. (2015). *Restrictive Imputation of Incomplete Survey Data*. PhD thesis, Utrecht University.

Vink, G., Frank, L. E., Pannekoek, J., and Van Buuren, S. (2014). Predictive mean matching imputation of semicontinuous variables. *Statistica Neerlandica*, 68(1):61–90.

Vink, G., Lazendic, G., and Van Buuren, S. (2015). Partioned predictive mean matching as a large data multilevel imputation technique. *Psychological Test and Assessment Modeling*, 57(4):577–594.

Vink, G. and Van Buuren, S. (2013). Multiple imputation of squared terms. *Sociological Methods & Research*, 42(4):598–607.

Vink, G. and Van Buuren, S. (2014). Pooling multiple imputations when the sample happens to be the population. *arXiv:1409.8542*.

Visscher, T. L. S., Viet, A. L., Kroesbergen, H. T., and Seidell, J. C. (2006). Underreporting of BMI in adults and its effect on obesity prevalence estimations in the period 1998 to 2001. *Obesity*, 14(11):2054–2063.

Von Hippel, P. T. (2007). Regression with missing *y*'s: An improved strategy for analyzing multiply imputed data. *Sociological Methodology*, 37(1):83–117.

Von Hippel, P. T. (2009). How to impute interactions, squares, and other transformed variables. *Sociological Methodology*, 39(1):265–291.

Von Hippel, P. T. (2013). Should a normal imputation model be modified to impute skewed variables? *Sociological Methods & Research*, 42(1):105–138.

Von Hippel, P. T. (2018). How many imputations do you need? A two-stage calculation using a quadratic rule. *Sociological Methods & Research*, doi.org/10.1177/0049124117747303.

Vroomen, J. M., Eekhout, I., Dijkgraaf, M. G., Van Hout, H., De Rooij, S. E., Heymans, M. W., and Bosmans, J. E. (2016). Multiple imputation strategies for zero-inflated cost data in economic evaluations: Which method works best. *The European Journal of Health Economics*, 17(8):939–950.

Wagstaff, D. A. and Harel, O. (2011). A closer examination of three small-sample approximations to the multiple-imputation degrees of freedom. *Stata Journal*, 11(3):403–419.

Waljee, A. K., Mukherjee, A., Singal, A. G., Zhang, Y., Warren, J., Balis, U., Marrero, J., Zhu, J., and Higgins, P. D. R. (2013). Comparison of imputation methods for missing laboratory data in medicine. *BMJ open*, 3(8):e002847.

Wallace, M. L., Anderson, S. J., and Mazumdar, S. (2010). A stochastic multiple imputation algorithm for missing covariate data in tree-structured survival analysis. *Statistics in Medicine*, 29(29):3004–3016.

Walls, T. A. and Schafer, J. L., editors (2006). *Models for Intensive Longitudinal Data*. Oxford University Press, Oxford.

Wang, N. and Robins, J. M. (1998). Large-sample theory for parametric multiple imputation procedures. *Biometrika*, 85(4):935–948.

Wang, Q. and Dinse, G. E. (2010). Linear regression analysis of survival data with missing censoring indicators. *Lifetime Data Analysis*, 17(2):256–279.

Wang, Y. J. and Kuo, K.-L. (2010). Compatibility of discrete conditional distributions with structural zeros. *Journal of Multivariate Analysis*, 101(1):191–199.

Wei, G. C. G. and Tanner, M. A. (1991). Applications of multiple imputation to the analysis of censored regression data. *Biometrics*, 47(4):1297–1309.

Weisberg, H. I. (2010). *Bias and causation: Models and judgment for valid comparisons*. John Wiley & Sons, Hoboken, NJ.

White, I. R. and Carlin, J. B. (2010). Bias and efficiency of multiple imputation compared with complete-case analysis for missing covariate values. *Statistics in Medicine*, 29(28):2920–2931.

White, I. R., Daniel, R., and Royston, P. (2010). Avoiding bias due to perfect prediction in multiple imputation of incomplete categorical variables. *Computational Statistics & Data Analysis*, 54(10):2267–2275.

White, I. R., Horton, N. J., Carpenter, J. R., and Pocock, S. J. (2011a). Strategy for intention to treat analysis in randomised trials with missing outcome data. *British Medical Journal*, 342:d40.

White, I. R. and Royston, P. (2009). Imputing missing covariate values for the Cox model. *Statistics in Medicine*, 28(15):1982–1998.

White, I. R., Royston, P., and Wood, A. M. (2011b). Multiple imputation using chained equations: Issues and guidance for practice. *Statistics in Medicine*, 30(4):377–399.

White, I. R. and Thompson, S. G. (2005). Adjusting for partially missing baseline measurements in randomized trials. *Statistics in Medicine*, 24(7):993–1007.

Wickham, H. and Grolemund, G. (2017). *R for Data Science*. O'Reilly Media, Inc., Sebastopol, CA.

Willett, J. B. (1989). Some results on reliability for the longitudinal measurement of change: Implications for the design of studies of individual growth. *Educational and Psychological Measurement*, 49:587–602.

Wood, A. M., White, I. R., and Royston, P. (2008). How should variable selection be performed with multiply imputed data? *Statistics in Medicine*, 27(17):3227–3246.

Wood, A. M., White, I. R., and Thompson, S. G. (2004). Are missing outcome data adequately handled? A review of published randomized controlled trials in major medical journals. *Clinical Trials*, 1(4):368–376.

Wu, L. (2010). *Mixed Effects Models for Complex Data*. Chapman & Hall /CRC, Boca Raton, FL.

Wu, W., Jia, F., and Enders, C. K. (2015). A comparison of imputation strategies for ordinal missing data on Likert scale variables. *Multivariate Behavioral Research*, 50(5):484–503.

Yang, X., Belin, T. R., and Boscardin, W. J. (2005). Imputation and variable selection in linear regression models with missing covariates. *Biometrics*, 61(2):498–506.

Yao, Y., Chen, S.-C., and Wang, S.-H. (2014). On compatibility of discrete full conditional distributions: A graphical representation approach. *Journal of Multivariate Analysis*, 124:1–9.

Yates, F. (1933). The analysis of replicated experiments when the field results are incomplete. *Empirical Journal of Experimental Agriculture*, 1(2):129–142.

Yu, L.-M., Burton, A., and Rivero-Arias, O. (2007). Evaluation of software for multiple imputation of semi-continuous data. *Statistical Methods in Medical Research*, 16(3):243–258.

Yu, M., Reiter, J. P., Zhu, L., Liu, B., Cronin, K. A., and Feuer, E. J. (2017). Protecting confidentiality in cancer registry data with geographic identifiers. *American Journal of Epidemiology*, 186(1):83–91.

Yucel, R. M. (2008). Multiple imputation inference for multivariate multilevel continuous data with ignorable non-response. *Philosophical Transactions of the Royal Society A*, 366(1874):2389–2403.

Yucel, R. M. (2011). Random covariances and mixed-effects models for imputing multivariate multilevel continuous data. *Statistical Modelling*, 11(4):351–370.

Yucel, R. M. (2017). Impact of the non-distinctness and non-ignorability on the inference by multiple imputation in multivariate multilevel data: A simulation assessment. *Journal of Statistical Computation and Simulation*, 87(9):1813–1826.

Yucel, R. M., He, Y., and Zaslavsky, A. M. (2008). Using calibration to improve rounding in imputation. *The American Statistician*, 62(2):125–129.

Yucel, R. M. and Zaslavsky, A. M. (2005). Imputation of binary treatment variables with measurement error in administrative data. *Journal of the American Statistical Association*, 100(472):1123–1132.

Yusuf, S., Zucker, D., Passamani, E., Peduzzi, P., Takaro, T., Fisher, L., Kennedy, J. W., Davis, K., Killip, T., and Norris, R. (1994). Effect of coronary artery bypass graft surgery on survival: overview of 10-year results from randomised trials by the coronary artery bypass graft surgery trialists collaboration. *The Lancet*, 344(8922):563–570.

Yuval, N. (2014). *Sapiens*. Random House, New York.

Zhang, Q. and Wang, L. (2017). Moderation analysis with missing data in the predictors. *Psychological Methods*, 22(4):649–666.

Zhang, W., Zhang, Y., Chaloner, K., and Stapleton, J. T. (2009). Imputation methods for doubly censored HIV data. *Journal of Statistical Computation and Simulation*, 79(10):1245–1257.

Zhao, E. and Yucel, R. M. (2009). Performance of sequential imputation method in multilevel applications. In *ASA 2009 Proceedings of the Survey Research Methods Section*, pages 2800–2810, Alexandria, VA.

Zhao, J. H. and Schafer, J. L. (2016). ***pan***: *Multiple Imputation for Multivariate Panel or Clustered Data*. R package version 1.4.

Zhao, Y. and Long, Q. (2017). Variable selection in the presence of missing data: Imputation-based methods. *Wiley Interdisciplinary Reviews: Computational Statistics*, 9(5).

Zhou, X. H., Zhou, C., Lui, D., and Ding, X. (2014). *Applied Missing Data Analysis in the Health Sciences*. John Wiley & Sons, Chichester, UK.

Zhu, J. (2016). *Assessment and Improvement of a Sequential Regression Multivariate Imputation Algorithm*. PhD thesis, University of Michigan.

Zhu, J. and Raghunathan, T. E. (2015). Convergence properties of a sequential regression multiple imputation algorithm. *Journal of the American Statistical Association*, 110(511):1112–1124.

Author index

Subject index